UNEP

Sudan

Post-Conflict Environmental Assessment

This report by the United Nations Environment Programme was made possible
by the generous contributions of the Governments of Sweden and the United Kingdom

Table of contents

Foreword

The peace deal signed in Nairobi by the Sudanese government and the Sudan People's Liberation Movement on 9 January 2005 put an end to more than two decades of civil war in the country. The United Nations family in Nairobi is proud to have played a lead role in the conclusion of the peace process by hosting an exceptional meeting of the United Nations Security Council in November 2004, which facilitated negotiations that led to a Comprehensive Peace Agreement being reached in early 2005.

For most of Sudan, it is now time to focus on recovery, reconstruction and development. In this context, the Government of National Unity and the Government of Southern Sudan requested UNEP to conduct an environmental assessment of the country in order to evaluate the state of Sudan's environment and identify the key environmental challenges ahead. This report presents the findings of the fieldwork, analysis and extensive consultations that were carried out between December 2005 and March 2007, and contains:

- an overview of the environment of Sudan and the assessment process;

- analysis and recommendations for the major cross-cutting issues of climate change, desertification, conflict, and population displacement; and

- analysis and recommendations for key environmental issues in nine different sectors (urban/health, industry, agriculture, forestry, water, wildlife, marine environment, law and foreign aid).

Sudan will not benefit fully from the tangible dividends of peace as long as conflict rages on in Darfur. Despite the signing of a peace agreement in May 2006, violence and insecurity continue to prevail in the region. The United Nations, through its Secretary-General, has designated the resolution of the crisis in Darfur as a main priority, and it is hoped that the findings and recommendations presented in this UNEP report will contribute to this goal.

Indeed, UNEP's investigation has shown clearly that peace and people's livelihoods in Darfur as well as in the rest of Sudan are inextricably linked to the environmental challenge. Just as environmental degradation can contribute to the triggering and perpetuation of conflict, the sustainable management of natural resources can provide the basis for long-term stability, sustainable livelihoods, and development. It is now critical that both national and local leadership prioritize environmental awareness and opportunities for the sustainable management of natural resources in Sudan.

We wish to sincerely thank the Governments of Sweden and the United Kingdom for their generous financial support, which enabled UNEP to carry out this assessment, organize two environmental workshops for national delegates in Sudan in 2006, and publish this report.

In addition, this assessment would not have been possible without the support of our colleagues in the UN Sudan Country Team, including those in sister agencies such as UNDP, UNICEF, FAO, UNHCR, WFP and OCHA. The Ministries of Environment of the Government of National Unity and the Government of Southern Sudan were also active partners in the assessment process, providing both information and support. We hope that UNEP can remain a long-term partner of the Sudanese authorities and people as they address the environmental challenges ahead.

Achim Steiner
United Nations Under-Secretary-General
Executive Director
of the United Nations Environment Programme

Executive summary

Introduction

In January 2005, the Sudanese Government and the Sudan People's Liberation Army signed a Comprehensive Peace Agreement, putting an end to twenty-two years of continuous civil war. With peace and a fast-growing economy fueled by its emerging oil industry, most of the country can now focus on recovery and development.

Sudan, however, faces a number of challenges. Among these are critical environmental issues, including land degradation, deforestation and the impacts of climate change, that threaten the Sudanese people's prospects for long-term peace, food security and sustainable development. In addition, complex but clear linkages exist between environmental problems and the ongoing conflict in Darfur, as well as other historical and current conflicts in Sudan.

Post-conflict environmental assessment

With a view to gaining a comprehensive understanding of the current state of the environment in Sudan and catalysing action to address the country's key environmental problems, the Government of National Unity (GONU) and Government of Southern Sudan (GOSS) requested the United Nations Environment Programme (UNEP) to conduct a post-conflict environmental assessment of Sudan. The goal of the UNEP assessment was accordingly to develop a solid technical basis for medium-term corrective action in the field of environmental protection and sustainable development.

Assessment process

The post-conflict environmental assessment process for Sudan began in late 2005. Following an initial appraisal and scoping study, fieldwork was carried out between January and August 2006. Different teams of experts spent a total of approximately 150 days in the field, on ten separate field missions, each lasting one to four weeks. Consultation with local and international stakeholders formed a large and continuous part of UNEP's assessment work, with the total number of interviewees estimated to be over two thousand. Parties consulted include representatives of federal, state and local governments, NGOs, academic and research institutions, international agencies, community leaders, farmers, pastoralists, foresters and businesspeople.

The UNEP team on mission in Northern state. Different teams of experts spent 150 days in the field, on ten separate field missions, each lasting one to four weeks

Figure E.1 General map of Sudan

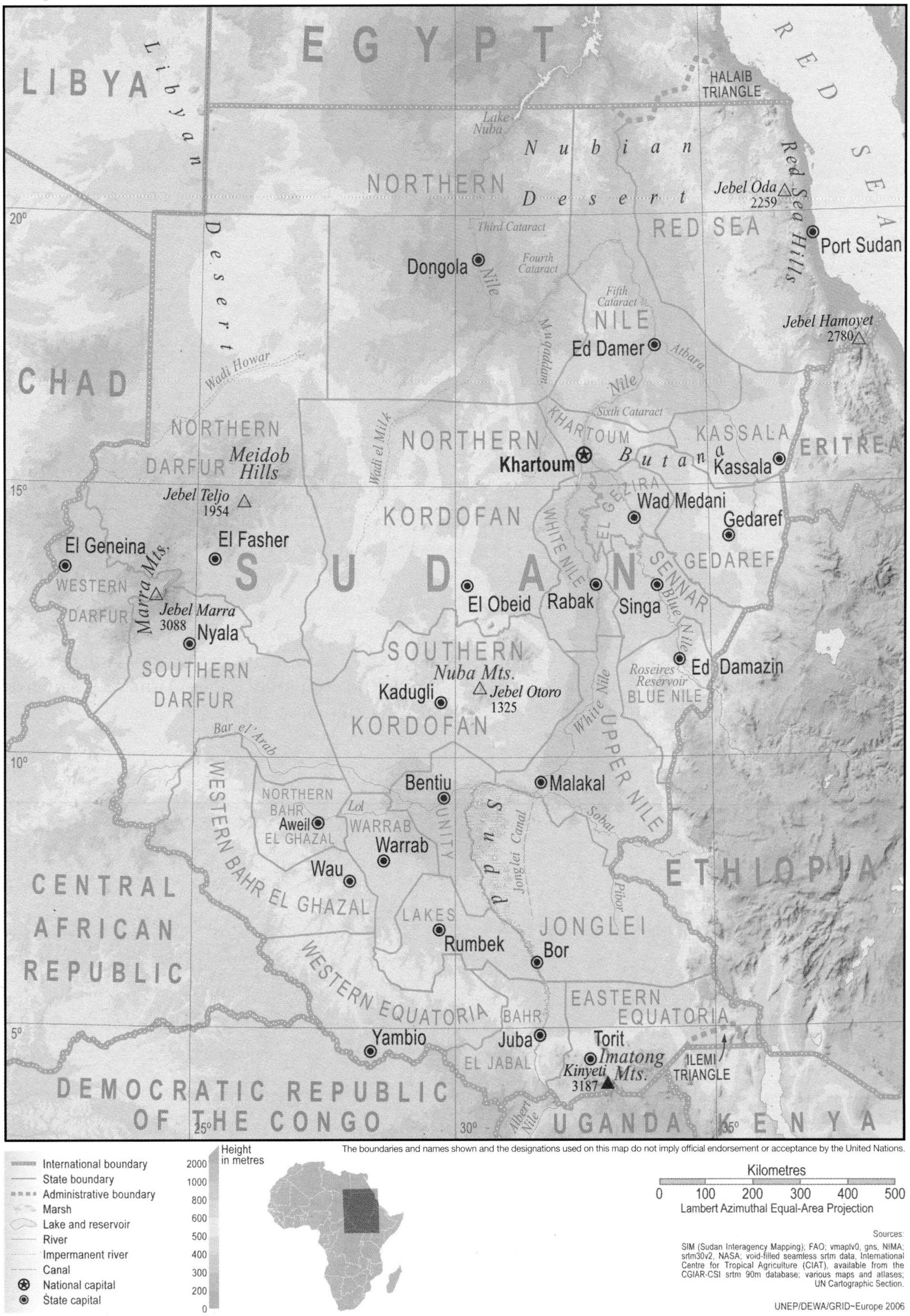

The UNEP team interviews a group of local men in Umm al Jawasir, in Northern state. Community hearings and consultations were a critical component of UNEP's assessment work

The assessment team was comprised of a core UNEP team and a large number of national and international partners who collaborated in a range of roles. These partnerships were crucial to the project's success, as they enabled the fieldwork, ensured that the study matched local issues and needs, and contributed to national endorsement of the assessment's outcomes. UNEP also worked closely with the Government of National Unity and the Government of Southern Sudan, and specific efforts were made to align UNEP activities with a government initiative known as the National Plan for Environmental Management.

Summary of the findings

The assessment identified a number of critical environmental issues that are closely linked to the country's social and political challenges.

Strong linkages between environment and conflict: a key issue in the Darfur crisis

The linkages between conflict and environment in Sudan are twofold. On one hand, the country's long history of conflict has had significant impacts on its environment. Indirect impacts such as population displacement, lack of governance, conflict-related resource exploitation and under-

investment in sustainable development have been the most severe consequences to date.

On the other hand, environmental issues have been and continue to be contributing causes of conflict. Competition over oil and gas reserves, Nile waters and timber, as well as land use issues related to agricultural land, are important causative factors in the instigation and perpetuation of conflict in Sudan. Confrontations over rangeland and rain-fed agricultural land in the drier parts of the country are a particularly striking manifestation of the connection between natural resource scarcity and violent conflict. In all cases, however, environmental factors are intertwined with a range of other social, political and economic issues.

UNEP's analysis indicates that there is a very strong link between land degradation, desertification and conflict in Darfur. Northern Darfur – where exponential population growth and related environmental stress have created the conditions for conflicts to be triggered and sustained by political, tribal or ethnic differences – can be considered a tragic example of the social breakdown that can result from ecological collapse. Long-term peace in the region will not be possible unless these underlying and closely linked environmental and livelihood issues are resolved.

**Population displacement: significant
environmental impacts**

With over five million internally displaced persons
(IDPs) and international refugees, Sudan has the
largest population of displaced persons in the world
today. In Darfur, internal displacement has occurred
at an unprecedented rate since 2003, with some 2.4
million people affected. This massive population
displacement has been accompanied by significant
human suffering and environmental damage. Areas
around the larger camps – particularly in Darfur
– are severely degraded, and the lack of controls and
solutions has led to human rights abuses, conflicts
over resources and food insecurity. Although this is
not a new phenomenon, the scale of displacement
and the particular vulnerability of the dry northern
Sudanese environment may make this the most
significant case of its type worldwide.

In addition, the large-scale return of southern
Sudanese to their homeland following the cessation
of the civil war is likely to result in a further wave
of environmental degradation in some of the more
fragile return areas.

*Cattle in poor condition on overgrazed land near
El Geneina, Western Darfur. Intense competition
over declining natural resources is a contributing
cause of the ongoing conflict in the region*

*Desertification and the associated loss of
agricultural land are not an inevitable and
unstoppable process. Good management
practices can sustain agriculture even in
seemingly arid and hostile environments,
as in this dune belt in Northern Kordofan*

**Desertification and regional climate change:
contributing to poverty and conflict**

An estimated 50 to 200 km southward shift of
the boundary between semi-desert and desert
has occurred since rainfall and vegetation records
were first held in the 1930s. This boundary
is expected to continue to move southwards
due to declining precipitation. The remaining
semi-desert and low rainfall savannah on sand,
which represent some 25 percent of Sudan's
agricultural land, are at considerable risk of
further desertification. This is forecast to lead to
a significant drop (approximately 20 percent) in
food production. In addition, there is mounting
evidence that the decline in precipitation due to
regional climate change has been a significant
stress factor on pastoralist societies – particularly
in Darfur and Kordofan – and has thereby
contributed to conflict.

Natural disasters: increasing vulnerability and impacts

Sudan has suffered a number of long and devastating droughts in the past decades, which have undermined food security and are strongly linked to human displacement and related conflicts. The vulnerability to drought is exacerbated by the tendency to maximize livestock herd sizes rather than quality, and by the lack of secure water sources such as deep boreholes that can be relied on during short dry spells.

Despite serious water shortages, floods are also common in Sudan. The most devastating occur on the Blue Nile, as a result of deforestation and overgrazing in the river's upper catchment. One of the main impacts of watershed degradation and associated flooding is severe riverbank erosion in the narrow but fertile Nile riverine strip.

Agriculture: severe land degradation due to demographic pressure and poorly managed development

Agriculture, which is the largest economic sector in Sudan, is at the heart of some of the country's most serious and chronic environmental problems, including land degradation in its various forms, riverbank erosion, invasive species, pesticide mismanagement in the large irrigation schemes, and water pollution. Disorganized and poorly managed mechanized rain-fed agriculture, which covers an estimated area of 6.5 million hectares, has been particularly destructive, leading to large-scale forest clearance, loss of wildlife and severe land degradation.

In addition, an explosive growth in livestock numbers – from 28.6 million in 1961 to 134.6 million in 2004 – has resulted in widespread degradation of the rangelands. Inadequate rural land tenure, finally, is an underlying cause of many environmental problems and a major obstacle to sustainable land use, as farmers have little incentive to invest in and protect natural resources.

Forestry: a deforestation crisis in the drier regions, risks and opportunities in the south

Deforestation in Sudan is estimated to be occurring at a rate of over 0.84 percent per annum at the national level, and 1.87 percent per annum in UNEP case study areas. It is driven principally by energy needs and agricultural clearance. Between 1990 and 2005, the country lost 11.6 percent of its forest cover, or approximately 8,835,000

The most serious and common natural disaster facing the population of Sudan is drought. Rural communities such as this village in Khartoum state have faced waves of drought since the 1970s, which have exacerbated rural poverty and precipitated large-scale displacement to the northern cities

Abandoned degraded agricultural land in a former irrigation scheme near Tandelti in Northern Kordofan

hectares. At the regional level, two-thirds of the forests in north, central and eastern Sudan disappeared between 1972 and 2001. In Darfur, a third of the forest cover was lost between 1973 and 2006. Southern Sudan is estimated to have lost 40 percent of its forests since independence and deforestation is ongoing, particularly around major towns. Extrapolation of deforestation rates indicate that forest cover could reduce by over 10 percent per decade. In areas under extreme pressure, UNEP estimates that total loss could occur within the next 10 years.

These negative trends demonstrate that this valuable resource upon which the rural population and a large part of the urban population depend completely for energy is seriously threatened. The growing use of fuelwood for brick-making in all parts of Sudan is an additional cause for concern. In Darfur, for instance, brick-making provides a livelihood for many IDP camp residents, but also contributes to severe localized deforestation. If it were properly managed, however, the forestry sector could represent a significant opportunity for economic development and sustainable north-south trade.

A mango orchard in Juba, Central Equatoria. The combination of higher rainfall and lower population and development pressure results in Sudan's remaining forest cover being concentrated in the southern half of the country

The rusting wreckage of the Jonglei canal excavator lies in the unfinished main channel. This failed venture illustrates the risks associated with developing large-scale projects in socially and environmentally sensitive areas without local support

Dams and water projects: major impacts and conflict linkages

UNEP considers the principal and most important environmental issue in the water resource sector in Sudan to be the ongoing or planned construction of over twenty large dams. While its electrical output is expected to bring major benefits to the country, the Merowe dam epitomizes environmental and social concerns over the country's ambitious dam-building programme. Although it is the first dam project in Sudan to have included an environmental impact assessment, the process did not meet international standards, and would have benefited from more transparency and public consultation. Major environmental problems associated with the Merowe dam include silt loss for flood recession agriculture, dam sedimentation and severe riverbank erosion due to intensive flow release within short time periods.

In addition, the active storage capacity of all of Sudan's existing dam reservoirs (with the exception of Jebel Aulia) is seriously affected by sediment deposition. Dams have also caused major degradation of downstream habitats, particularly of the maya wetlands on the Blue Nile and of the riparian *dom* palm forests in the lower Atbara river.

The infamous Jonglei canal engineering mega-project, which started in the 1970s, was closely linked to the start of the north-south civil war. As it was not completed, its anticipated major impacts on the Sudd wetlands never came to pass. The unfinished canal bed, which does not connect to any major water bodies or watercourses, now acts only as a giant ditch and embankment hindering wildlife migrations. Nevertheless, lessons learnt from this project should be carefully studied and applied to existing efforts in peacebuilding between north and south, especially as economic motivations for the project still exist, including from international partners.

Urban issues and environmental health: rapid and chaotic urbanization and chronic waste and sanitation issues

Uncontrolled sprawl, chronic solid waste management problems and the lack of wastewater treatment are the leading environmental problems facing Sudan's urban centres. The explosive growth of the capital Khartoum continues relentlessly, with 64 percent of the country's urban population residing in the area. The larger towns of Southern Sudan are also experiencing very rapid growth fueled by the return of formerly displaced persons, estimated at 300,000 by end of 2006. In Darfur, the majority of the two million displaced are found on the fringes of urban centres, whose size in some cases has increased by over 200 percent in the last three years.

Sewage treatment is grossly inadequate in all of Sudan's cities, and solid waste management practices throughout the country are uniformly poor. In the majority of cases, garbage of all types accumulates close to its point of origin and is periodically burnt. These shortcomings in environmental sanitation are directly reflected in the elevated incidence of waterborne diseases, which make up 80 percent of reported diseases in the country.

Waste pickers at the main Khartoum landfill site. Waste management is problematic throughout Sudan.

Industrial pollution: a growing problem and a key issue for the emerging oil industry

Environmental governance of industry was virtually non-existent until 2000, and the effects of this are clearly visible today. While the situation has improved over the last few years, UNEP has found that major challenges remain in the areas of project development and impact assessment, improving the operation of older and government-managed facilities, and influencing the policies and management approach at the higher levels of government.

Due to the relatively limited industrial development in Sudan to date, environmental damage has so far been moderate, but the situation could worsen rapidly as the country embarks on an oil-financed development boom. The release of effluent from factories and the disposal of produced water associated with crude oil extraction are issues of particular concern, as industrial wastewater treatment facilities are lacking even in Khartoum. Industrial effluent is typically released into the domestic sewage system, where there is one.

The release of industrial effluent from older factories lacking wastewater treatment facilities is an issue of particular concern

The all-women State Environment Council Secretariat in Gedaref state. The CPA and Interim Constitution devolve extensive responsibility to state governments in the area of environmental governance. State-level structures, however, remain under-funded and in need of substantial investment

Other issues include air emissions, and hazardous and solid waste disposal. While UNEP observed generally substandard environmental performance at most industrial sites, there were exceptional cases of responsible environmental stewardship at selected oil, sugar and cement facilities visited.

Wildlife and protected areas: depleted bio-diversity with some internationally significant areas and wildlife populations remaining

The past few decades have witnessed a major assault on wildlife and their habitats. In northern and central Sudan, the greatest damage has been inflicted by habitat destruction and fragmentation from farming and deforestation. Larger wildlife have essentially disappeared and are now mostly confined to core protected areas and remote desert regions. In the south, uncontrolled and unsustainable hunting has decimated wildlife populations and caused the local eradication of many of the larger species, such as elephant, rhino, buffalo, giraffe, eland and zebra. Nonetheless, Sudan's remaining wildlife populations, including very large herds of white-eared kob and tiang antelope, are internationally significant.

Approximately fifty sites throughout Sudan – covering 10 and 15 percent of the areas of the north and south respectively – are listed as having some form of legal protection. In practice, however, the level of protection afforded to these areas has ranged from slight to negligible, and several exist only on paper today. Many of these important areas are located in regions affected by conflict and have hence suffered from a long-term absence of the rule of law. With three exceptions (Dinder, Sanganeb and Dongonab Bay National Parks), the data on wildlife and protected areas is currently insufficient to allow for the development of adequate management plans.

Marine environment: a largely intact ecosystem under threat

UNEP found the Sudanese marine and coastal environment to be in relatively good condition overall. Its coral reefs are the best preserved ecosystems in the country. However, the economic and shipping boom focused on Port Sudan and the oil export facilities may rapidly change the environmental situation for the worse. Steady degradation is ongoing in the developed strip from Port Sudan to Suakin, and the symptoms of overgrazing and land degradation are as omnipresent on the coast as elsewhere in dryland Sudan. Mangrove stands, for example, are currently under severe pressure along the entire coastline. Pollution from land-based sources and the risk of oil spills are further issues of concern.

Environmental governance: historically weak, now at a crossroads

By granting the Government of Southern Sudan and the states extensive and explicit responsibility in the area of environment and natural resources management, the CPA and new Interim Constitutions have significantly changed the framework for environmental governance in Sudan and helped create the conditions for reform.

At the national level, the country faces many challenges to meet its international obligations, as set out in the treaties and conventions it has signed over the last thirty years. Although the technical skill and level of knowledge in the environmental sector are high and some legislation is already in place, regulatory authorities have critical structural problems, and are under-resourced.

In Southern Sudan, environmental governance is in its infancy, but the early signs are positive. High-level political and cross-sector support is visible, and UNEP considers the new structures to be relatively suited to the task.

Environment and international aid: reduced environmental impact of relief operations and improved UN response to environmental issues necessary

The environmental assessment of the international aid programme in Sudan raised a number of issues that need to be resolved to avoid inadvertently doing harm through the provision of aid, and to improve the effectiveness of aid expenditure in the environmental sector. UNEP's analysis indicates that while most aid projects in Sudan do not cause significant harm to the environment, a few clearly do and the overall diffused impact of the programme is very significant.

One major and highly complex issue is the environmental impact of the provision of food and other emergency aid to some 15 percent of the population, and the projected impact of the various options for shifting back from aid dependence to autonomous and sustainable livelihoods. Indeed, the country is presently caught in a vicious circle of food aid dependence, agricultural underdevelopment and environmental degradation. Under current

The coral reefs of the Red Sea coast are the best preserved ecosystems of Sudan

circumstances, if aid were reduced to encourage a return to agriculture, the result in some areas would be food insecurity and an intensification of land degradation, leading to the high likelihood of failure and secondary displacement.

The integration of environmental considerations into the current UN programme in Sudan needs to be significantly improved. In addition, the environment-related expenditure that does occur – while acknowledged and welcome – suffers from a range of management problems that reduce its effectiveness. Priorities for the UN and its partners in this field are improved coordination and environmental mainstreaming to ensure that international assistance 'does no harm' to Sudan's environment, and 'builds back better'

Recommendations

1. **Invest in environmental management to support lasting peace in Darfur, and to avoid local conflict over natural resources elsewhere in Sudan.** Because environmental degradation and resource scarcity are among the root causes of the current conflict in Darfur, practical measures to alleviate such problems should be considered vital tools for conflict prevention and peacebuilding. Climate change adaptation measures and ecologically sustainable rural development are needed in Darfur and elsewhere to cope with changing environmental conditions and to avoid clashes over declining natural resources.

A group of southern Sudanese travels down the White Nile aboard a ferry, returning to the homeland after years of displacement due to the civil war. A massive return process is currently underway for the four million people displaced during the conflict

A food aid delivery awaits distribution at Port Sudan. Fifteen percent of Sudan's population depends on international food aid for survival

2. **Build capacity at all levels of government and improve legislation to ensure that reconstruction and economic development do not intensify environmental pressures and threaten the livelihoods of present and future generations.** The new governance context provides a rare opportunity to truly embed the principles of sustainable development and best practices in environmental management into the governance architecture in Sudan.

3. **National and regional governments should assume increasing responsibility for investment in the environment and sustainable development.** The injection of oil revenue has greatly improved the financial resources of both the Government of National Unity and the Government of Southern Sudan, enabling them to translate reform into action.

4. **All UN relief and development projects in Sudan should integrate environmental considerations in order to improve the effectiveness of the UN country programme.** Better coordination and environmental mainstreaming are necessary to ensure that international assistance 'does no harm' to Sudan's environment.

The way forward and the UNEP Sudan country programme

This report's 85 detailed recommendations include individual cost and time estimates, and nominate responsible parties for implementation. While they envisage a central and coordinating role for the environment ministries of GONU and GOSS, the wholehearted support and participation of many other government ministries and authorities, as well as several UN agencies, are also needed. The total cost of the recommendations is USD 120 million with expenditure spread over five years. UNEP considers that the majority should be financed by GONU and GOSS, with the balance provided by the international community.

For its part, UNEP plans to establish a Sudan country programme for the period of at least 2007-2009, and stands ready to assist the Government of Sudan and international partners in the implementation of these recommendations.

Introduction

Introduction

1.1　Background

In January 2005, after more than two decades of devastating civil war, the Sudanese central government in Khartoum and the Sudan People's Liberation Army in the south signed a historic Comprehensive Peace Agreement. This landmark achievement – which was followed by the adoption of an Interim Constitution – brought peace to most of the country for the first time in a generation.

Now, thanks to the rapid development of its oil industry, Sudan is one of the fastest-growing economies in Africa. Direct investment and international aid are starting to flow into the country on a large scale, and some parts of Sudan are undergoing brisk development.

As it focuses on recovery and development, however, the country faces a number of key challenges. Chief among them are several critical environmental issues – such as land degradation, deforestation and the impacts of climate change – that threaten Sudan's prospects for long-term peace, food security and sustainable development.

Recent tensions in north-south border regions have highlighted several environmental issues that constitute potential flashpoints for renewed conflict, including the environmental impacts of the oil industry and the management of the country's water resources.

In Darfur, where violence and insecurity continue to prevail despite the signing of a peace agreement in May 2006, complex but clear linkages exist between environmental problems and the ongoing conflict. Indeed, climate change, land degradation and the resulting competition over scarce natural resources are among the root causes as well as the consequences of the violence and grave humanitarian situation in the region.

Natural resource management and rehabilitation, therefore, are not only fundamental prerequisites to peacebuilding in Darfur and the rest of Sudan – they must become a national priority if the country is to achieve long-term social stability and prosperity.

With a view to obtaining a comprehensive understanding of the current state of the environment in Sudan, and catalysing action to address the country's key environmental problems, the Government of National Unity (GONU) and Government of Southern Sudan (GOSS) requested the United Nations Environment Programme (UNEP) to conduct a post-conflict environmental assessment of Sudan. The present report is the principal product of the resulting national-scale assessment project, managed by UNEP over the period November 2005 to January 2007.

1.2　Objectives

Goal and objectives

The goal of the UNEP post-conflict environmental assessment for Sudan was to develop a solid technical basis for medium-term (1-5 years) corrective action in the field of environmental protection and sustainable development. This goal was expanded into five objectives:

1. Provide neutral and objective information on the most critical environmental problems facing the country, and on the potential risks to human health, livelihoods and ecosystem services;

2. Recommend strategic priorities for sustainable resource management and identify the actors, timelines and costs necessary for implementation;

3. Facilitate the development of national environmental policy and strengthen the capacity for national environmental governance;

4. Raise awareness and catalyse financial support for environmental projects by national authorities, UN actors, NGOs and donors; and

5. Integrate environmental issues into the recovery and reconstruction process.

This report aims to present the post-conflict environmental issues for Sudan in a single concise document accessible to a wide audience of non-experts. A number of detailed studies were prepared in parallel to provide the technical basis for this PCEA report. Access to the technical report series and further information on Sudan's environment can be obtained from the UNEP Sudan website at http://sudanreport.unep.ch.

Links to the UN country team in Sudan and international UN processes

This report is designed to fit within the United Nations country- and global-level frameworks for Sudan. At the country level, this study aims to assist the UN family to integrate or 'mainstream' environmental issues into the UN programme for Sudan, according to the framework provided by the UN Country Team Forum, the annual UN Sudan Work Plan process, and the Sudan National and Darfur Joint Assessment Missions.

At the global level, this report is designed to link with ongoing UN reform processes, which focus on issues such as aid effectiveness, improved coordination and better integration of cross-cutting issues like the environment.

A new and developing theme at the global level – addressed by such high-level bodies as the High-level Panel on System-wide Coherence in the Areas of Development, Humanitarian Assistance and the Environment – is the recognition that environmental degradation has become a major contributor to food insecurity, conflict and vulnerability to natural disasters. It could be argued that this evident in Sudan today.

1.3   Assessment scope

The geographical scope of UNEP's survey extended to all states of the Republic of Sudan, the coastline, and to territorial seas.

The assessment's technical scope was developed in two stages – an initial broad scan was followed by a targeted study focused on identified key themes. The final twelve themes, as reflected in the chapters of this report were: natural disasters and desertification; conflict and peace-building; population displacement; urban environment and environmental health; industry; agriculture; forest resources; freshwater resources; wildlife and protected area management; marine environments and resources; environmental governance and awareness; and international aid.

To ensure linkages to the some of the major humanitarian and governance issues the UN and partners are attempting to address in Sudan,

UNEP's assessment work also included the following six cross-cutting topics:

1. Capacity-building: to build national capacity during the process by maximizing the use of government counterparts and technical experts;

2. Engagement with local partners: to link the UNEP process with existing and new local initiatives for environmental assessment and management;

3. Livelihoods and food security: to explicitly link the observed environmental issues with their impact on the poor, particularly on the rural poor;

4. Gender: to link environmental issues and impacts with gender, as issues such as water and firewood scarcity have a disproportionately negative impact on women;

5. Peacebuilding: to analyse the linkages between conflict and environment in order to assist ongoing conflict prevention and resolution efforts; and

6. Aid effectiveness: to critically assess the success of what has been attempted so far in this sector and design a more effective response to the environmental issues identified.

1.4   Methodology

Assessment process

The post-conflict environmental assessment process for Sudan commenced in earnest in late 2005. The major components of this process were:

- an initial appraisal and scoping study;
- consultation;
- desk studies;
- fieldwork;
- remote sensing;
- analysis; and
- development of the recommendations and reporting.

The fieldwork and consultation process are described in more detail below.

Fieldwork

UNEP's fieldwork was carried out between January and August 2006. Different teams of experts spent a total of approximately 150 days in the field, on ten separate field missions, each lasting one to four weeks. The states covered and the timing of each mission are set out below, while the locations visited and field trip routes are shown in Figure 1.1.

Table 1.   UNEP field missions in Sudan

Timing	States visited
February 2006	Northern and Red Sea states, and the coastline
March 2006	Northern and Southern Kordofan
March 2006	Institutional assessment in Juba
April 2006	Khartoum, Kassala, Gedaref, El Gezira, White Nile, and Blue Nile states
May 2006	Central Equatoria (Bahr el Jabal) and Jonglei states
May 2006	Institutional assessment in Khartoum
June 2006	Northern, Western and Southern Darfur
July 2006	Lakes, Northern and Western Bahr el Ghazal, and Upper Nile states
July 2006	Central Equatoria (Bahr el Jabal) state and the town of Yei
August 2006	Northern state

The total distance travelled was in the order of 12,000 km. The average fieldwork day included three to five stakeholder meetings of varying formality; the total number of interviewees is estimated to have been over two thousand.

Constraints and acknowledged gaps in assessment coverage

The two major constraints encountered in the course of the assessment were security risks posed by ongoing military action and fieldwork logistics in Southern Sudan and Darfur. Lesser but nonetheless significant limitations included minefields and the lack of environmental data due to extended periods of conflict.

UNEP considers the technical and geographical scope of the fieldwork to be adequate for the purposes of this assessment. Given the size of Sudan, however, and the security and other constraints detailed above, it was not possible to survey all regions thoroughly. The following areas received only limited coverage:

In the relatively undeveloped areas of Southern Sudan and Darfur, distances are great and roads are poor. In the wet season, mud and flooded stream crossings preclude road travel and restrict aircraft landings in many locations

Figure 1.1   UNEP fieldwork routes

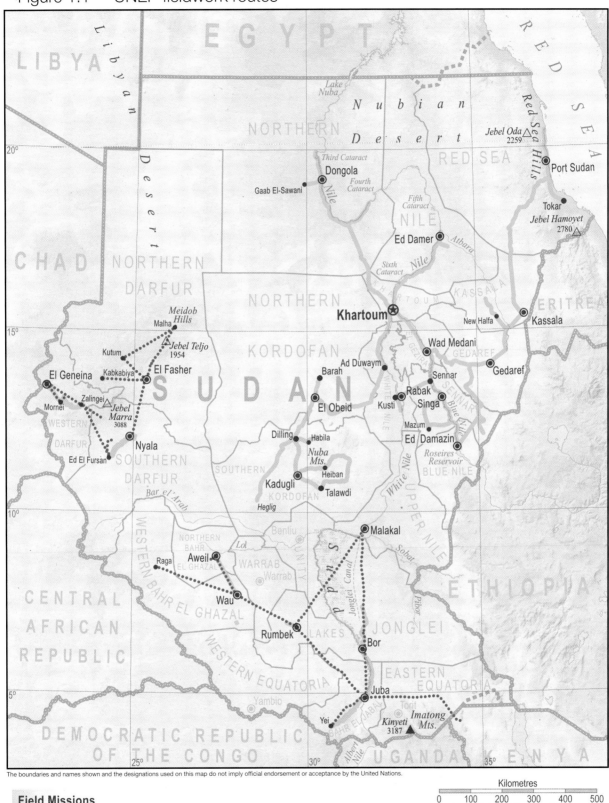

The boundaries and names shown and the designations used on this map do not imply official endorsement or acceptance by the United Nations.

Field Missions

•••••• Travel by air

——— Travel by road

UNEP/DEWA/GRID~Europe 2006

Kilometres

0   100   200   300   400   500

Lambert Azimuthal Equal-Area Projection

Sources:
SIM (Sudan Interagency Mapping); FAO; vmaplv0, gns, NIMA; srtm30v2, NASA; void-filled seamless srtm data, International Centre for Tropical Agriculture (CIAT), available from the CGIAR-CSI srtm 90m database; various maps and atlases; UN Cartographic Section.

- Abyei, Unity state and Upper Nile (oilfields in particular);

- Eastern Equatoria (particularly the Imatong ranges and the dry plains in the far east);

- Western Equatoria (the tropical rainforest in particular);

- the Jebel Marra plateau in Darfur;

- the far south of Southern Darfur, west of Western Bahr el Ghazal; and

- the Eastern Front region on the border of Kassala and Red Sea state.

UN helicopters were the only viable method of transportation in many parts of Darfur and Southern Sudan

UNEP link to national institutions and processes

In order to maximize local engagement in the assessment process and its outcomes, UNEP worked closely with the Government of National Unity (GONU) and the Government of Southern Sudan (GOSS) throughout 2006. Specific efforts were made to align UNEP activities with a government initiative known as the National Plan for Environmental Management (NPEM).

In practical terms, UNEP provided technical and financial support for two major environmental workshops in 2006, one held in Khartoum in July and the other in Juba in November. At these events, technical papers were presented and national delegates discussed and debated regional and national environmental issues.

The draft report consultation process also allowed for UNEP material to be integrated into NPEM documents as they were being developed.

Stakeholders consultation meetings were organized in early 2007 by the Ministry of Environment and Physical Development in Khartoum to discuss and review the draft UNEP post-conflict environmental assessment report

GONU and GOSS report review and endorsement

UNEP engaged the Government of National Unity and the Government of Southern Sudan in a formal process of draft document review. While it incorporates the agreed solutions and wording from that process, this final report is, however, first and foremost an independent UNEP report, with endorsement from the GONU and GOSS.

1.5 Assessment team and consultations

The assessment team was comprised of a core UNEP team and a large number of national and international partners who contributed in a range of roles. The full list of contributors is presented in Appendix III and summarized below:

- UNEP Post-Conflict and Disaster management Branch (core team including seconded individual consultants);

- UNEP Regional Office for Africa and UNEP GRID;

- other UN agencies, including UNOPS, UNDP, WFP, FAO, UNHCR, UNICEF, OCHA, and DSS;

- UN Mission in Sudan;

- African Union Mission in Sudan;

- USAID and the European Commission;

- Government of National Unity Ministry of Environment and Physical Development, including the Secretariat of the Higher Council for Environment and Natural Resources;

- Government of National Unity Ministries of Agriculture and Forestry; of Energy and Mining; and of Irrigation and Water Resources;

- Government of National Unity Remote Sensing Authority and Forests National Corporation;

- Government of Southern Sudan Ministry of Environment, Wildlife Conservation and Tourism;

- Government of Southern Sudan Ministry of Agriculture and Forestry;

- Sudanese Environmental Conservation Society;

- Boma Wildlife Training Centre;

- Kagelu Forestry Training Centre;

- World Agroforestry Centre (ICRAF);

- Rift Valley Institute; and

- Nile Basin Initiative.

Consultation with local stakeholders formed a large and continuous part of UNEP's assessment work, as here in the small village of Mireir, Southern Darfur

The UNEP team discusses a local agricultural project with men from the village of Um Belut, Southern Darfur

These partnerships were absolutely crucial to the project's success, as they facilitated the fieldwork, ensured that the study matched local issues and needs, and contributed to national endorsement of the assessment's outcomes.

Consultations

Consultation with local and international stakeholders formed a large and continuous part of UNEP's assessment work. The list of parties consulted, which is provided in Appendix III, included representatives of federal, state and local governments, non-governmental agencies, academic institutions, international agencies, local residents, agriculturists, pastoralists, foresters and business people.

Key partners in the process were the two counterpart ministries for UNEP, the Government of National Unity's Ministry of Environment and Physical Development, located in Khartoum, and the Government of Southern Sudan's Ministry of Environment, Wildlife Conservation and Tourism, located in Juba. These counterparts accompanied UNEP staff on several of the field missions and provided the main link to other branches of their respective governments.

1.6   Report structure

This report has four main sections:

1. An introduction providing the details of the assessment process;

2. A 'country context' chapter offering general background information on Sudan;

3. Twelve thematic assessment chapters, each in a common format:

   • introduction and assessment activities;
   • overview of the sector or theme;
   • overview of the environmental impacts and issues related to the theme;
   • discussion of the individual impacts and issues; and
   • theme-specific conclusions and detailed recommendations;

4. A conclusion presenting a summary of findings and recommendations, and a discussion of the general way forward.

The twelve thematic chapters are grouped and sequenced according to the type of issue under discussion, as follows:

Cross-cutting issues

Chapter 3 - Natural disasters and
desertification;
Chapter 4 - Conflict and environment;
Chapter 5 - Population displacement;

Sectoral issues

Chapter 6 - Urban environment and
environmental health;
Chapter 7 - Industry;
Chapter 8 - Agriculture;
Chapter 9 - Forest resources;
Chapter 10 - Freshwater resources;
Chapter 11 - Wildlife and protected area
management;
Chapter 12 - Marine environments and
resources;

Institutional response to the issues

Chapter 13 - Environmental governance and
awareness;
Chapter 14 - International aid.

Recommendation format

In each thematic chapter, recommendations are
provided in the following standard format:

- **Numbering:** All recommendations are
numbered to aid collation and tracking;

- **Description:** A one- to four-line description
of the recommendation, including a note
on the scope applicable to the stated cost, if
appropriate;

- **Category (CA):** One of seven categories of
response the recommendation pertains to, as
set out below;

- **Primary beneficiary (PB):** The party
considered by UNEP to be the main target
or recipient of the project's benefits. Note
that in many cases, projects have a large
number of direct and indirect beneficiaries,
and that many of the benefits will derive

*The post-conflict assessment process also included photography and filming: over
35 hours of footage and 5,000 photographs were taken*

from subsequent work done by the primary beneficiary. This is particularly the case in governance or capacity-building projects directed at a specific government sector;

- **United Nations partner (UNP):** The UN agency considered by UNEP to be most suitable to be the primary partner to the beneficiary in the implementation of the project. In the absence of a clear nominee, UNEP remains the default (although a default role is not preferred for a number of reasons). The partner role may range from monitoring only to full involvement through the provision of advice, services and equipment;

- **Cost estimate (CE):** The estimated cost for all parties combined (beneficiary and partners) to implement the recommendation. Note that many governance recommendations will result in laws, policies and plans that will have a major economic impact. This follow-on cost is not included in the estimate. All costs are in USD million, in divisions of USD 100,000; and

- **Duration (DU):** The estimated time required for completion of the project from scoping to close-out. Recommendations are given in the range of one to five years.

The recommendations have been divided into seven categories of response to align with UN and donor agency structures and strategies for assistance to Sudan, as follows:

1. **Governance and rule of law (GROL)** covers the areas of policy development, planning and legislation. In some case, this entails the reform of existing structures, policies, plans and laws;

2. **Technical assistance (TA)** covers the provision of expert advice and technical services, with the objective of addressing an immediate need;

3. **Capacity-building (CB)** covers all topics where the main objective is to improve the ability of the beneficiary to fulfill its mandate, through activities such as mentoring, training and providing equipment and support services. Capacity-building logically follows technical assistance;

A UNEP expert interviews Chadian refugees in Um Shalaya camp, in Western Darfur

A UNEP expert documenting the mission

4. **Government investment (GI)** covers a range of subjects for which UNEP considered that all the factors needed to resolve the issue were generally already in place, except for sufficient funding by the host government. This category thus applied mainly to areas where local technical and human capacity were rated as relatively high and solutions were already devised, but lack of funding prevented the responsible party from fulfilling its mandate.

5. **Awareness-raising (AR)** covers all topics where the main objective is to expose a wide audience to the concepts and issues of environment and sustainable development (focusing on those specific to Sudan). This includes activities such as environmental education, stakeholder briefings, media releases and document distribution.

6. **Assessment (AS)** covers all forms of proposed follow-up assessments and related studies warranted by UNEP. This includes specific studies on subjects and regions that UNEP was not able to include adequately in the scope of this national report due to cost, time and document size constraints.

7. **Practical action (PA):** the majority of the above categories of recommendations focus on building human resources and generating outputs in the form of legislation, policies, plans and other documents. UNEP believes that a certain percentage of projects in the environmental sector should also include or consist of practical action, in order to provide and promote the visible and concrete benefits of good environmental governance and awareness. Such practical projects could include tree-planting, waste clean-up and sustainable building construction. This report strongly emphasizes demonstration projects to catalyse positive change on a larger scale.

Country Context

Country context

2.1 Introduction

Introduction to the national context

The Republic of Sudan is the largest country in Africa. Its highly diverse landscape ranges from desert to tropical forest, and its abundant natural resources include oil, timber, extensive agricultural land, and marine and inland fisheries. The country is also culturally diverse, as it bridges the Islamic culture of North Africa with the largely Christian south, and comprises hundreds of distinct tribal and ethnic groups.

Unfortunately, Sudan has long been plagued by civil war and regional conflict. In the fifty years since achieving independence, the country as a whole has been at peace for only eleven years (1972-1983). While a historic peace agreement was reached for Southern Sudan in 2005, conflict rages on in Darfur. Adding to the burden of war, Sudan has experienced several severe droughts in the past thirty years, and food production in many regions has dropped at the same time as the population has increased.

The combined impacts of conflict and food insecurity have caused over five million Sudanese to be both internally and internationally displaced into camps and urban fringes, and over five million to receive international food aid [2.1, 2.2].

Introduction to the international context

The international community currently provides Sudan with over USD 2 billion per annum in aid, through humanitarian crisis response programmes, recovery and development programmes, and peacekeeping operations. This major investment is delivered through a number of organizations, including the Sudanese Government, donor country governments, the UN family of agencies and the World Bank, bilateral agencies, and national and international non-governmental organizations [2.1, 2.2].

The objectives of this vast and complex programme of assistance are threefold: 1) to prevent, contain and resolve conflict, 2) to save human lives and reduce suffering, and 3) to assist sustainable development. In practical terms, this translates into the achievement and maintenance of peace agreements, and positive numerical indicators in poverty reduction and sustainable development as provided by the UN Millennium Development Goals.

2.2 Society

Population

A detailed national census has never been carried out for all of Sudan; all population figures must therefore be regarded as broad estimates that are rapidly made obsolete by a swelling population with a growth rate estimated to exceed 2.6 percent [2.1, 2.3, 2.4, 2.5, 2.6]. In addition, all detailed data collection to date has excluded Southern Sudan, whose population is broadly estimated at 7-10 million [2.1, 2.7]. Taking these limitations into account, the population of Sudan in 2006 could be estimated to be between 35 and 40 million, with approximately 70 percent living in rural areas, and the other 30 percent living in the capital Khartoum and the country's six other largest cities: Port Sudan, Kassala, Omdurman, El Obeid, Wad Medani, Gedaref and Juba [2.8].

Farmers in Mornei, Western Darfur. The majority of Sudanese live in rural areas and depend on agriculture for their livelihood

Figure 2.1   Sudan population density

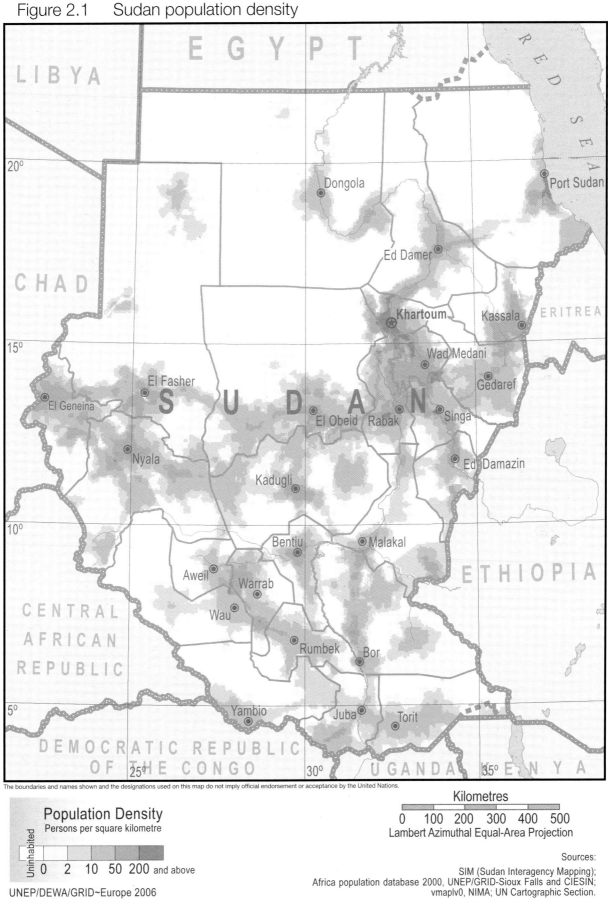

The boundaries and names shown and the designations used on this map do not imply official endorsement or acceptance by the United Nations.

Population Density
Persons per square kilometre

Uninhabited

0   2   10   50   200   and above

UNEP/DEWA/GRID~Europe 2006

Kilometres

0   100   200   300   400   500

Lambert Azimuthal Equal-Area Projection

Sources:
SIM (Sudan Interagency Mapping);
Africa population database 2000, UNEP/GRID-Sioux Falls and CIESIN;
vmaplv0, NIMA; UN Cartographic Section.

Development status

Sudan is rated as a least developed country by UNCTAD, and this is reflected in the most recent Millennium Development Goals Report, Human Development Report and related figures for the country.

It should be noted that these national and regional figures mask very wide regional variations, as wealth and development are concentrated in urban areas and northern states.

Gum arabic farmer from the Jawama'a tribe in El Darota, Northern Kordofan

Table 2.   Development context for Sudan

Indicators	Value	Year
Population size (million)	37	2006
Population growth rate (%)	2.6	1998-2003
Life expectancy at birth (years)	56.5	2004
GDP per capita (USD)	640	2005
Prevalence of HIV/AIDS in adult population (age 15-49) (%)	1.6	2003
Contraceptive prevalence (women age 14-49)	7	2004
Population with access to improved water supply (%)	70	2004
Population with access to improved sanitation (%)	34	2004
Population undernourished (%)	26	2000
Percentage of malnourished children under five (%)	27	2003
Infant mortality rate (per 1,000 live births)	62	2004
Children immunized against measles (%)	50	2000
Gross enrolment rate in primary education (%)	59.6	2004
Youth literacy rate (age 15-24) (%)	60.9	2004
Ratio of girls to boys in primary education (%)	88	2000
Under five mortality rate (per 1,000 live births)	90	2005
Birth attended by skilled health staff (%)	57	2004
Maternal mortality rate (per 100,000 live births)	590	2000
Fixed lines and mobile telephone subscribers (per 1,000)	69	2003

Table 3.   Key socio-economic indicators for Southern Sudan

Indicators	Value	Year
Population size (million)	7,514	2003
Refugees or internally displaced persons (million)	4.8	2002
Population growth rate (%)	2.9	2001
Life expectancy at birth (years)	42	2001
GNP per capita (USD)	< 90	2002
Percentage of population earning less than one USD a day (%)	> 90	2000
Prevalence of HIV/AIDS in adult population age 15-49 (%)	2.6	2001
Population without access to drinking water (%)	73	2001
Adult literacy rate (%)	24	2001
Net enrolment ratio in primary education (%)	20	2000
Ratio of girls to boys in primary education (%)	36	2000
Under five mortality rate (per 1,000 live births)	250	2001
Maternal mortality rate (per 100,000 live births)	1,700	2000

Dinka tribe children in the town of Bor, Jonglei state

Ethnicity and religion

Sudan comprises hundreds of ethnic and tribal divisions and language groups, with two major distinct cultures: Arab and Black African. Arab populations generally live in the northern states, which cover most of Sudan's territory and include most of the country's largest urban centres. The Black African culture has its heartland in the south but extends north into Blue Nile state, the Nuba mountains region, and the three Darfur states. In addition, several million internally displaced people, mainly from the south, have relocated to the cities and agricultural regions in the north and centre of the country.

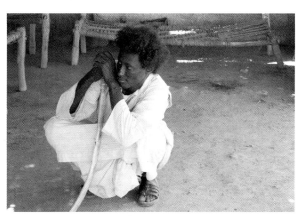

Beja tribesman in Gadamayai, Red Sea state

Most of the estimated 25-30 million Sudanese living in the northern regions are Arabic-speaking Muslims, though traditional, non-Arabic mother tongues are also widely used. Among these are several distinct tribal groups: the Kababish of Northern Kordofan, a camel-raising people; the Ja'alin and Shaigiyya groups of settled tribes along the rivers; the semi-nomadic Baggara of Kordofan and Darfur; the Hamitic Beja in the Red Sea area and the Nubians of the northern Nile areas, some of whom have been resettled on the Atbara river; as well as the Negroid Nuba of Southern Kordofan, the Fur in the western reaches of the country, and the Funj in southern Blue Nile state [2.12].

The southern states, with a population of around 7-10 million, are home to many tribal groups and many more languages than are used in the north. Though some practice indigenous traditional beliefs, southern Sudanese are largely Christian. The Dinka – whose population is estimated at more than one million – is the largest of the many Black African tribes. Along with the Shilluk and the Nuer, they are 'Nilotic' tribes. The Azande are 'Bantus'; the Moro and the Madi, who live in the west, are 'Sudanic', while the Acholi and Otuho, who live in the extreme south, are 'Nilo-hamites'.

History

Sudanese civilization dates back to at least 3000 BC [2.12]. It long concentrated along the northern reaches of the Nile river, the area that came to be known as Nubia. The region's three principal kingdoms were converted to Coptic Christianity by missionaries in the 6th century AD. These Black Christian kingdoms coexisted with their Muslim Arab neighbours in Egypt for centuries, until the influx of Arab immigrants brought about their collapse in the 13th to 15th centuries. Sudan was then partly converted to Islam.

By 1874, Egypt had conquered all of Sudan and encouraged British interference in the region. This aroused Muslim opposition and led to the revolt of the Mahdi, who captured Khartoum in 1885 and established a Muslim theocracy that lasted until 1898, when their forces were defeated by the British in the Battle of Omdurman. The country was then run jointly as the Anglo-Egyptian Sudan, a 'condominium' in which the British were the dominant partner. When Egypt became a British protectorate in 1914, Sudan was taken under British rule until it achieved independence in 1956 [2.12].

The recent history of Sudan has been marked by turmoil, with several periods of conflict and a series of natural disasters leading to massive population displacement. Civil strife began with the Torit mutiny in 1955 and intensified until 1962, by which time the south was effectively at war with the north. This situation lasted until 1972. A fragile peace then prevailed for eleven years, but from 1983, the war was more or less continuous until January 2005, when it was officially ended by the signing of a Comprehensive Peace Agreement (CPA) between the Sudanese Government based in Khartoum and the Sudan People's Liberation Movement (SPLM) and allies in the south.

Low-level conflict, which had been ongoing in Darfur for a generation, developed into a new regional civil war in 2003. The war continues today, despite the signing of the Darfur Peace Agreement in 2006. Low-level conflict also took place in eastern Sudan from the 1990s, though a provisional peace agreement was concluded in October 2006.

A detailed account of historical and current conflicts in Sudan is provided in Chapter 4.

2.3 Governance and economy

Governance structure

In accordance with the provisions of the 2005 peace agreement, Sudan is now ruled by a central government, the Government of National Unity (GONU), headed by the President, Omar Hassan Ahmed El Bashir, and the First Vice-President, Salva Kiir Mayardit. The First Vice-President is also the leader of the SPLM and the President of the new Government of Southern Sudan (GOSS), which has substantial regional autonomy. This structure will stay in effect until 2011, at which time Southern Sudan may choose through a referendum either to remain an autonomous region or to become independent.

Sudan is divided into twenty-five states. Each has its own state government and a measure of executive and legislative authority. The GOSS administers ten states. Two states, Blue Nile and Southern Kordofan, as well as part of a third state (the Abyei region), are geographically part of the north, but have historical, tribal and ethnic links to the south [2.12]. A compromise was reached for these three areas in the peace agreement. The nation's capital Khartoum is subject to a special regime that differs from the rest of the north: as the peace accord states that Khartoum 'shall be a symbol of national unity and reflect the diversity of Sudan', it is administered by an eight-member cabinet composed of four members from the National Congress Party (NCP), two members of the Sudan People's Liberation Movement (SPLM) and two from other northern parties. While Sharia (Islamic law) continues to be the legal system in the north, non-Muslims – mainly Southerners – are exempt from it.

The governance system in Sudan has been severely affected by the four decades of instability the country has undergone. Developing governance and the rule of law is accordingly one of the major challenges set out in the UN and Partners Work Plan for 2007 [2.1].

A detailed discussion of Sudan's governance structures is provided in Chapter 13.

Figure 2.2   Sudan political map

The boundaries and names shown and the designations used on this map do not imply official endorsement or acceptance by the United Nations.

 **Three Areas**

UNEP/DEWA/GRID~Europe 2006

Kilometres

0   100   200   300   400   500

Lambert Azimuthal Equal-Area Projection

Sources:
SIM; vmaplv0, NIMA; various maps; UN Cartographic Section.

Economy

Despite relatively abundant natural resources, Sudan is currently a very poor country due to underdevelopment, conflict and political instability. In 2004, the gross domestic product per person was estimated at USD 740 (using Purchasing Power Parity figures), as compared to USD 3,806 and USD 1,248 for neighbouring Egypt and Kenya respectively.

While the production and export of oil are growing significantly in importance, Sudan's primary resources are agricultural. Sorghum is the country's principal food crop, and livestock, cotton, sesame, peanuts and gum arabic are its major agricultural exports. Sudan, however, remains a net importer of food and a major recipient of food aid.

Industrial development, which consists of agricultural processing and various light industries located in Khartoum North, is limited in Sudan. The country is reputed to have great mineral resources but the real extent of these is unknown.

Extensive petroleum exploration began in the mid-1970s and export began in 1999. Sudan's current production is approximately 500,000 barrels per day, and it is expected that the oil industry will soon rival agriculture in importance.

While Sudan remains poor overall, an 11.8 percent growth of the GDP is forecast for 2007 [2.3] and parts of the country have recently started

A sandstorm in Khartoum in May 2006. Sand and dust storms are common throughout northern and central Sudan

to experience rapid development. The present oil-financed economic and construction boom is focused on Khartoum, Port Sudan and a limited number of mega-projects such as the Merowe dam. Most of the major projects are managed and partly financed by foreign investors and multinational firms, including Middle Eastern and Asian companies.

Sudan's industrial sector, including its oil industry, is discussed in more detail in Chapter 7.

2.4 Climate

Average monthly temperatures in Sudan vary between 26°C and 36°C. The hottest areas, where temperatures regularly exceed 40°C, are found in the northern part of the country.

The dominant characteristic of Sudan's climate is a very wide geographical variation in rainfall [2.15]. In the north, annual precipitation ranges from close to zero near the border with Egypt, to approximately 200 mm around the capital, Khartoum. Sand and dust storms that can cover vast regions and last for days at a time are a defining feature of this low rainfall belt.

Spate irrigation crops in the Tokar delta, Red Sea state. Agriculture is the largest economic sector in Sudan

A camel herder in Northern state. The northernmost third of Sudan has a desert climate

In central Sudan, a division of seasons can be observed:

- winter or dry season (December-February);

- advancing monsoon season (March-May); and

- retreating monsoon season (October-November).

Just south of Khartoum, annual precipitation rarely exceeds 700 mm. In addition, precipitation is relatively erratic, with a combination of short- and long-term droughts, and periods of heavy rainfall.

The extreme south-west is almost equatorial: the dry season is very short and falls in between two peak rainy seasons, and annual precipitation can exceed 1,600 mm.

The issue of climatic variability and its link to environmental problems is covered in more detail in Chapter 3.

Figure 2.3   Sudan average annual temperature

The boundaries and names shown and the designations used on this map do not imply official endorsement or acceptance by the United Nations.

Mean Annual Temperature
(1961-1990)

20 22 24 26 28 30 °C

UNEP/DEWA/GRID~Europe 2006

Kilometres

0   100   200   300   400   500
Lambert Azimuthal Equal-Area Projection

Sources:
IPCC and CRU; SIM (Sudan Interagency Mapping); vmaplv0, NIMA;
UN Cartographic Section.

Figure 2.4 Sudan average annual precipitation

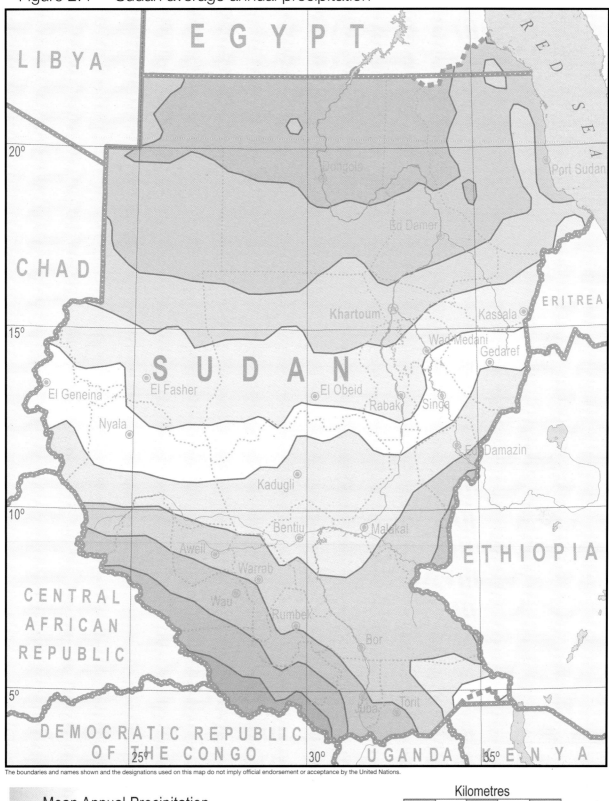

The boundaries and names shown and the designations used on this map do not imply official endorsement or acceptance by the United Nations.

Mean Annual Precipitation
(1961-1990)

0 25 100 200 400 600 800 1000 1200 1400 1600 mm per year

UNEP/DEWA/GRID~Europe 2006

Kilometres

0 100 200 300 400 500
Lambert Azimuthal Equal-Area Projection

Sources:
IPCC and CRU;SIM (Sudan Interagency Mapping);
vmaplv0, NIMA; UN Cartographic Section.

2.5 Geography and vegetation zones

A large and geographically diverse country

With an area of 2.5 million km², Sudan is the largest country in Africa. Its territory crosses over 18 degrees of latitude, which results in an extremely diverse environment ranging from arid desert in the north to tropical forests in the south. Sudan is bordered by ten countries: Egypt, Eritrea, Ethiopia, Kenya, Uganda, the Democratic Republic of Congo, the Central African Republic, Chad and Libya.

The majority of Sudan is very flat, with extensive plains in an altitude range of 300 to 600 m above sea level. Isolated mountain ranges are found across the country, including the Red Sea hills in the far north-east, the Jebel Marra plateau in the west, the Nuba mountains in the centre, and the Imatong mountains in the south-east. The average elevation of these mountains is 1,000 m above sea level, but the highest point is Mount Kinyeti in the Imatong range, which reaches 3,187 m.

The dominant river system in Sudan is the Nile, whose basin extends over 77 percent of the country. The river's two main tributaries, the Blue and White Nile, flow into Sudan from Ethiopia and Uganda respectively, and meet in Khartoum before flowing north into Egypt. In an otherwise arid terrain, the Nile plays a crucial role in the country's various ecosystems. Sudan also has over 750 km of coastline and territorial waters in the Red Sea, which include an archipelago of small islands.

Twenty-nine percent of Sudan's total area is classified as desert, 19 percent as semi-desert, 27 percent as low rainfall savannah, 14 percent as high rainfall savannah, 10 percent as flood region (swamps and areas affected by floods) and less than one percent as true mountain vegetation [2.15]. Note that the precise figures in each class are highly dependent upon the classification system and date; the above are based on recent FAO figures.

Different regions and associated environmental issues

Due to its geographic and climatic diversity, environmental issues affecting Sudan differ radically across the country. To provide context for the issues under discussion in the following chapters, the most ecologically significant regions and geographic features of Sudan are briefly described below. From an environmental perspective, the most important regions and features are:

1. territorial seas;
2. the coastline and islands;
3. northern, central and south-eastern arid regions, including mountain ranges;
4. the central semi-arid region known as the Sahel belt;
5. the Marra plateau;
6. the Nuba mountains;
7. wetlands;
8. the southern clay plains;
9. savannah of various types based on rainfall and soil profile;
10. subtropical lowlands and the plateau in the extreme south of Sudan; and
11. the Imatong, Dongotona, Acholi and Jebel Gumbiri mountain ranges.

It should be noted that many different versions of ecological, soil, vegetation and livelihood zoning for Sudan are in circulation, for a range of purposes [2.15, 2.16, 2.17, 2.18]. The zones listed above and discussed in more detail below are a simplified blend of these classifications, with a focus on major variations between ecosystems.

Sandstorm in Northern Darfur

Figure 2.5   Sudan regional environments

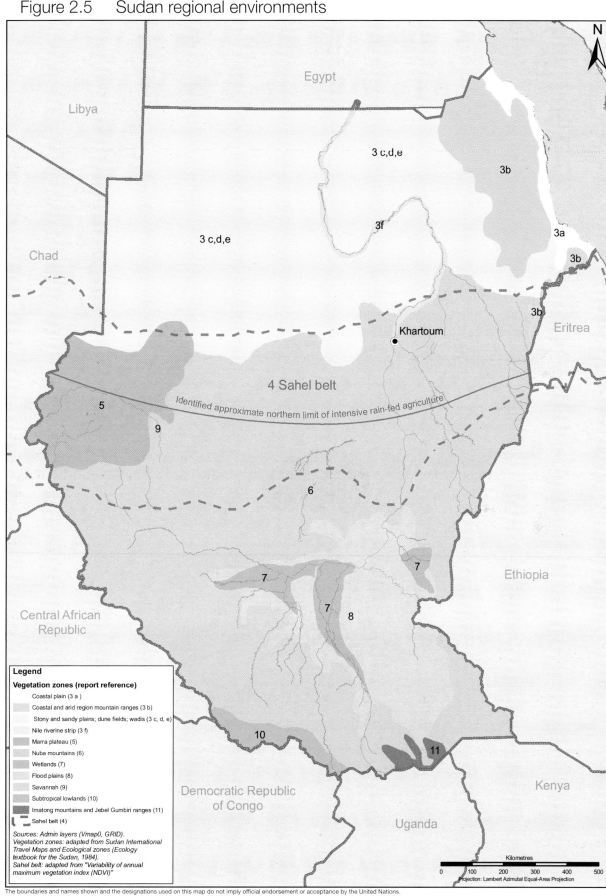

1. Territorial seas

The Sudanese Red Sea is famous for its attractive and mostly pristine habitats, and particularly for its coral reefs. The Red Sea is home to a variety of pelagic fish including tuna, but the overall fish density is relatively low due to limited nutrient input. Sudan's territorial waters host important populations of seabirds and turtles, as well as mammals like dugong, dolphins and whales.

Sudan is a member of the Regional Organization for the Conservation of the Environment of the Red Sea and the Gulf of Aden (PERSGA).

2. Coastline and islands

The coastline of Sudan on the Red Sea is approximately 750 km long, not including all the embayments and inlets. Numerous islands are scattered along the coast, the majority of which have no water or vegetation. The dominant coastal forms are silty beaches, rocky headlands and salt marshes. Fringing coral reefs are very common and water clarity is high due to the lack of sedimentation.

Average precipitation in the coastal areas is extremely low, ranging from 36 mm per year at Halaib to 164 mm per year at Suakin, so

A Manta ray in Sanganeb Marine National Park

that the desert extends all the way to the tide mark. The only exception is the Tokar delta, which receives substantial run-off from seasonal streams originating in the Ethiopian and Eritrean highlands. The islands and most of the coastline are relatively undisturbed and host important feeding and nesting sites for a variety of seabirds.

Barren headland with fringing reef 100 km north of Port Sudan

© RED SEA ENTERPRISES, PORT SUDAN

The coral reefs fringing the Sudanese coastline and islands are generally in excellent condition

A salt marsh 40 km south of Port Sudan. Offshore, seagrass beds support various forms of marine life

3. Northern and south-eastern arid regions

The majority of Sudan can be classified as arid land, with approximately 29 percent classified as true desert (less than 90 mm of rain per year). Four of the northern states are located within the Sahara desert and its margins. A small area in the extreme south-east of the country (the Toposa region) is also semi-arid.

The common features of the northern deserts are extreme temperatures, very low rainfall and, as a result, sparse vegetation. Within this pattern, variations are due to nuances in precipitation, geology, topography, and isolated riverine regions. Important sub-regions within the northern deserts include:

a. The coastal plain. This gently sloping plain, which is some 56 km wide in the south near Tokar and approximately 24 km wide near the Egyptian border, is intersected by spurs of the adjacent mountain ranges and *wadis* (intermittently flowing rivers). A notable feature is the Tokar delta, which has sufficient groundwater and seasonal flooding to support intensive agriculture.

The coastal plain 10 km south of Suakin

b. Coastal and arid region mountain ranges. The coastal mountain range runs virtually uninterrupted along the entire coastline, with peak elevations generally in the order of 1,100 m. Mountains also extend along the Eritrean and Ethiopian borders, where they form the western edge of the Ethiopian plateau. The coastal and other hyper-arid regional mountain environments are characterized by very thin or absent soil cover and negligible vegetation, except in alluvial valleys and isolated oases.

Coastal Jebel

Figure 2.6   Coastal plain and mountain range

The boundaries and names shown and the designations used on this map do not imply official endorsement or acceptance by the United Nations.

c. Stony and sandy plains. The majority of deserts in Sudan are stony and sandy plains, which represent areas of wind erosion. In the most extreme cases, soil cover is completely absent over large areas.

d. Dune fields. Sand dunes occur across most of the Sahara and Sudanese deserts, although their types and density vary significantly from region to region. The largest dune fields are found in the north-west, in Northern state. Dunes can be mobile or immobile/fixed; the former present major threats to agricultural land in arid regions.

Stony plain 60 km north of Port Sudan

Mobile dune in Northern state

e. *Wadis*. *Wadis* or *khors* (generally dry seasonal watercourses) are ecological hotspots within desert and semi-desert environments. Drainage and infiltration from seasonal rainfall events concentrate beneath the dry stream beds, and support trees and short-lived grasses, in addition to higher densities of the more drought-resistant shrub species.

f. The Nile riverine strip. The waters of the Nile have sustained civilizations in the arid regions of Egypt and Sudan since the development of agriculture over 10,000 years ago. The annual wet season flow surge results in regular flooding and sediment deposition on a narrow strip along nearly the entire length of the Nile, in an otherwise very arid environment. The width of the cultivated and heavily developed strip has been expanded by irrigation schemes, but outside of these areas, it is generally no more than two kilometres wide.

With the exception of the Nile riverine strip and the coastal plain, the desert regions of Sudan are relatively undeveloped, as the land can only support low-intensity pastoralism and isolated oasis communities.

4. The central semi-arid region: the Sahel belt

The Sahel, which extends from Senegal eastward to Sudan, forms a narrow transitional band between the arid Sahara to the north and the humid savannah to the south. With eight to eleven dry months per year, it has an approximate annual precipitation of 300-600 mm. As the bulk of agriculture in Sudan is practised within and to the south of the Sahel belt, most of the original landscape has been altered: the majority of central Sudan, where rain-fed and irrigated agriculture predominate, is now covered by flat and open fields with limited tree cover.

In its natural state, the Sahel belt is characterized by baobab and acacia trees, and sparse grass cover. Since the late 20th century, it has been subjected to desertification and soil erosion caused by natural climate change, as well as overgrazing and farming. The countries of the Sahel zone also suffered devastating droughts and famine in the early 1970s, and again in the 1980s. Apart from long-term droughts, the Sahel is prone to highly variable rainfall, with associated problems for livestock- and crop-rearing.

Nile riverine agriculture, Northern state. A narrow strip of irrigated land on either side of the main Nile in the desert regions supports up to three crops a year

Figure 2.7   Nile riverine strip

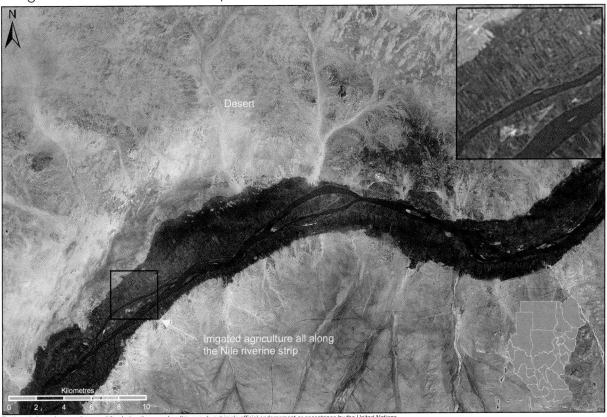

Desert

Irrigated agriculture all along
the Nile riverine strip

Kilometres

0   2   4   6   8   10

The boundaries and names shown and the designations used on this map do not imply official endorsement or acceptance by the United Nations.

Figure 2.8   Sahel belt and Gezira irrigation scheme

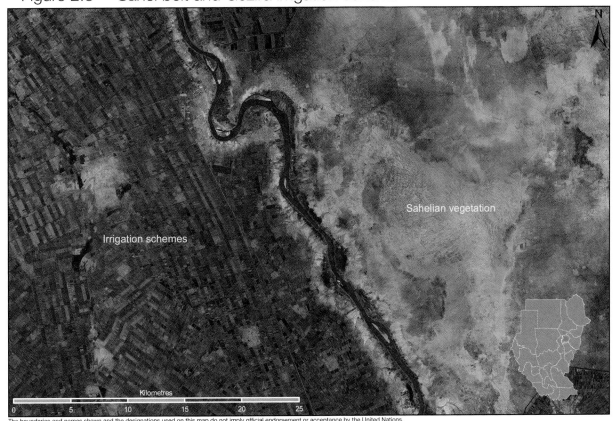

Sahelian vegetation

Irrigation schemes

Kilometres

0   5   10   15   20   25

The boundaries and names shown and the designations used on this map do not imply official endorsement or acceptance by the United Nations.

A baobab tree in the Sahel during the dry season, Northern Kordofan

Accurately mapping and defining the Sahel in Sudan is problematic due to the limited records available and the changing climate. Accordingly, UNEP has used three different indicators for the Sahel belt and the associated limits of rain-fed agriculture:

- historical rainfall records converted to annual average contours for 300-600 mm;

- the approximate northern limit of intensive rain-fed agriculture as indicated by UNEP

Figure 2.9   Jebel Marra and Sahel belt

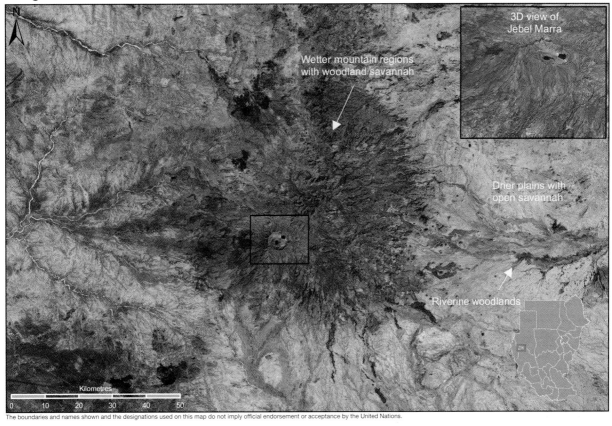

The boundaries and names shown and the designations used on this map do not imply official endorsement or acceptance by the United Nations.

Bushland and wadi on the southern limit of the Sahel, Southern Darfur

analysis of Landsat images dating from 2000 to 2005 (note that scattered rain-fed agriculture and pastoralism occur well north of this line); and

• a measure of annual rainfall and associated vegetation variability recorded by satellite images (analysis by the Vulnerability Analysis and Mapping Unit, WFP-Khartoum), using an annual change rate of 15 percent or more for the period 1982-2003 [2.11].

5. The Marra plateau

The Marra plateau is a rugged volcanic range that occupies approximately 80,000 km² in central Darfur, with an average altitude of 1,500 m and a maximum elevation of 3,088 m at Jebel Marra. The higher and more southerly parts of the plateau have a wetter microclimate (over 600 mm of rain per year) than the surrounding area, which is relatively arid with erratic rainfall. The plateau originally had extensive woodlands, which have been partly removed for agricultural development.

6. The Nuba mountains

The Nuba mountains are a set of widely spaced small mountains located in the state of Southern Kordofan. Their average altitude is 900 m with a maximum elevation of 1,326 m at Jebel Heiban. They are relatively steep-sided, with extensive hinterlands and a wetter microclimate that results in higher-density forest coverage than the surrounding savannah.

7. Wetlands

Permanent wetlands make up approximately five percent of the area of Southern Sudan, while a much greater area, both north and south, is seasonally flooded. The largest wetlands and flood plains are all linked to the Nile tributaries that traverse the southern plains. The largest wetland is the Sudd, which is formed by the White Nile in very flat topography between the towns of Bor and Malakal. Covering more than 30,000 km², the Sudd comprises multiple channels, lakes and swamps, with a maze of thick emergent aquatic vegetation.

Fringing swamps on the White Nile, Jonglei state

In the south, the wetlands are essentially undeveloped and represent a safe haven for wildlife, including migratory birds.

Villages perched on steep hillsides in the Nuba mountains, Southern Kordofan

Figure 2.10   Sudd wetland and flood plains

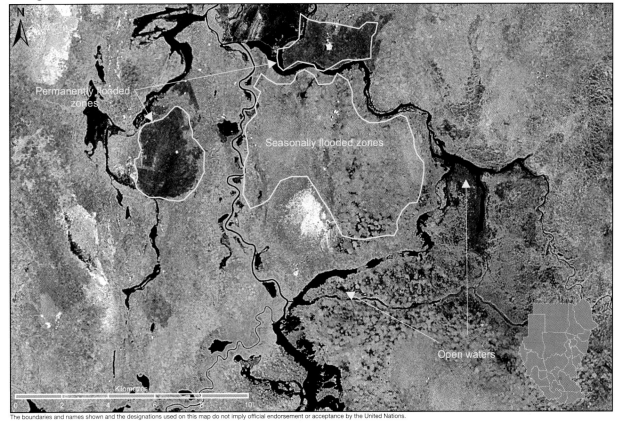

The boundaries and names shown and the designations used on this map do not imply official endorsement or acceptance by the United Nations.

Mongalla gazelles grazing in the tall grass of the clay plains in Padak district, Jonglei state

8. Flood plains

Much of central and south Sudan is covered by sediment deposited in the Nile basin and known locally as 'black cotton' soil. Due to its high clay content, the soil in these areas retains water in the wet season to form very soft and virtually impassable shallow flood plains. In the dry season, the water disappears from all but a few swamps, waterholes and tributaries, and the clay shrinks

and cracks. These areas are relatively fertile but difficult to cultivate.

The geographic border between flood plains and the drier Sahel belt is somewhat arbitrary in the clay soil regions, as even the dry areas flood easily during high rainfall events. The boundary between flood plains and wetlands is also often arbitrary, as many parts of Southern Sudan consist of a network of seasonally variable wetlands interlacing multiple small flood plains.

White-backed vultures resting on the new grass of the seasonally flooded 'toic' in Padak district

9. Savannah

Large areas of central and south Sudan are considered to be savannah, classified as low-density woodland, mixed scrub and grassland. Within this broad class, the density and proportions of the three vegetation types vary significantly according to regional climates, soil types, topography and the influence of deliberate seasonal burning, which tends to favour the development of grasslands.

10. Subtropical lowlands

The extreme south and south-west of Sudan can be classified as subtropical. This is reflected in the vegetation, which changes relatively abruptly from savannah to semi-tropical forest in the region south and south-west of Juba.

The land bordering the Democratic Republic of Congo in the south-west rises to form a continuous low range known as the Ironstone hills. These hills also form the boundary between the Nile and Congo watersheds. The region supports intensive agriculture and some forestry, but is otherwise undeveloped.

High rainfall woodland savannah in Bor district, Jonglei state

11. The Imatong, Dongotona, Acholi and Jebel Gumbiri mountain ranges

The Imatong, Dongotona and Acholi mountain ranges flank the White Nile in the extreme south of Southern Sudan. Their average altitude is 900 m, with a peak elevation of 3,187 m at Mount Kinyeti, which is the highest point in Sudan. They are characterized by steep slopes and high rainfall, resulting in dense forest and high-yield agriculture. The Jebel Gumbiri mountains, further west, support extensive teak plantations.

High rainfall woodland savannah with a small seasonal wetland in Wau district, Western Bahr el Ghazal

Figure 2.11   Imatong, Dongotona and Acholi mountain ranges

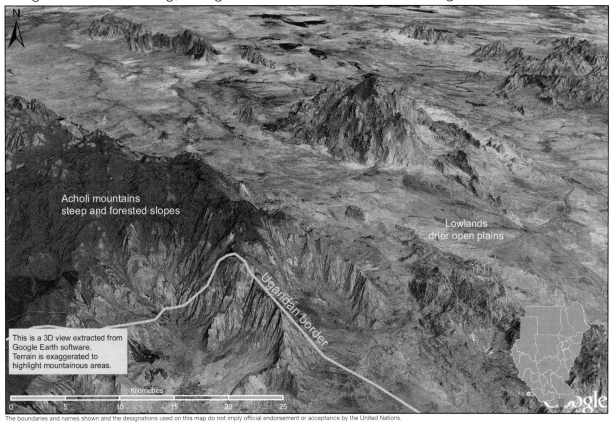

The boundaries and names shown and the designations used on this map do not imply official endorsement or acceptance by the United Nations.

Subtropical vegetation and red ironstone soil in Yei district, Central Equatoria

Natural Disasters and Desertification

A Beja nomad village in Kassala state. Climate change and desertification threaten the livelihoods of millions of Sudanese living on the edge of the dry Sahel belt.

Natural disasters and desertification

3.1 Introduction and assessment activities

Introduction

Natural disasters in the contrasting forms of drought and flooding have historically occurred frequently in Sudan, and have contributed significantly to population displacement and the underdevelopment of the country. A silent and even greater disaster is the ongoing process of desertification, driven by climate change, drought, and the impact of human activities.

In Sudan, desertification is clearly linked to conflict, as there are strong indications that the hardship caused to pastoralist societies by desertification is one of the underlying causes of the current war in Darfur.

Given the severity of the impact of such events and processes, there is a clear and urgent need for improved climate analysis, disaster prediction and risk reduction for Sudan in general, and for Darfur

in particular. The current and forecast impact of desertification, especially, is poorly understood, and major efforts are required to investigate, anticipate and correct this phenomenon.

This chapter discusses the key linkages between natural disasters, desertification and the environment, as well as options for mitigating both the risk of disasters occurring and their impact when they do occur.

Assessment activities

UNEP's work on climate change and natural disasters in Sudan was part of the larger investigation of the agricultural, forestry and water resource sectors; fieldwork details are accordingly provided in Chapters 8, 9 and 10 respectively.

Though relatively little background literature can be found on flooding in Sudan, a significant body of documentation is available on drought. In addition, a detailed and authoritative project on climate in Sudan was completed in 2003 with the assistance of the UN Framework Convention on Climate Change (UNFCCC) [3.1]. The final reports from this project provide much of the technical basis for the country-specific climate change work presented in this chapter.

Rainfall in the Sahel commonly falls in short torrential bursts, resulting in extensive but short-lived flooding

Even though 2006 was a relatively 'good' year, this small dam in Western Darfur dried up completely. Rain only falls during four months of the year, so surface reserves do not last through the dry season

3.2   Water shortages

Sudan suffers from a chronic shortage of freshwater overall. In addition, water distribution is extremely unequal, with major regional, seasonal and annual variations. Underlying this variability is a creeping trend towards generally drier conditions.

Annual climate variability and drought

Insufficient and highly variable annual precipitation is a defining feature of the climate of most of Sudan. A variability analysis of rainfall records from 1961 to 1990 in Northern and Southern Kordofan found that annual precipitation ranged from 350 to 850 mm, with an average annual variation of 65 percent in the northern parts of Northern Kordofan and 15 percent in the southern parts of Southern Kordofan [3.1].

Annual variability and relative scarcity of rainfall – in the north of Sudan in particular – have a dominant effect on agriculture and food security, and are strongly linked to displacement and related conflicts. Drought events also change the environment, as dry spells kill otherwise long-lived trees and result in a general reduction of the vegetation cover, leaving land more vulnerable to overgrazing and erosion.

Together with other countries in the Sahel belt, Sudan has suffered a number of long and devastating droughts in the past decades. All regions have been affected, but the worst impacts have been felt in the central and northern states, particularly in Northern Kordofan, Northern state, Northern and Western Darfur, and Red Sea and White Nile states. The most severe drought occurred in 1980-1984, and was accompanied by widespread displacement and localized famine. Localized and less severe droughts (affecting between one and five states) were also recorded in 1967-1973, 1987, 1989, 1990, 1991, 1993 and 2000 [3.1].

Isolated drought years generally have little permanent effect on the environment. In the case of central Sudan, however, the eighteen recorded years of drought within the last half-century are certain to have had a major influence on the vegetation profile and soil conditions seen in 2006.

Recent research has indicated that the most likely cause of these historical droughts was a medium-term (years) change in ocean temperature, rather than local factors such as overgrazing [3.2]. Therefore, the potential for such droughts to occur again remains.

Long-term regional rainfall reduction

In addition to and separately from the variation in precipitation noted above, there is mounting evidence of long-term regional climate change in several parts of the country. This is witnessed by a very irregular but marked decline in rainfall, for which the clearest indications are again found in Kordofan and Darfur states.

Table 4 below summarizes the long-term trends noted, as indicated by thirty-year moving averages of annual precipitation for three locations in Darfur.

Precipitation records have been kept in Darfur since 1917. However, there are still only three continuously monitored stations for an area of over 0.8 million km². The data below shows an overall trend of declining rainfall, with the most marked decrease on the northern edge of the Sahel in Northern Darfur. Since records began, the ten-year moving average for El Fasher has declined from 300 mm per annum to approximately 200 mm, while the last time rainfall exceeded 400 mm was in 1953 [3.3].

The scale of historical climate change as recorded in Northern Darfur is almost unprecedented: the reduction in rainfall has turned millions of hectares of already marginal semi-desert grazing land into desert. The impact of climate change is considered to be directly related to the conflict in the region, as desertification has added significantly to the stress on the livelihoods of pastoralist societies, forcing them to move south to find pasture.

A more detailed discussion of linkages between climate change and conflict in Darfur is provided in Chapter 4.

The foundations of an abandoned village on the steep hills of the northern limits of the Jebel Marra plateau, Northern Darfur. Evidence of abandonment of rural land can be found all along the northern edge of the Sahel

Climate change model predictions provide grim warnings for dryland Sudan

The Sudan climate change study conducted in 2003 provides a solid technical basis for discussion. Moreover, a range of very recent regional studies, as well as a number of additional assessments of the potential impacts of climate change, indicate good agreement with earlier work. Following is a concise summary of this work, to set the context for the findings of UNEP's assessment.

Table 4.   Long-term rainfall reduction in Darfur

Rain gauge location	Average annual rainfall (mm) 1946 - 1975	Average annual rainfall (mm) 1976 - 2005	Reduction (-)	Percentage
El Fasher, Northern Darfur	272.36	178.90	- 93.46	- 34 %
Nyala, Southern Darfur	448.71	376.50	- 72.21	- 16 %
El Geneina, Western Darfur	564.20	427.70	- 136.50	- 24 %

There is generally no clear edge to the desert, but in this case in Northern Darfur, the boundary between the overgrazed sandy rangeland and the threatened rain-fed agricultural zone is quite marked

The 2003 study selected Northern and Southern Kordofan for detailed analysis; all the results presented thus relate to those areas only. A 'baseline climate' was determined using rainfall and temperature data from 1961 to 1990. A range of global warming scenarios were then modelled to predict changes in temperature and rainfall from the baseline to the years 2030 and 2060.

The climate model results indicated a 0.5 to 1.5°C rise in the average annual temperature and an approximate five percent drop in rainfall, though results varied across the study area. These findings were then used to project the scale of potential changes in crop yields for sorghum, millet and gum arabic.

The final results are alarming: the crop models show a major and potentially disastrous decline in crop production for Northern Kordofan and lesser but significant drops further south. For example, the modelled sorghum production in the region of El Obeid is predicted to drop by 70 percent, from 495 kg/hectare to 150 kg/hectare.

These dramatic findings are due principally to the fact that the region is situated on the fringes of the Sahara desert and on the northern limit of viability for rain-fed crop production, where even small increases in temperature and minor reductions in precipitation could tip the balance towards desert-like conditions.

Other climate models covering all of Africa generally predict similar problems, although there are some major differences in predicted annual rainfall [3.4, 3.5]. One model, which focused on changes in the growing season, predicted that in the Sahel belt, growing seasons would reduce and the percentage of failed harvests would increase [3.6]. The scale of the change varies from region to region, but in Darfur it is predicted to be in the order of 5 to 20 percent from 2000 levels by 2020.

Summary: history and modelling combine for a downward forecast

Historical data, anecdotal field reports and modelling all point to the same general trend. Overall, rainfall is becoming increasingly scarce and/or unreliable in Sudan's Sahel belt, and this trend is likely to continue. On this basis alone, large tracts of the Sahel will be severely impacted by declining food productivity over the next generation and beyond.

Settlements like Malka in Northern Darfur are already on the margins of survival; a small reduction in rainfall could suffice to render large parts of the semi-arid desert fringe unviable. Land degradation is clearly visible as large swathes of bare red subsoil

3.3 Desertification: Sudan's greatest environmental problem

Desertification, as defined in the UN Convention to Combat Desertification, is the degradation of land in arid, semi-arid and dry sub-humid areas caused by climatic change and human activities.

In northern Sudan, there is high awareness of the issue of desertification within the academic community, and historical evidence of a number of attempts to quantify and/or limit the extent of the problem since at least the 1950s [3.7]. As early as 1953, a landmark study discussed several of the sources of the problem (such as overgrazing), as well as its implications (long-term damage and reductions in productivity) [3.8].

UNEP considers that three compounding desertification processes are underway in Sudan, which are relatively difficult to distinguish, separate and quantify on the ground:

1. Climate-based conversion of land types from semi-desert to desert. The scale and duration of the reduction in rainfall noted above is sufficient to have changed the natural environment, irrespective of human influence. This type of change occurs as a regional process, where less drought-resistant vegetation gradually dies off or fails to reproduce, resulting in a lower-density mix of different species. In a shift as rapid as that observed in Northern Darfur and Northern Kordofan, this is manifest first and foremost in the widespread death of trees during drought events, which are not followed by recovery. This has been the case for *Acacia senegal*, the tree that produces gum arabic (see Case Study 8.2), for example. The limited figures available indicate a southward shift in desert climate of approximately 100 km over 40 years [3.7].

2. Degradation of existing desert environments, including *wadi*s and oases. At least 29 percent of Sudan is already true desert. Within this large area, however, are hundreds of smaller wetter regions

Fuelwood vendors in Red Sea state. Deforestation is a major cause of land degradation in desert environments. Tree cover is concentrated in seasonal wadis, where it helps retain soil that would otherwise be swept away by wind and flash floods

Figure 3.1   Desertification in Bara district, Northern Kordofan

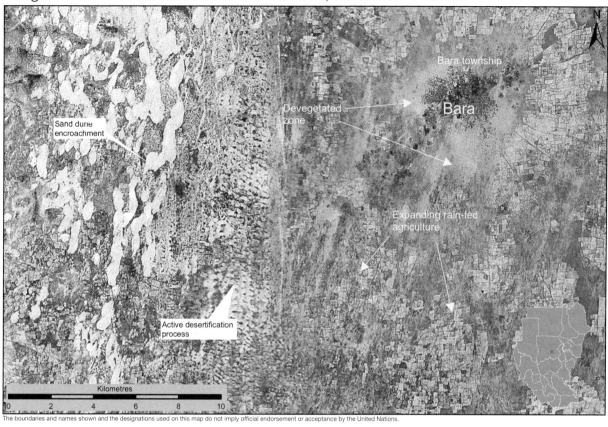

The boundaries and names shown and the designations used on this map do not imply official endorsement or acceptance by the United Nations.

These date palms are submerged by shifting sands. Farmers have attempted to hold back the sands by building walls around the trees, but these will eventually be submerged as well

resulting from localized rainfall catchments, rivers and groundwater flows. Virtually all such areas inspected by UNEP were found to be moderately to severely degraded, principally due to deforestation, overgrazing and erosion.

3. Conversion of land types from semi-desert to desert by human action. Over-exploitation of semi-desert environments through deforestation, overgrazing and cultivation results in habitat conversion to desert, even though rainfall may still be sufficient to support semi-desert vegetation. In Sudan, a particular problem has been the conversion of dry and fragile rangelands into traditional and mechanized cropland. A detailed analysis of these processes is provided in Chapter 8.

Regional differences in soil types and topography also play a part in this complex three-pronged process. The soil in the north and west of Sudan, for instance, is sandy and prone to water and wind erosion, while the south and east have more resistant clay soil. In addition, mountain ranges such as the Jebel Marra plateau form high rainfall watersheds in otherwise arid areas.

To summarize, there is sufficient disseminated evidence to support the following findings:

- Moderate to severe land degradation is ongoing in the desert and semi-arid regions that cover the northern half of Sudan;

- A 50 to 200 km southward shift of the boundary between desert and semi-desert has occurred since rainfall and vegetation records began in the 1930s. This shift, however, has not been well quantified and is based largely on anecdotal evidence and small-scale studies;

- The desert and semi-desert boundaries are expected to continue to shift southwards due to declining precipitation/reliability of precipitation;

- Most of the remaining semi-arid and low rainfall savannah on sand, representing approximately 25 percent of Sudan's agricultural land, is at considerable risk of further desertification, to the extent that food production in these regions will at minimum plateau, and more

likely continue to drop significantly (i.e. up to 20 percent or more); and

- Modelled predictions of a future 70 percent drop in food production in Northern Kordofan have actually already taken place on a smaller scale and on a short-term and local basis, due to reduced rainfall and ongoing land degradation and abandonment. This trend is expected to worsen with time and the predicted result is that in the absence of major changes in agricultural patterns, food insecurity will only increase in these regions.

The area at greatest risk is the Sahel belt, as shown in Figure 2.5. It includes the conflict-affected parts of Darfur, the previously drought-stricken parts of Northern Kordofan and Khartoum states, and conflict- and drought-stricken Kassala state.

Much of the evidence for the above findings is piecemeal, anecdotal and/or based on site-specific data. The limited numerical data available does validate the anecdotal findings, but further solid and comprehensive analysis is clearly needed.

A thin tree belt prevents a dune from overwhelming irrigated fields in Northern state

The fields' survival is threatened by uncontrolled cutting in the nearby protective tree belt

This abandoned field within a collapsed irrigation scheme in Khartoum state previously supported low density rangeland. It is now barren and its remaining topsoil is being blown away

3.4   Water damage

Flooding

Despite serious water shortages, floods are common in Sudan. The two predominant types of floods are localized floods caused by exceptionally heavy rains and run-off (flash floods), and widespread floods caused by overflow of the Nile and its tributaries.

Severe flash floods were recorded in 1962-1965, 1978-1979, 1988, 1994, 1998, 1999 [3.1] and 2006. This last flood was directly observed by UNEP in the field. Though generally short in duration, these events can cause major damage to villages and urban and agricultural areas located in catchment and drainage zones.

Nile floods usually originate from heavy rainfall in the (now largely deforested) catchment areas of the Ethiopian mountains, which causes unpredictable surges in the flow of the Blue Nile. The sequence of severe Nile floods – which were recorded in 1878, 1946, 1988, 1994, 1998 and 2006 – clearly shows that the frequency of flooding has increased dramatically over the last twenty years.

Riverbank erosion

Riverbank erosion is a natural phenomenon in Sudan that can, in extreme cases, be characterized as a local disaster due to its social and environmental

On the main Nile in Northern state. One of the causes of riverbank erosion is the increased frequency of sand dune migration into the Nile, as the rapid influx of sand alters the riverflow, resulting in downstream erosion as well as sediment deposition

impacts. This problem is most acute on the main Nile downstream from Khartoum, where peak wet season flows and river channel changes result in very rapid removal of land from riverside terraces.

The destruction witnessed by UNEP field teams is impressive. For example, an estimated 17 percent of Ganati (1,420 ha), 25 percent of El Zouma (200 ha) and 30 percent of El Ghaba (1,215 ha) cooperative societies in Northern state have been swept away in flood peaks [3.9]. Moreover, bank erosion leads to sedimentation problems elsewhere.

Flash flooding 20 km north of Khartoum, September 2006

The submerged Sunut Forest wetland in the metropolitan Khartoum area, August 2006. The flooding of the Nile is an annual natural event

Farmers in Northern state watch as the date palms on which their livelihoods depend are washed away by riverbank erosion

3.5   Disaster risk reduction and the mitigation of desertification

The potential to predict and limit impacts

The past thirty years have seen major developments in the field of disaster prediction and risk reduction. It is now generally recognized that while the natural phenomena causing disasters are in most cases beyond human control, the vulnerability (of affected communities) is generally a result of human activity. This is particularly clear in Sudan.

Drought. The vulnerability to drought is partly related to social and development factors such as the tendency to maximize herd sizes rather than herd quality, and the lack of secure water resources such as deep boreholes which can be relied upon during short-term droughts.

Completely degraded rangeland in Northern Darfur. This area immediately outside a large IDP camp has been subject to a combination of long-term overgrazing and fodder gathering, with topsoil largely removed and virtually no remnant vegetation or seed stock

As a result of overgrazing, the thin topsoil of this rangeland near El Geneina in Western Darfur is being eroded by wind and water

Desertification. While climate-related desertification cannot be easily addressed, desertification due to human activity can be limited through appropriate land use planning and regulation, to avoid over-exploitation of fragile semi-desert regions.

Flooding. The increase in Blue Nile flooding is considered to result partly from deforestation and overgrazing in the Ethiopian highlands. Besides, the impact of floods in Khartoum state is generally highest in the slums and IDP camps located in low-lying areas previously left unoccupied as they are known by locals to be flood-prone.

Riverbank erosion. While adjustments in river morphology are a natural phenomenon, human action in altering stream discharge and sediment loads has played a significant role in accelerating the process. The main impacts include watershed degradation from deforestation, overgrazing and poor farming practices that increase stream turbidity, and the effects of dams on the Blue Nile and Atbara rivers. The removal of riverbank vegetation through fires or grazing further aggravates the problem, as it weakens the banks' ability to withstand the erosive power of flood peaks. In this context, UNEP anticipates that pulsed water released from the new Merowe dam will become a major cause of downstream riverbank erosion on the main Nile (see Case Study 10.1).

Action required in addition to more studies and plans

Reducing the vulnerability of communities to natural disasters is the core principle of disaster risk reduction. Environmental protection is one component of an integrated response to the issue. For Sudan, this translates into the need for practical risk-reduction measures, such as better rangeland management to create a buffer capacity to deal with periodic droughts, or catchment protection to mitigate flood risk.

There are already numerous policies, strategy papers and small-scale projects aimed at tackling drought and desertification in Sudan [3.7], and similar work is commencing on flood risk reduction. These positive early steps should be supported with substantial follow-up actions.

3.6 Conclusions and recommendations

Conclusion

Conflict, displacement and food insecurity are three of the most pressing issues facing Sudan, and the main reasons for the current international humanitarian aid effort. Natural and partly man-made disasters such as drought, desertification and floods are major contributing causes to these problems.

For the Government of Sudan, tackling these issues will require a major investment in improving natural resources management, as well as the elaboration of new policies for the sustainable use of natural resources. Investment by the international community is also warranted as part of the shift from humanitarian relief to sustainable development assistance.

The role of vegetation in controlling desertification is exemplified in this photograph of degraded rangeland in Khartoum state. The clump of grass has been grazed but its roots still retain the underlying soil, while surrounding soil has been removed by wind erosion

Riverbank erosion removed the supports of this irrigation pump intake system within months of its installation, and threatens to destroy it completely. Without mitigatory measures, the site is not suitable for such a project

Background to the recommendations

Rather than establish major investment programmes focused solely on natural disasters and desertification, it is recommended that these issues be integrated into development and food security programmes at the national level. Accordingly, many recommendations relevant to this topic are spread throughout specific sector chapters, including agriculture, forestry, water resources and environmental governance (Chapters 8, 9, 10 and 13 respectively). In this chapter, recommendations are limited to data collection, analysis and coordination.

Because the areas of disaster risk reduction, desertification and adaptation to climate change in Sudan could benefit greatly from better data, robust analysis and improved data accessibility, investing in science is a main theme for these recommendations. A second theme is awareness-raising, as alarming findings such as those expressed in climate change work to date should be validated and widely communicated to promote a national response to these challenges.

Finally, international assistance should play a strong role in the fields of climate change adaptation and disaster risk reduction, as these are global issues for which extensive expertise and financial resources are available to help countries like Sudan.

Recommendations for the Government of National Unity

R3.1   Invest in national weather and drought forecasting services, including in measures to increase data collection and existing data accessibility, and provide improved early warning of drought episodes. This work should tie into existing international early warning and forecasting programmes, such as the US-based Famine Early Warning System.

CA: GI; PB: GONU MAF; UNP: UNEP; CE: 3M; DU: 5 years, ongoing

R3.2   Undertake a major study to truly quantify desertification in Sudan. This should include a combination of fieldwork and remote sensing on both local and national scales.

CA: GI; PB: GONU MAF; UNP: UNEP; CE: 0.5M; DU: 2 years

R3.3   Validate and disseminate climate change findings together with desertification findings. The results of the two studies should be used as the benchmark for land use planning in the dryland states of Sudan.

CA: AS; PB: GONU MAF; UNP: UNEP; CE: 0.5M; DU: 2 years

Conflict and the Environment

The African Union Mission in Sudan (AMIS) military escort for UNEP fieldwork near El Geneina, Western Darfur. Intense competition over declining natural resources is one of the underlying causes of the ongoing conflict.

Conflict and the Environment

4.1 Introduction and assessment activities

Introduction

Sudan has been wracked by civil war and regional strife for most of the past fifty years, and at the time of finalizing this report, in June 2007, a major conflict rages on in Darfur. At the same time, Sudan suffers from a number of severe environmental problems, both within and outside current and historical conflict-affected areas. UNEP's assessment has found that the connections between conflict and environment in Sudan are both complex and pervasive: while many of the conflicts have been initiated partly by tension over the use of shared natural resources, those same resources have often been damaged by conflict.

This chapter is divided into three main sections:

1. **a conflict overview**, presenting a summary of the history of recent conflicts in Sudan;

2. **an overview of the role of natural resources** in the instigation and continuation of historical and current conflicts, listing the major resources of concern and focusing specifically on conflicts involving rangelands and rain-fed agricultural land; and

3. **a brief environmental impact assessment of the various conflicts**, evaluating the direct and indirect impacts of conflict on Sudan's environment.

Chronic environmental problems are covered in other chapters, though it should be noted that at the local level, the boundary between chronic and conflict-related environmental issues is often unclear.

Assessment activities

The assessment of conflict-related issues was an integral part of fieldwork throughout the country. In addition, UNEP carried out a number of specific activities, including:

- walkover inspections of destroyed military equipment in Juba, Bor and Padak, in Southern Sudan;

Visible remnants of war: abandoned armoured vehicles in Juba, Jonglei state, Southern Sudan

- viewing of unexploded ordnance (UXO) and mined areas (where walkovers were not possible) in Juba, Yei, Malakal and the Nuba mountains;

- walkover inspections of burnt and destroyed villages and forests east of El Geneina in Western Darfur, and low flyovers in other conflict-affected parts of Darfur;

- viewing of weaponry held by various armed parties throughout Sudan;

- interviews with de-mining and military experts within Sudan; and

- interviews with conflict-affected communities in Darfur, Southern Kordofan and Southern Sudan.

These activities were considered sufficient to obtain an overview of the direct impacts of conflict and related issues for most of Sudan, though UNEP was not able to carry out sufficient fieldwork in Darfur to allow for a full analysis. Moreover, UNEP chose not to investigate in detail the social and political aspects of conflicts in Sudan, focusing instead on their environmental dimension.

4.2   Overview of conflicts in Sudan

A complex mosaic

Conflicts have directly affected over 60 percent of the country for the last 50 years, and hence greatly influenced its development [4.1, 4.2]. Understanding Sudan's complex mosaic of conflicts is an essential first step in establishing the linkages between conflict and environment in the region. This section accordingly provides a brief summary of the chronology and geography of the various confrontations, together with a short account of the tactics and weaponry used. A thorough review of social and political factors might be taken into consideration in a comprehensive conflict analysis, but is outside the scope of this environmental assessment.

Tribal and small-scale conflicts

Tribal and small-scale conflicts fought only with small arms have occurred continuously throughout the history of Sudan [4.3]. No part of the country has been exempt from such clashes, but they have been concentrated in the south, west and east of the country for the last thirty years. Their causes are generally poorly recorded, but include disputes over cattle theft, access to water and grazing, and local politics [4.3]. Many – though not all – of the large-scale conflicts in Sudan have a connection to tribal friction.

The major conflicts

The majority of large-scale conflicts in Sudan have been long-term (five years or more) confrontations between forces aligned with the central Sudanese government based in Khartoum and an array of anti-government forces. The government side has comprised conventional army and air forces, and allied local militias. The opposition has consisted of local militias which – in the case of the Sudan People's Liberation Army (SPLA) in Southern Sudan – evolved into a united resistance army with a parallel governance and administration structure (the Sudan People's Liberation Movement or SPLM).

Major conflicts have at times extended over as much as 60 percent of the territory of Sudan, principally in the ten southern states, but also in the west (all three Darfur states), the centre (Blue Nile and Southern Kordofan states), the east (Kassala state) and the north-east (Red Sea state). In total, over 15 million people have been directly affected, not including the approximately six million people currently still impacted in Darfur. Total conflict-related casualties are unknown, but estimated by a range of sources to be in the range of two to three million [4.4].

Although the government forces' weaponry has included tanks and heavy artillery, most military confrontations have been fought mainly with light weapons such as AK47 assault rifles. The opposition forces' armament has been generally light, with a small number of tanks and other heavy weapons. Only government forces have had airpower.

Landmines have been used widely in most major conflicts. Minefields have been abandoned without marking or extraction and are mostly unmapped. As a result, Sudan now suffers from a severe landmine legacy which continues to cause civilian casualties. It should be noted that there are no reports of extensive use of landmines in the ongoing war in Darfur.

Figure 4.1   Conflicts in Sudan: 1957–2006

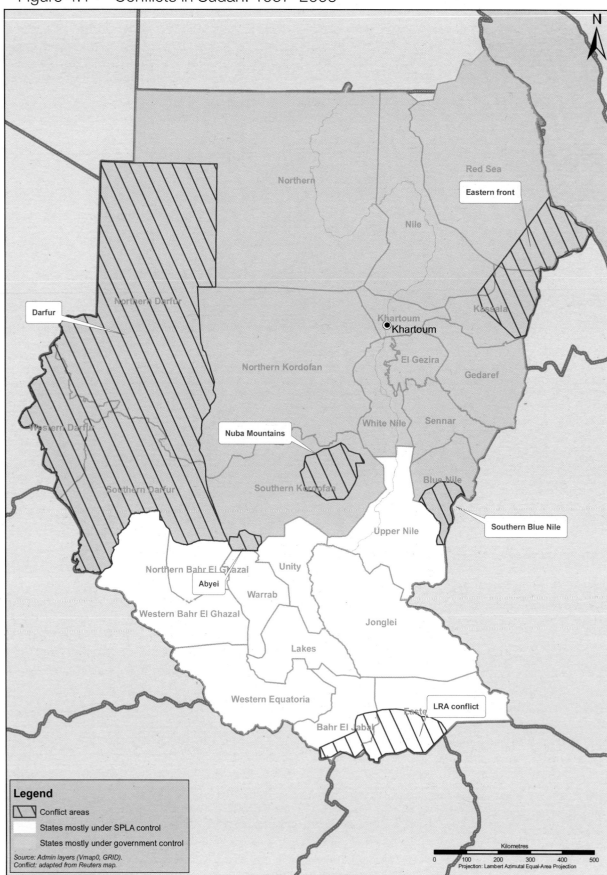

A destroyed village and badly eroded land seen from the air in Northern Darfur

There is no firm field or documented evidence of any unconventional weapons (chemical, nuclear or biological) ever being held or used in Sudan. Some local communities reported that drinking water wells had been poisoned in Darfur, but in the absence of detail and opportunity for inspection, UNEP did not investigate this issue further.

The history and current status of each of the major conflict areas is briefly described below. The geographical extent of the various conflicts as interpreted by UNEP is shown in Figure 4.1.

Darfur

Fighting in Darfur has occurred intermittently for at least thirty years. Until 2003, it was mostly confined to a series of partly connected tribal and local conflicts [4.5]. In early 2003, these hostilities escalated into a full-scale military confrontation in all three Darfur states, which also frequently spills into neighbouring Chad and the Central African Republic.

The ongoing Darfur conflict is characterized by a 'scorched earth' campaign carried out by militias over large areas, resulting in a significant number of civilian deaths, the widespread destruction of villages and forests, and the displacement of victims into camps for protection, food and water. Over two million people are currently displaced, and casualties are estimated by a range of sources to be between 200,000 and 500,000 [4.6].

A downed fighter-bomber near Padak, Jonglei state

Southern Sudan

In the fifty years since Sudan's independence, the south has experienced only eleven years of peace. During most of the civil war, the central Sudanese government held a number of major towns and launched air attacks and dry-season ground offensives into the surrounding countryside. The opposition forces, the Sudan People's Liberation Army (SPLA) and their allies, fought guerrilla actions, besieged towns and conducted ground offensives in both wet and dry seasons. Most of the countryside, however, saw little or no military activity. Frontlines with prolonged, active fighting were confined to northern-central border regions and besieged towns. The fiercest fighting took place in the 1990s, with frontlines changing constantly and several towns being taken many times.

The conflict extended to areas in central Sudan, such as Abyei district, Blue Nile and the Nuba mountains in Southern Kordofan. Known as the 'Three Areas', these regions retain a high level of political uncertainty today. Small-scale conflict due to the Ugandan militia the Lord's Resistance Army (LRA) has also occurred intermittently in the far south even after the signing of the Comprehensive Peace Agreement in January 2005, and some instability persists in other border regions, particularly in Upper Nile.

Nuba mountains

The Nuba mountains were a SPLA stronghold in the 1990s. The SPLA held the forested regions and steeper terrain, while the open ground and surrounding plains were largely occupied by government forces. The area saw extensive fighting and aerial bombardment [4.7].

Kassala state - Eastern front

The region bordering Eritrea in Kassala state was a stronghold of the Beja people, who were allied with the SPLA. Conflict flared up in the 1990s, but a separate peace agreement between the central government and eastern forces – known as the Eastern Sudan Peace Agreement – was concluded in October 2006.

Red Sea state - Eritrean conflict

The Tokar region in Red Sea state was affected by low-level conflict between Sudan and Eritrea and local allied groups for twelve years, beginning in 1992. Hostilities ceased completely only with the signing of the CPA in early 2005.

The ongoing LRA conflict

Traditionally based in northern Uganda, directly south of the Sudan's Eastern Equatoria state, the Lord's Resistance Army (LRA) has fought against the Ugandan armed forces for over twenty years. In 2005 and 2006, the conflict spread to Southern Sudan and the Democratic Republic of Congo. As of June 2007, a ceasefire is in effect but peace negotiations have stalled and sporadic conflict is ongoing.

4.3   Analysis of the role of natural resources as a contributing cause of conflict in Sudan

It is acknowledged that there are many factors that contribute to conflict in Sudan that have little or no link to the environment or natural resources. These include political, religious, ethnic, tribal and clan divisions, economic factors, land tenure deficiencies and historical feuds. In addition, where environment and natural resource management issues are important, they are generally **contributing** factors only, not the sole cause for tension.

The Nuba mountains were the scene of sustained fighting in the 1990s

The conflict on the Eastern front was fought in the barren hills of Kassala state, near Ethiopia

As noted previously, 'non-environmental' factors have been excluded from detailed examination in this assessment to allow for a tighter focus on the environmental dimensions of conflict. Also excluded is any analysis of the subsequent behaviour of the conflicting parties, except where it is directly relevant to the environment, as is the case for the targeted destruction of natural resources.

Four natural resources closely linked to conflict in Sudan

In Sudan, four categories of natural resources are particularly linked to conflict as contributing causes:

1. oil and gas reserves;
2. Nile waters;
3. hardwood timber; and
4. rangeland and rain-fed agricultural land (and associated water points).

Potential conflicts over oil, Nile waters and hardwood timber are national-scale issues. Tensions over rangeland and rain-fed agricultural land are primarily local, but have the potential to escalate and exacerbate other sources of conflict to the extent of becoming national-scale issues, as is presently the case in Darfur.

The linkages between these resources/land uses and conflict are discussed below; the fourth category is examined in more detail in a separate section, as it has strong ties to the ongoing conflict in Darfur.

Note that groundwater (on a regional scale), wildlife, freshwater fisheries and all types of marine resources are excluded from this list of important contributing causes, as there is no evidence that they have been major factors in instigating conflict in Sudan to date.

Competition over oil and gas reserves

Though the major north-south conflict started well before oil was discovered in central Sudan, competition for ownership and shares in the benefits of the country's oil and gas reserves was a driving force for the conflict and remains a source of political tension today [4.4]. This is, however, considered to be primarily an economic, political and social issue, and is hence not addressed in detail in this report.

Of more relevance to UNEP, in this context, are the environmental impacts of the oil industry and their potential to catalyse conflict in the future. Consultations in central and south Sudan revealed widespread and intense dissatisfaction with the oil industry's environmental performance, coupled with the above-mentioned general concerns about ownership and benefit-sharing. In summary, the population in the vicinity of the oilfields said they felt subjected to all of the downsides of the presence of the oil industry (including its environmental impacts) without receiving a share in the benefits. Experience from other countries, such as Nigeria, shows that the root causes for this type of resentment must be addressed in order to avoid long-term instability and conflict at the local level. Part of the solution is to improve the environmental performance of the industry.

The environmental aspects of this issue are covered in a more detailed assessment of the oil industry in Chapter 7.

Camels graze in a destroyed village in Western Darfur. The trees have been cut for fuelwood and to provide the animals with fodder. Fighting over grazing land has been ongoing in Darfur since 1920 at least

Water is the most precious natural resource in the drier regions. Goats, cattle and camels all use this crowded water point in Southern Kordofan

Conflict over water rights and benefits from the Nile

Competition for the benefits accrued from the use of surface water was also an important contributing factor of the civil war, as illustrated by the Jonglei canal project (see Case Study 10.2), which was a cause as well as a victim of the conflict that flared up in Southern Sudan in 1983. The significance of this issue has not declined over time and tensions over attempts to re-start the project are still high.

However, a number of institutional safeguards are likely to prevent a re-instigation of conflict over water rights alone at the state and federal level. First, as a high profile and easily identifiable issue, it receives significant attention from GONU and GOSS leadership, as well as international assistance in the form of programmes like the Nile Basin Initiative. Second, major projects such as new dams or canals require both large investments and long periods of time, and this development process (in its modern form at least) has a range of built-in safeguards to identify and mitigate the risk of conflict. Water issues are covered in more detail in Chapter 10.

Timber and the war economy

While there is no indication that timber has been a major contributing cause of the instigation of conflict in Sudan, there is clear evidence that revenue from hardwood timber sales helped sustain the north-south civil war. Timber became part of the war economy, and there are now signs that this process is being repeated with charcoal in Darfur. Overall however, the timber-conflict linkage in Sudan is considered to be mainly an environmental impact issue (rather than a conflict catalyst). This is discussed in more detail in the next section, and in Chapter 9.

Local conflicts over rangeland and rain-fed agricultural land

Local clashes over rangeland and rain-fed agricultural land have occurred throughout Sudan's recorded history. In the absence of demographic and environmental change, such conflicts would generally be considered a social, political or economic issue and not warrant an assessment purely on environmental grounds. However, environmental issues like desertification, land degradation and climate change are becoming major factors in these conflicts. This topic is addressed in more detail in the following section.

Low quality degraded rangeland in Northern Darfur. To survive in these regions, pastoralists must travel across agricultural areas to find water and fodder for their herds, which commonly leads to conflict

4.4 Environmental linkages to local conflicts over rangeland and rain-fed agricultural land

Introduction and limits to the observed linkages

It is important to note that while environmental problems affect rangeland and rain-fed agricultural land across virtually all of Sudan, they are clearly and strongly linked to conflict in a minority of cases and regions only. These linkages do exist, but their significance and geographic scale should not be exaggerated.

That said, there is substantial evidence of a strong link between the recent occurrence of local conflict and environmental degradation of rangeland and rain-fed agricultural land in the drier parts of Sudan.

The actors of conflict at the local level: three major competing and conflicting groups

The rural ethnic and livelihood structures of Sudan are so complex and area-specific that any summary of the issue of resource competition on a national scale is by definition a gross simplification. For instance, traditional pastoralist and agricultural societies in Sudan are not always clearly separated: in many areas, societies (families, clans and even whole tribes) practice a mixture of crop-growing and animal-rearing. Nonetheless, there are some relatively clear boundaries – defined as much by livelihoods as by any other factor – between different tribes, clans and ethnic groups.

For the purposes of this discussion, UNEP has classified the hundreds of distinct rural social units present in the current and historical conflict regions into three major groups, based on livelihood strategies:

Unexploded ordnance partially buried in a pit outside Juba, Jonglei state

This mined road in Jonglei state has not been used by vehicles for a decade, but locals still walk along it to collect firewood and access farm plots

1. predominantly sedentary crop-rearing societies/tribes;

2. predominantly nomadic (transhumant) livestock-rearing societies/tribes; and

3. owners of and workers on mechanized agricultural schemes.

All three groups depend on rainfall for their livelihood. The other major rural group is comprised of farmers using river and groundwater for irrigation. To date, however, irrigated agriculture has not been a major factor in local conflicts in Sudan.

Most of the recorded local conflicts are within and between the first two groups: pastoralists and agriculturalists fighting over access to land and water. The third group, the mechanized farming lobby, is generally not directly involved in conflict, but has played a very strong role in precipitating it in some states, through uncontrolled land take from the other two groups. In the Nuba mountains and in Blue Nile state, combatants reported that the expansion of mechanized agricultural schemes onto their land had precipitated the fighting, which had then escalated and coalesced with the major north-south political conflict [4.7, 4.8, 4.9].

The historical background: a tradition of local conflict and resolution

Violent conflict resulting partly from competition over agricultural and grazing land is a worldwide and age-old phenomenon. In Sudan – and particularly in Darfur and Kordofan – there is an extensive history of local clashes associated with this issue [4.3, 4.5, 4.10, 4.11]. A 2003 study on the causes of conflict in Darfur from 1930 to 2000, for example, indicates that competition for pastoral land and water has been a driving force behind the majority of local confrontations for the last 70 years (see Table 5).

Table 5.   Causes of local conflicts in Darfur from 1930 to 2000 [4.3, 4.5]

No.	Tribal groups involved	Year	Main cause of conflict
1	Kababish, Kawahla, Berti and Medoub	1932	Grazing and water rights
2	Kababish, Medoub and Zyadiya	1957	Grazing and water rights
3	Rezeigat, Baggara and Maalia	1968	Local politics of administration
4	Rezeigat, Baggara and Dinka	1975	Grazing and water rights
5	Beni Helba, Zyadiya and Mahriya	1976	Grazing and water rights
6	Northern Rezeigat (Abbala) and Dago	1976	Grazing and water rights
7	N Rezeigat (Abbala) and Bargo	1978	Grazing and water rights
8	N Rezeigat and Gimir	1978	Grazing and water rights
9	N Rezeigat and Fur	1980	Grazing and water rights
10	N Rezeigat (Abbala) and Bargo	1980	Grazing and water rights
11	Taaisha and Salamat	1980	Local politics of administration
12	Kababish, Berti and Ziyadiya	1981	Grazing and water rights
13	Rezeigat, Baggara and Dinka	1981	Grazing and water rights
14	N Rezeigat and Beni Helba	1982	Grazing and water rights
15	Kababish, Kawahla, Berti and Medoub	1982	Grazing and water rights
16	Rezeigat and Mysseriya	1983	Grazing and water rights
17	Kababish, Berti and Medoub	1984	Grazing and water rights
18	Rezeigat and Mysseriya	1984	Grazing and water rights
19	Gimir and Fallata (Fulani)	1987	Administrative boundaries
20	Kababish, Kawahla, Berti and Medoub	1987	Grazing and water rights
21	Fur and Bidayat	1989	Armed robberies
22	Arab and Fur	1989	Grazing, cross-boundary politics
23	Zaghawa and Gimir	1990	Administrative boundaries
24	Zaghawa and Gimir	1990	Administrative boundaries
25	Taaisha and Gimir	1990	Land
26	Bargo and Rezeigat	1990	Grazing and water rights
27	Zaghawa and Maalia	1991	Land
28	Zaghawa and Marareit	1991	Grazing and water rights
29	Zaghawa and Beni Hussein	1991	Grazing and water rights
30	Zaghawa, Mima and Birgid	1991	Grazing and water rights
31	Zaghawa and Birgid	1991	Grazing and water rights
32	Zaghawa and Birgid	1991	Grazing and water rights
33	Fur and Turgum	1991	Land
34	Zaghawa and Arab	1994	Grazing and water rights
35	Zaghawa Sudan and Zaghawa Chad	1994	Power and politics
36	Masalit and Arab	1996	Grazing, administration
37	Zaghawa and Rezeigat	1997	Local politics
38	Kababish Arabs and Midoub	1997	Grazing and water rights
39	Masalit and Arab	1996	Grazing, administration
40	Zaghawa and Gimir	1999	Grazing, administration
41	Fur and Arab	2000	Grazing, politics, armed robberies

The scorched earth tactics used by militias in Darfur include cutting and burning trees in a haphazard manner

Until 1970, there is also a well-documented history of local resolution for such conflicts, through established mediation and dispute resolution mechanisms. Since then, however, legal reforms have essentially destroyed many of these traditional structures and processes, and failed to provide a viable substitute. In addition, the last thirty years have seen an influx of small arms into the region, with the unfortunate result that local conflicts today are both much more violent and more difficult to contain and mediate.

Theories of natural resource scarcity and application to local conflict in Sudan

Academic research and the discourse on the role of natural resource scarcity as a driver of conflict have developed significantly over the last decade [4.12, 4.13, 4.14, 4.15]. In light of the ongoing Darfur crisis, Sudan is a prime example of the importance, complexity and political sensitivity of this topic. The following analysis borrows heavily from the language and concepts used by leading researchers in this field.

As a basis for discussion, the environmentally significant factors that contribute to conflict related to rangeland and rain-fed agricultural land have been divided into four groups:

- supply: factors affecting the available resources;
- demand: factors affecting the demand for resources;
- land use: changes affecting the way remaining resources are shared; and
- institutional and development factors.

While all the purely environmental factors are 'supply' issues, they have to be put into the context of 'demand' and 'institution-specific' factors.

During the major north-south conflict, the town of Wau in Western Bahr el Ghazal was a centre for the logging and regional export of teak. The trade was effectively halted by the closing of the rail link; only a small-scale local teak trade subsists today

Supply – an unreliable and dwindling resource

The noted environmental issues affecting agriculture in Sudan all result in a dwindling supply of natural resources:

- **Desertification, soil erosion and soil exhaustion** (depletion of nutrients and compaction) lower agricultural productivity and, in the worst cases, take land out of use for the long term. This has been well documented but poorly quantified in Sudan (see Chapters 3 and 8);

- **Deforestation,** particularly in the drylands, has resulted in a near permanent loss of resources including seasonal forage for pastoralists and natural fertilizer/soil recovery services for farmers. Deforestation rates in the areas studied by UNEP average 1.87 percent per annum (see Chapters 8 and 9);

- **Historical climate change** has reduced productivity in some areas due to a decline in rainfall. A major and long-term drop in precipitation (30 percent over 80 years) has been recorded in Northern Darfur, for example. The implications of such a decline on dry rangeland quality are obvious (see Chapter 3); and

- **Forecast climate change** is expected to further reduce productivity due to declining rainfall and increased variability, particularly in the Sahel belt. A drop in productivity of up to 70 percent is forecast for the most vulnerable areas (see Chapter 3).

Ever increasing demands on resources

The demand for natural resources in Sudan is uniformly increasing, due to the following factors:

- **Human population growth** is the underlying driver of increased demand for natural resources. Sudan has an overall growth rate of over 2.6 percent per annum, masking much higher localized rates. In central Darfur, for example, government statistics indicate a regional population (linear) growth rate of 12 percent per annum, from 3 persons/km² in 1956 to 18 persons/km² in 2003 [4.16]. These growth rates are indicative of large-scale in-migration, in this case mainly from the north and possibly due to environmental factors such as desertification; and

- **Livestock population and growth rates**; government officials and academics have tracked the population increase of livestock since the 1960s. In northern and central Sudan alone, it is estimated to have increased by over 400 percent between 1961 and 2004 (see Chapter 8) [4.17].

Land use changes – a dwindling share of resources for pastoralists

The horizontal expansion of agriculture into areas that were previously either rangeland or forest has been a well recognized trend for the last four decades. The northwards expansion of rain-fed agriculture into marginal areas historically only used for grazing has been particularly damaging. Three examples from the recent UNEP-ICRAF [4.18] study of land use changes illustrate a major reduction in rangeland areas due to expanding agriculture (see Chapters 8 and 9):

- In Ed Damazin, Blue Nile state, agricultural land (mainly mechanized), increased from 42 to 77 percent between 1972 and 1999, while rangeland effectively disappeared, dropping from 8.3 to 0.1 percent;

- In the El Obeid region of Northern Kordofan, rain-fed agricultural land increased by 57.6 percent between 1973 and 1999, while rangeland decreased by 33.8 percent and wooded pasture by 27 percent; and

- In the Um Chelluta region of Southern Darfur, rain-fed agricultural land increased by 138 percent between 1973 and 2000, while rangeland and closed woodland decreased by 56 and 32 percent respectively.

In addition to the loss of grazing land, agricultural expansion has also blocked livestock migratory routes between many of the widely separated dry and wet season pastures, and between the herds and daily watering points. A further complication is that sedentary farmers are increasingly raising their own livestock, and are hence less willing to give grazing rights to nomads in transit [4.19] (see Chapter 8 for a more detailed discussion of these issues).

Institutional factors – failing to rectify the issues

Agricultural institutions and environmental governance in Sudan are discussed in detail in Chapters 8 and 13 respectively. In summary, the rural environment has been impacted by a combination of ill-fated reform and development programmes, as well as legal reforms and failures in environmental governance. One key issue is the difficulty of developing and applying a practical, just and stable system of rural land tenure in an ethnically complex society of intermingled sedentary farmers and transhumants/nomads. This has not been achieved in Sudan so far.

A lack of development and livelihood options

Outside of the main urban areas, Sudan remains very poor and underdeveloped. Rural populations consequently have very few options to solve these agricultural crises, as solutions like agricultural development, improvements in pasture and stock quality, and using working capital to cover short-term needs and alternative employment are simply not available [4.19].

The net result – disappearing livelihoods for dryland pastoralist societies

The clear trend that emerges when these various elements are pieced together is that of a **significant long-term increase in livestock density on rangelands that are reducing in total area, accessibility and quality**. In environmental terms, the observed net result is overgrazing and land degradation. In social terms, the reported consequence for pastoralist societies is an effectively permanent loss of livelihoods and entrenched poverty.

Pastoralist societies in Sudan have always been relatively vulnerable to losing their livelihoods due to erratic rainfall, but the above-noted combination of factors has propelled many pastoralists into a negative spiral of poverty, displacement, and in the worst cases, conflict. Their coping strategies, which have been well documented [4.16, 4.19], include:

- Abandoning pastoralism as a livelihood in favour of sedentary agriculture, or displacement to cities;

- Increasing or varying the extent of annual herd movements where possible, with a general trend towards a permanently more southerly migration;

- Maximizing herd sizes as an insurance measure (assisted by the provision of water points and veterinary services);

- Changing herd composition, replacing camels by small animals, mainly sheep, in response to the curtailment of long-distance migration;

- Competing directly with other grazers for preferred areas of higher productivity (**entailing a conflict risk**);

- Moving and grazing livestock on cropland without consent (**entailing a conflict risk**); and

- Reducing competition by forcing other pastoralists and agriculturalists off previously shared land (**as a last resort - the proactive conflict scenario**).

Variations of all of these strategies can be observed throughout Sudan, particularly in the drier regions.

Displaced populations settle on the outskirts of existing towns, as seen here in El Fasher, Northern Darfur, where the new settlement is distinguished by white plastic sheeting. These new arrivals add to the environmental burden on the surrounding desert environment

Camel herders from the Shanabla tribe at a water point in El Tooj, Southern Kordofan. The southward migration of camel herders is a harbinger of renewed conflict in the Nuba mountains

CS 4.1   The southward migration of camel herders into the Nuba mountains and subsequent resource competition

The Nuba mountains region in Southern Kordofan provides an example of the increase in natural resource competition and local conflict that results from the combination of agricultural expansion, land degradation and the southward migration of pastoralists.

At the start of the civil war in the 1980s, cattle-herding pastoralists from the Hawazma Baggara tribe started penetrating deeper into the Nuba mountains in search of water and pasture for their cattle, due to the loss of grazing land to mechanized agriculture and drought. The rivalry that ensued with the indigenous Nuba tribe, who practised a combination of sedentary farming and cattle-rearing, contributed to the outbreak of large-scale armed conflict. Meanwhile, as some of the dry season pastures around Talodi were off-limits during the conflict years, the Hawazma had to remain in their wet season grazing lands in Northern Kordofan, exerting greater pressure on the vegetation there.

In 2006, UNEP observed the return of Hawazma Baggara to their former grazing camps in conflict zones in Southern Kordofan, for example near Atmoor. UNEP also witnessed the presence of the camel-herding Shanabla tribe in the midst of thick woodland savannah at El Tooj (now reportedly reaching up to lakes Keilak and Abiad).

This new southward migration of camel herders constitutes an indicator of livestock overcrowding and rangeland degradation in Northern Kordofan, and is a harbinger of further conflict with the Nuba. At Farandala in SPLM-controlled territory, the Nuba expressed concern over the widespread mutilation of trees due to heavy lopping by the Shanabla to feed their camels, and warned of 'restarting the war' if this did not cease.

Conclusions on the role of environmental issues in conflicts over rangeland and rain-fed agricultural land

Pastoralist societies have been at the centre of local conflicts in Sudan throughout recorded history. The most significant problems have occurred and continue to occur in the drier central regions, which are also the regions with the largest livestock populations, and under the most severe environmental stress.

As there are many factors in play – most of which are not related to the environment – land degradation does not appear to be the dominant causative factor in local conflicts. It is, however, a very important element, which is growing in significance and is a critical issue for the long-term resolution of the Darfur crisis. The key cause for concern is the **historical, ongoing and forecast shrinkage and degradation of remaining rangelands in the northern part of the Sahel belt.**

Much of the evidence for UNEP's analysis is anecdotal and qualitative; it has been gathered through desk study work, satellite images and interviews of rural societies across Sudan. The consistency and convergence of reports from a range of sources lend credibility to this analysis, although further research is clearly needed, with a particular emphasis on improved quantification of the highlighted issues and moving beyond analysis to search for viable long-term solutions.

A conference on the topic of environmental degradation and conflict in Darfur was held in Khartoum in 2004. The proceedings [4.20] illustrated the depth of local understanding of the issue. Given the situation observed in 2007, however, UNEP must conclude that this high-quality awareness-raising exercise was unfortunately apparently not transformed into lasting action.

4.5 Assessment of the environmental impacts of conflict

Introduction

This section approaches the linkages between conflict and environment from the reverse angle to the above analysis, by examining **if and how armed conflict has resulted in negative or positive impacts on the environment in Sudan.** Direct impacts, indirect impacts and key conflict-related issues are identified and discussed in this chapter. Detailed discussion and recommendations on the various environmental issues of concern (e.g. deforestation) are referred to the corresponding sector chapter.

Definitions and impact listings

The following definitions are used for direct, indirect and secondary environmental impacts of conflict in Sudan:

- **Direct impacts** are those arising directly and solely from military action;

- **Indirect and secondary impacts** are all impacts that can be credibly sourced in whole or in part to the conflicts and the associated war economy, excluding the direct impacts.

On this basis, UNEP has developed the following list of impacts for discussion:

Direct impacts include:
- landmines and explosive remnants of war (ERW);
- destroyed target-related impacts;
- defensive works; and
- targeted natural resource destruction.

Indirect and secondary impacts include:
- environmental impacts related to population displacement;
- natural resource looting and war economy resource extraction;
- environmental governance and information vacuum; and
- funding crises, arrested development and conservation programmes.

Direct impacts

Landmines and explosive remnants of war

Landmines and other explosive remnants of war (ERW) are a major problem in Sudan. Thirty-two percent of the country is estimated to be affected [4.4], with the greatest concentration in Southern Sudan (see Case Study 4.2). As many as twenty-one of the country's twenty-five states may be impacted, although the true extent of Sudan's landmine problem remains unknown, as a comprehensive survey of the issue has not been undertaken to date.

In 1983, southern military forces sabotaged these generators powering the Jonglei canal excavator. Plans to restart the giant water project constitute a major potential flashpoint for renewed conflict

The reported and registered number of landmine casualties over the past five years totals 2,200, though again, no systematic data collection and verification mechanism exists. In addition, there is no data at all on animal casualties from mines in Sudan, but these are expected to be much higher than the human casualty rate. The impacts of landmines on wildlife would only be significant (at the ecosystem level) if individual losses affected locally threatened populations of key species.

The potential impacts of landmines and ERW can be divided into chemical and physical categories. Conventional explosives, such as TNT and RDX, found in artillery shells and mines are highly toxic and slow to degrade. While they present an acute toxic hazard if ingested, the toxic risk is considered insignificant compared to the risk of injury from explosion.

Apart from human casualties, another major impact of landmines is impeded access to large areas for people and their livestock. In Sudan, access to some areas has been reduced for decades, as they have remained mined or suspected as such since the beginning of the conflict.

In all but the driest areas, the result of reduced access has been the relatively unimpeded growth of vegetation. UNEP fieldwork, in the Nuba mountains in particular, revealed extensive areas of woodland regrowth in suspected minefields. Such regrowth can have a beneficial effect on the affected areas and associated wildlife populations, but the flow of benefits to people is usually reduced, as they cannot safely extract resources (e.g. water, fuelwood, fodder) from these sites. Despite the risks, however, UNEP teams witnessed people walking, herding cattle and gathering fuel in clearly marked minefields.

The dumping of waste on minefields and on top of unexploded ordnance creates a major safety problem (top); unexploded ordnance is loosely stacked and scattered across the area (bottom)

CS 4.2   Unexploded ordnance, minefields and deforestation at Jebel Kujur, Juba district

The Jebel Kujur massif near the city of Juba in the state of Central Equatoria (Bahr el Jabal) clearly illustrates the localized but severe impacts of conflict affecting many urban centres in Southern Sudan, as well as the environmental governance challenges facing the new government.

During the 1983-2005 conflict, Juba was a garrison town for the central government military, and was continuously under siege and frequently attacked by SPLA forces. The town itself still shows extensive scarring, and overgrown entrenchments, minefields and scattered unexploded ordnance are visible on the fringes. Deforestation and soil erosion are severe, particularly at Jebel Kujur, which originally supported a dense forest cover. A quarry is also operating at one end of the range.

In late 2006, clean-up was ongoing, but there were still minefields and areas of stacked ordnance in the foothills of Jebel Kujur. Despite the obvious risks, cattle grazing, scrap recovery and waste dumping were routinely taking place in these areas. Plastic waste was being dumped directly on top of unexploded artillery shells and rocket-propelled grenades, creating obvious serious hazards for site users and greatly increasing the future cost of de-mining and rehabilitation.

The removal of explosive remnants of war (ERW) from Jebel Kujur is a difficult but short-term activity. The greater challenges are sustainable solutions for waste management for the growing city and reforestation of the massif.

Destroyed target-related impacts

Target-related impacts refer to the effects on the environment of direct military action on targets, irrespective of the method. The physical destruction of the environment from conventional weaponry (bombs, artillery shells and mortars) principally takes the form of cratering, and damaged or destroyed buildings, trees, and industrial facilities.

Though cratering has been reported by de-mining staff in Southern Sudan, there is no indication that more than a few hectares are affected at each conflict location. Similarly, the destruction of trees by direct military action is considered negligible compared to other causes of deforestation in Sudan. No lasting environmental damage is expected either from the destruction of buildings, apart from the generation of inert solid waste as rubble.

The single most significant industrial target in conflicts to date is the Jonglei canal excavator, which was sabotaged 40 km north of Padak in Jonglei state. The rusting excavator is currently used as a nesting site by eagles and is home to several beehives. UNEP experts inspected the excavator and its surroundings, and concluded that its direct environmental impact was negligible.

Neither the oilfields in the south, nor the transfer pipeline to Port Sudan were ever successfully attacked to the extent that significant environmental damage ensued.

UNEP concludes that the absence of vulnerable industrial targets in historical conflict zones has prevented any major environmental contamination from chemical spillage, and that other target-related impacts have been insignificant in environmental terms.

Defensive works

Major defensive works such as trench networks and bunkers were noticeably absent throughout the country, but de-mining staff in Southern Sudan reported that limited defence works could be found on the outskirts of besieged garrison towns.

Southern communities gave consistent reports of government forces clearing trees from the periphery of the garrison towns to deny cover to attacking forces. UNEP site inspections of the outskirts of Juba, Malakal and Aweil certainly indicated that deforestation has occurred, but it was not possible to attribute this solely to defensive works, as several other causes of deforestation were also evident at these locations (see Chapter 9).

In many rural areas of Southern Sudan, the only direct and lasting evidence of the conflict is scattered steel scrap, such as this grenade fragment outside Juba, Jonglei state

Targeted natural resource destruction

In Darfur, the deliberate targeting of vital natural resource-related infrastructure, such as rural water pumps, has been well documented by NGOs and inspection reports from the African Union Mission in Sudan (AMIS)[4.21]. Local populations in Darfur have also reported many instances of deliberate natural resource destruction by raiding militia, whose principal targets are trees, crops and pastures. Crops and pastures are burned and trees are cut. UNEP directly observed evidence of destructive tree-cutting in destroyed and deserted villages east of El Geneina in Western Darfur (see Case Study 4.3). Aid workers have reported similar targeted tree-cutting in other parts of Darfur.

Given the lack of quantifiable data on field conditions in Darfur, it is not possible to estimate the significance of this phenomenon. UNEP can only state that it is occurring and that it will add to the deforestation problem in the region (see Chapter 8).

Indirect and secondary environmental impacts of conflict

The environmental impacts of population displacement

After civilian deaths and injuries, the most significant effect of conflict on the population of Sudan has been displacement – people fleeing conflict zones seeking security. An estimated five million people (7 to 12 percent of the estimated total population of Sudan) have been displaced to date, and less than one million have returned. The number of displaced is rising due to the continuing conflict in Darfur. The great majority of the displaced have come from rural areas and migrated to camps on the outskirts of towns and cities. Over two million have relocated to the capital city, Khartoum.

The severe and complex environmental consequences of displacement include:

- deforestation in camp areas;
- devegetation in camps areas;
- unsustainable groundwater extraction in camps;
- water pollution in camp areas;
- uncontrolled urban slum growth;
- the development of a 'relief economy' which can locally exacerbate demand for natural resources;
- fallow area regeneration and invasive weed expansion; and
- return- and recovery-related deforestation.

Not all displacement in Sudan is due to conflict. Drought and economic factors are also major contributing causes. For this reason, the environmental impacts of all the different types of displacement are separately discussed in Chapter 5.

Looting of natural resources - war economy resource extraction

Natural resource looting is defined as the uncontrolled and often illegal extraction of natural resources that commonly occurs during extended conflicts. In this context, natural resources are often badly impacted and also have a role in sustaining the conflict.

In Sudan, the resources in question are timber (lumber and charcoal), ivory and bushmeat. Although oil is a contested natural resource in Sudan, it is excluded from this discussion as UNEP found no evidence of significant uncontrolled, concealed or illegal extraction. The potential and actual environmental impacts of the oil industry are covered in Chapter 7.

The looting of timber occurred on both sides in the north-south conflict. The most significant extraction concerned high value timber in Southern Sudan and fuelwood for charcoal in the Nuba mountains.

In Southern Sudan, UNEP received consistent verbal reports, backed by literature [4.22], of extraction and export (regional and international) of plantation teak and natural mahogany by government as well as SPLA forces and associated militias, though extraction was limited on both sides to areas within their respective control and close to transportation corridors. Northern government forces extracted timber from Wau, exporting it north via the rail link, and from Juba and other Nile towns, exporting by barge. The SPLA exported plantation teak southwards, from the Equatoria states to Uganda.

An abandoned grinding stone in the former village of Hashaba, south-east of El Geneina, destroyed in the conflict

CS 4.3   Targeted natural resources destruction in Western Darfur

One of the defining impacts of the current conflict in Darfur has been the displacement of people from rural areas, and the destruction of villages and surrounding land by militias. During its field mission in June 2006, the UNEP assessment team, under armed escort from African Union forces, visited some of the areas south-east of El Geneina in Western Darfur. The mission found that the outlying villages had been damaged to the extent that hardly any evidence of their former existence remained. In addition to the demolition of infrastructure, the trees within village limits had been systematically cut down.

These observations from the areas around El Geneina were consistent with anecdotal information collected through interviews with IDPs in the camps of Northern, Western, and Southern Darfur.

While some trees may have been felled to provide fodder for livestock or to be sold for firewood in IDP camps, there is evidence that some were undoubtedly cut down maliciously. This is the case for mango trees, for instance, as their leaves are inedible for livestock. From a military perspective, destroying trees severs the former community's links to the land and reduces the likelihood of resettlement. The environmental consequences of the loss of tree cover include a net deficit of biomass available to the soil, as well as the loss of the trees' ability to fix nitrogen. Both result in a decrease in soil fertility.

UNEP has no data or basis on which to quantify the extent of this reported trade. It is clear, however, that it has come to an end or has at least been significantly reduced. For the northern forces, trade has been stopped by the closure of the Wau rail link and the demobilization of northern garrisons from the south, while the SPLA's extractions have been curtailed by the newly formed Government of Southern Sudan's 2005 ban on timber exports and customs controls on border roads.

In the Nuba mountains, UNEP field teams observed charcoal for sale at military checkpoints, indicating that the military may still play a role in this business in the area.

Both UNEP teams and the follow-up Darfur Joint Assessment Mission field teams found an active lumber industry in central Darfur, in historical as well as current conflict areas. While it was not possible to determine who the main actors in this trade were, it was clear that some uncontrolled logging linked to the conflict was occurring.

The elephant population in Southern Sudan was decimated during the north-south conflict. While it is likely that much of the ivory was shipped to Khartoum, which is the centre of ivory carving in the region, there is no firm evidence to identify the main actors of elephant poaching and ivory transportation. Note that while rhinoceros horn was undoubtedly a poaching target in Southern Sudan during the early stages of the conflict, this trade has stopped due to the virtual extinction of rhino in the region.

Though UNEP did not find proof of an ongoing widespread commercial bushmeat trade, local people in Southern Sudan reported that both sides in the north-south conflict had taken bushmeat to feed their forces, with the result that the larger edible mammals such as buffalo, giraffe, zebra and eland are locally extinct throughout much of the south.

In sum, the looting of natural resources has undoubtedly occurred in Sudan and has caused significant damage. However, the signing of the Comprehensive Peace Agreement has reduced the scale of such activities, though looting remains an issue for Darfur, and to some extent for the Nuba mountains.

Environmental governance and information vacuum

Conflict zones generally suffer from a lack of stable governance and limited observance of the rule of law. In environmental terms, this results in a complete lack of environmental protection as well as impunity for those, military or otherwise, who extract or process natural resources in an uncontrolled manner or cause other forms of environmental damage.

Conflict zones are also usually inaccessible for science-based data collection. In the case of Sudan, conflict-related security constraints have denied the environmental science community access to at least half of the country for over two decades. As a result, the true status of much of Sudan's environmental resources remains unknown or open to speculation, limiting rational decision-making for resource management and conservation.

Funding crises - arrested development and conservation programmes

Extended and major conflicts drain national resources and can lead to isolation from the international community. Decades of war in Sudan have helped ensure that it remain one of the world's poorest countries. Political issues have also constrained the flow of international knowledge and assistance to Sudan.

The result has been that conservation of the environment and the sustainable management of natural resources have not been regarded as priorities for Sudan since independence, and that even when they have been considered, they have generally not been sufficiently funded to bring about positive change.

The financial burden of virtually continuous warfare and the ensuing poverty can thus be considered as one of the root causes of the current state of the environment in Sudan.

Summary of the environmental impacts of conflict

The findings of UNEP's assessment of the environmental impacts of conflict in Sudan can be summarized as follows:

Direct impacts are overall minor:
- landmines and explosive remnants of war: significant;
- destroyed target-related impacts: not significant;
- defensive works: not significant; and
- targeted natural resource destruction: significant for Darfur, but currently not quantifiable.

Indirect and secondary impacts are major:
- environmental impacts related to population displacement: very significant;
- looting of natural resources: significant;
- environmental governance and information vacuum: significant; and
- funding crises: very significant.

These findings indicate that the way forward on environmental issues in post-conflict Sudan should not focus on the direct legacies of conflict (which are relatively minor). Attention should instead be paid to the indirect and secondary impact-related issues, as well as to chronic problems. This would be best achieved by integrating all of the issues into a holistic recovery programme rather than attempting to separate them on the basis of conflict linkages.

4.6 Conclusions and recommendations

Conclusion

The linkages between conflict and environment in Sudan are twofold. On one hand, the country's long history of conflict has had a significant impact on its environment. Indirect impacts such as population displacement, lack of governance, conflict-related resource exploitation and underinvestment in sustainable development have been the most severe consequences to date.

On the other hand, environmental issues have been and continue to be contributing causes of conflict. Competition over oil and gas reserves, Nile waters and timber, as well as land use issues related to agricultural land, are important causative factors in the instigation and perpetuation of conflict in Sudan. Confrontations over rangeland and rain-fed agricultural land in the drier parts of the country are a particularly striking manifestation of the connection between natural resource scarcity and violent conflict. In all cases, however, environmental factors are intertwined with a range of other social, political and economic issues.

UNEP's analysis indicates that there is a very strong link between land degradation, desertification and conflict in Darfur. Northern Darfur – where exponential population growth and related environmental stress have created the conditions for conflicts to be triggered and sustained by political, tribal or ethnic differences – can be considered a tragic example of the social breakdown that can result from ecological collapse. Long-term peace in the region will not be possible unless these underlying and closely linked environmental and livelihood issues are resolved.

Background to the recommendations

The analysis of the linkages between conflict and environment in Sudan has so far been largely confined to academic circles. In Sudan, only USAID has explicitly integrated peacebuilding into the design of its environmental programme in Southern Sudan [4.24]. It is important that this discussion be broadened to include the government and the United Nations. International peacekeeping initiatives and implementing organizations, such as the African Union Mission to Sudan (AMIS) and the United Nations Mission to Sudan (UNMIS), should particularly take this issue in account.

In addition to political solutions, practical measures to alleviate natural resource competition are urgently needed to help contain the current conflict and present a viable long-term solution for the development of rural Darfur. Elsewhere in Sudan, efforts should be focused first and foremost on identified environmental 'flashpoints', which are specific issues that constitute a potential trigger for the renewal of conflict. The most important of these is the environmental impact of the oil industry, but there are several others, including the charcoal industry in central Sudan, the potential for ivory poaching and the development of a timber mafia in Southern Sudan.

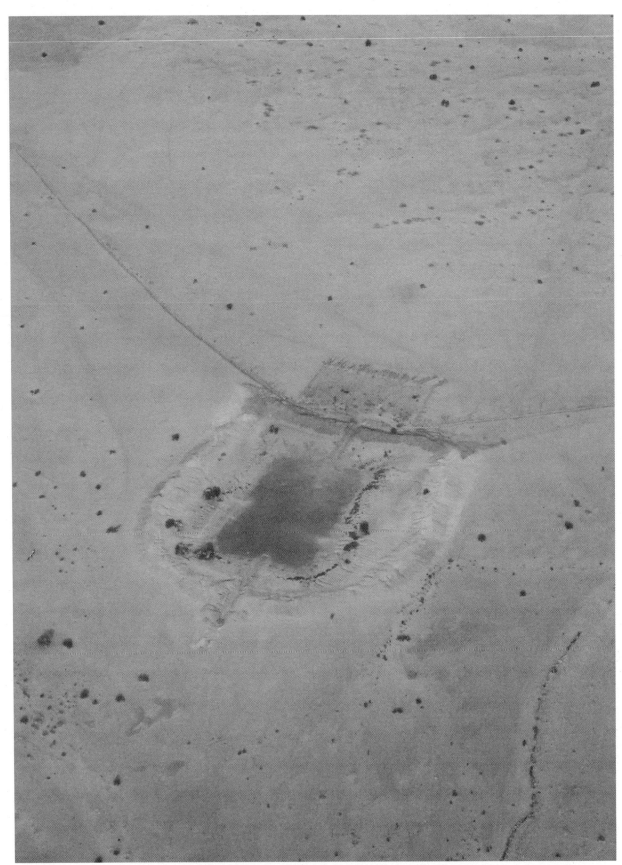

Parched and overgrazed land surrounding a dry livestock supply dam south of El Fasher, Northern Darfur, in June 2006. Environmental scarcity and degradation are two of the important contributing factors in the Darfur crisis

Possible measures – which are listed as recommendations in this and other chapters – include agricultural policy reform, developing the timber industry, and strengthening environmental governance. Such measures should be considered vital investments in conflict prevention and resolution rather than purely environmental conservation projects.

In summary, in the context of the CPA and the ongoing Darfur crisis, the attention of the environmental sector should be focused on the following three areas in order to assist peacebuilding and conflict resolution in Sudan:

1. reducing the environmental impact of the oil industry in central Sudan;

2. promoting more sustainable agriculture and pastoralism in dryland Sudan; and

3. providing information and technical assistance on environment-conflict issues to the national and international community working on peacebuilding and conflict resolution throughout Sudan, with an initial focus on Darfur.

Recommendations for the international community

R4.1 Bring the issue of environmental degradation and ecologically sustainable rural development to the forefront of peacebuilding activities in Sudan. This will entail a major awareness-raising exercise by UNEP and the international community in Sudan, and will need to be incorporated into response strategies for bodies such as the African Union, the UN Development Group (UNDG) and the UN Department of Peacekeeping Operations (UNDPKO).

CA: AW; PB: UNDPKO; UNP: UNEP; CE: 0.5M; DU: 1 year

R4.2 Bring natural resource assessment and management expertise into the existing peacebuilding and peacekeeping efforts in Sudan. UNEP or other organizations would provide technical assistance to the existing actors in this area for the south, east and Darfur, joining in the decision-making process. This should include significant direct support to governments and to both the African Union Mission to Sudan and the United Nations Mission to Sudan.

CA: TA; PB: UNMIS; UNP: UNEP and FAO; CE: 2M; DU: 3 years

R4.3 Conduct a specific environmental assessment for rural Darfur conflict regions as soon as security conditions and political stability permit. The major conflict which flared up in northern and central Darfur in September 2006 is expected to change and worsen the situation, in both humanitarian and environmental terms. An updated, detailed assessment focusing on land quality is needed to assist in the development of an appropriate recovery plan (when the time for recovery arrives). This very technical work would be used to supplement the existing body of largely qualitative work presented in the Darfur JAM interim report.

CA: AS; PB: UNMIS; UNP: UNEP and FAO; CE: 0.4M; DU: 1 year

Recommendations for the Government of National Unity

R4.4 Undertake strategic reform of the agricultural and pastoral sector. Without resolution of the underlying rural land use problems, the issue of the links between environmental degradation and conflict will remain insoluble. This recommendation is not costed as it is essentially an internal culture and strategic policy issue for GONU.

Population Displacement and the Environment

Sudan has the largest population of displaced persons in the world today. Nearly two million are in Darfur, in large settlements such as Abu Shouk IDP camp in El Fasher, Northern Darfur.

Population displacement and the environment

5.1 Introduction and assessment activities

Introduction

Over five million internally displaced persons (IDPs) and international refugees currently live in rural camps, informal settlements and urban slums in Sudan. This represents the largest population of displaced persons in the world today. Living conditions in these settlements are in many cases appalling: they are crowded and unsanitary, food and water are in short supply, insecurity is high, and livelihood opportunities are generally lacking. Some of these temporary settlements have existed for over twenty years with no improvement, and the conflict in Darfur is generating a new wave of displacement that is worsening the situation.

This massive population displacement has been accompanied by major environmental damage in the affected parts of the country. This is not a new phenomenon, but the scale of displacement and the particular vulnerability of the dry northern Sudanese environment may make this the most significant case of its type worldwide. Moreover, environmental degradation is also a contributing cause of displacement in Sudan, so that halting displacement will require concurrent action to halt environmental degradation.

Assessment activities

The assessment of displacement-related issues was included in UNEP's general fieldwork, which covered many of the areas where displacement had occurred and where returnees were expected. The environmental impact of displaced populations was a principal theme of the fieldwork in Darfur, while the impact of returnees on the rural environment was one of the main subjects of UNEP's work in Jonglei state. Locations visited include:

Dinka teenagers, who were raised in Ugandan refugee camps, wait to board the barge bringing them back to Bor district, Jonglei state

Figure 5.1   Displaced persons camps in Darfur

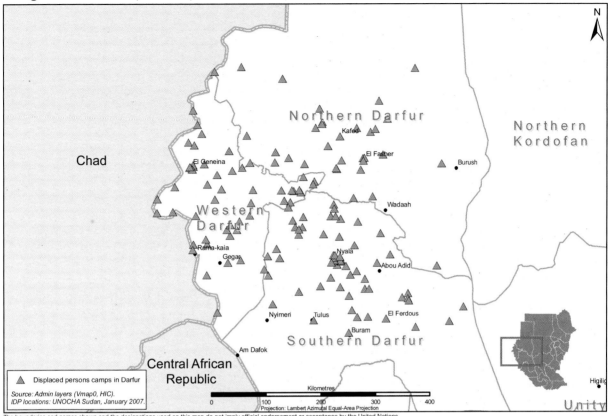

- IDP and refugee camps in Darfur: Mornei-Um Shalaya, Zalingei, Kalma, El Fasher-Abu Shouk, Kebkabiya, and Kutum-Kassab;

- villages on the outskirts of El Geneina, Western Darfur, destroyed and deserted as a result of conflict;

- IDP settlements in Port Sudan, Khartoum and Juba;

- Jonglei state way stations and return sites in Bor and Padak districts; and

- rural return sites in the Nuba mountains, Southern Kordofan.

Interviews with displaced persons took place at all of the above locations.

These displacement-specific activities were considered sufficient to obtain an overview of the issues, particularly for Darfur camps and for the Southern Sudan return process.

5.2   Overview of population displacement in Sudan

The world's largest displaced population

Over the past few decades, Sudan has witnessed more involuntary movement of people within and around its territory than any other country in the world. At the end of 2005, UNHCR estimated that some 700,000 Sudanese refugees lived outside the country [5.1]. Sudan has also offered asylum to a significant number of refugees from other countries in recent years, primarily from Chad, the Democratic Republic of Congo (DRC), Eritrea, Ethiopia, Somalia and Uganda.

Some of the highest numbers of refugees in the country were recorded during the 1990s: in 1993, for example, Sudan was host to some 745,000 refugees, the majority from Eritrea (57 percent), Chad (19 percent), and Ethiopia (2 percent) [5.2]. By the year 2000, the overall number had dropped to around 418,000 [5.3]. Estimates for 2005 indicate that approximately 147,000 refugees were officially recognized in Sudan [5.1]. The steady decline in numbers over this period is attributable to restored peace and security in neighbouring countries, and to a series of successful repatriation exercises.

The majority of refugees now seeking asylum in Sudan (77 percent) are Eritrean people [5.4] who live mainly in formal camps in the east. The influx of Eritrean refugees has been steady since 2003, as tension has increased in that country. In addition, there are 29,000 refugees from Uganda, DRC, Somalia, Ethiopia and other countries. With the exception of some 5,000 refugees from Chad, most live in Khartoum, Juba and other urban areas.

Besides hosting hundreds of thousands of refugees, Sudan has also generated more IDPs than any other country in the world – an estimated 5.4 million (see Table 6), or more than half the total IDP population on the continent [5.6, 5.5]. The International Displacement Monitoring Centre estimates that two million IDPs now live in Khartoum, most of whom have moved in with family members or set up squatter communities in neighbourhoods and fields around the capital. IDPs today account for 40 percent of Khartoum's total population [5.5]. In addition to squatter areas such as Soba Arradi, which hosts some 64,000 people, four official camps have been established to house IDPs: Omdurman es Salaam (120,000 people), Wad el Bashier (74,800 people), Mayo Farms (133,000 people) and Jebel Aulia (45,000 people).

Table 6. Location and number of internally displaced people in Sudan [5.5]

Location (state)	Number of IDPs
Khartoum	2,000,000
Northern	200,000
Red Sea	277,000
Kassala	76,000
Gedaref	42,000
Sennar	60,000
Blue Nile	235,000
White Nile	110,000
Upper Nile	95,000
Kordofan	189,000
Unity	135,000
Bahr el Ghazal	210,000
Equatoria	26,000
Greater Darfur	1,950,000
Total	**5,805,000**

Since 2003, internal displacement has occurred at an unprecedented rate in western Sudan. The Darfur crisis is reported to have affected some 2.4 million people, of whom 1.8 million are IDPs. Hundreds of thousands of people have already died, while conditions in many camps are far below international standards. In 2004, it was estimated that 465,000 households in Darfur would be in need of food assistance early in 2005 due to crop failure [5.7]. The same report noted that 90 percent of IDPs had lost their livestock, impeding income generation and water collection, and hindering return. Forty percent of the resident population had also lost their livestock.

The duration of displacement and the prospects for return

In Sudan as elsewhere, displaced populations return to their homelands if and when it is possible. For returns to take place on a large scale, however, a number of pre-conditions must be met:

- The original cause for displacement should have been removed, and physical security restored;

- Prospects for a livelihood in the homeland should be better than in the displaced location;

- Essential or important services (such as water, medical aid and schooling) should be available in the homeland and ideally equivalent to those in the displaced location;

- A practical means to travel back to the homeland safely (with possessions) should be available; and

- The return process must be sponsored or affordable for the displaced.

Because of these conditions, temporary displacements for any reason tend to turn into long-term processes or even permanent moves. Temporary settlements that exist for over a decade are not uncommon in Sudan. In Port Sudan, for instance, the UNEP team met IDPs from Northern Kordofan who had lived in informal settlements for twenty-three years and had no intention of returning to their homeland (see Case Study 5.1).

IDP camp residents told UNEP that they would rather remain in the Port Sudan area than return home, due to employment opportunities and improved education

CS 5.1   Fringe dwellers at Port Sudan: rural populations fleeing drought and seeking livelihoods in the cities

This informal settlement located in a *wadi* (seasonal riverbed) adjacent to the Port Sudan landfill is a typical example of uncontrolled urbanization triggered by natural causes. It houses over 500 families, the majority of which came from the El Obeid region in Northern Kordofan.

Interviewed residents stated that they had originally abandoned their farms due to extended drought and arrived in the region twenty-three years ago. The community was forcibly moved from a better site nine years ago by a land dispute and expanding urban development. The current site is seasonally flooded and has few amenities, aside from local schools and a water point installed by an aid project.

Despite the long-term nature of the settlement, all of the dwellings are temporary constructions. When asked about the potential for return to Northern Kordofan, the residents expressed no desire to do so, explaining that local employment and availability of schools were the determining factors in their decision to remain in Port Sudan. As the residents have no land tenure, however, they are at risk of being moved to even more distant fringes of Port Sudan as the city expands.

Large-scale returns of southern Sudanese currently in northern Sudan and in neighboring countries are now taking place but are expected to take several years to complete (see Figure 5.2). As of November 2006, over 17,000 refugees had returned to Southern Sudan through movements organized by UNHCR. An estimated total of 500,000 people returned to Southern Sudan, Abyei, and Southern Kordofan and Blue Nile states in 2006.

In Darfur, large camps appeared in 2003 and are presently increasing in population due to the intensification of the conflict in late 2006.

Figure 5.2   Forecast returns for Southern Sudan in 2007

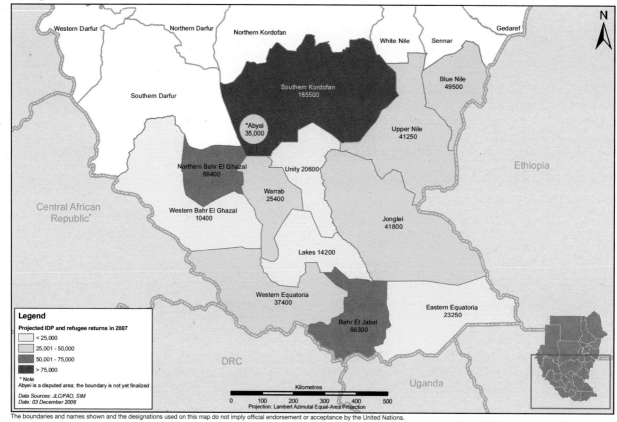

5.3 Overview of displacement-related environmental issues

Displacement-related environmental issues are widespread and often highly visible in major camps, settlements, urban slums and return areas. The most significant are:

1. environmental issues as a cause for displacement;

2. impacts related to the concentrations of people in camps or settlements:
    - deforestation and the fuelwood crisis in dry-land camp areas;
    - land degradation;
    - unsustainable groundwater extraction; and
    - water pollution;

3. other impacts related to the initial displacement;
    - uncontrolled urban and slum growth; and
    - fallow area regeneration (generally a positive impact);

4. the impacts of returnees and the environmental sustainability of rural returns; and

5. international environmental impacts.

5.4 Environment as one of three major causes of displacement in Sudan

There are three principal causes of displacement in Sudan:

- conflict-related insecurity and loss of livelihoods;
- natural and environmental causes: drought, desertification and flooding; and
- government-sponsored development schemes.

The principal cause of displacement has historically been the major conflicts that have afflicted Sudan since its independence. The second is natural disasters: drought, desertification and flooding, which are discussed in detail in Chapter 3. The third cause of displacement is government-sponsored development schemes, specifically mechanized rain-fed agricultural schemes, such as the Aswan dam and the new Merowe dam. In these cases, displacement takes the form of organized resettlements and land allocation for new agricultural schemes. The environmental impact of agricultural schemes and dams are covered in Chapters 8 and 10 respectively.

5.5 Environmental impacts of camps and settlements

Governance of settlements including environmental issues

The environmental impacts of camps in Sudan vary not only according to their physical location but also to their type (IDP or refugee camps), and to how long they have been in existence.

Oversight of refugee camps is the responsibility of the UN High Commissioner for Refugees (UNHCR) which, in turn, works with a government counterpart (the Commissioner for Refugees) and a range of other agencies and institutions, national and international, as required.

Responsibility for IDP camps is much less clear, particularly in Sudan, where some are run by the government and others by local authorities, militant groups, or international NGOs. Resources (funds, technical assistance and so forth) available to IDP and refugee camps also vary considerably. In general, IDP camps tend to have fewer relief resources than refugee camps. UNEP field teams encountered many families who deliberately elected to go to a refugee camp in preference to an IDP camp, because conditions were better in the former.

Environmental concerns have rarely – if ever – been a factor in the choice of sites for refugee or IDP camps in Sudan. No environmental assessment has ever been carried out prior to the site selection and establishment of any existing camp, nor is this a legal requirement.

A rapid environmental assessment conducted by OCHA at three camps in Darfur in 2001 highlighted another common concern which is addressed in this report: 'While the environment is an important factor in the Darfur crisis, there is no international agency with a specific mandate to consider or incorporate environmental issues into relief operations and peace efforts. This contrasts with the case for Darfur refugees in Chad, where UNHCR has a mandate to incorporate environmental issues into relief and return efforts' [5.8].

Deforestation and the fuelwood crisis in camp areas

One of the most significant environmental impacts of displaced population settlements is the severe deforestation that has occurred around the larger camps in the drier parts of the country.

Camp residents in Western Darfur cut wood chippings from a fallen tree for cooking fuel. The concentration of people into large settlements has also concentrated the demands on natural resources, resulting in severely deforested areas

This problem is related to the scale of the camps and to the standard of aid provision for displaced populations. Indeed, the level of assistance that displaced people receive in temporary settlements varies greatly. International refugees automatically qualify for assistance from UNHCR, while many IDPs do not. The assistance provided can include food aid, a water supply, basic sanitation facilities, tented accommodation or simply cover sheets and some basic household items.

What is virtually never provided is a source of energy for cooking food, boiling water or heating. In addition, when no formal accommodation is supplied, timber is needed to construct temporary dwellings. As a result, people living in camps and settlements are forced to find timber and fuelwood in the surrounding area. Livelihood strategies and the relief economy also play a role in the deforestation of camp areas: the collection of wood to fuel brick kilns, for example, is a major source of deforestation in a number of settlements in Darfur (see Case Study 5.2).

Deforestation is clearly visible around all major camp locations and can easily be detected by satellite in regions with otherwise good forest cover. In Nimule county on the border with Uganda, for instance, the illicit felling of trees for firewood and to clear land for slash-and-burn agriculture on the outskirts of a local IDP camp has resulted in the deforestation of a large area surrounding the camp (see Figure 5.3).

In drier regions, the effects are more difficult to detect but even more damaging. Much of northern and central Sudan is relatively dry, with low woodland density and slow growth rates. Tree cover is particularly sparse in Northern Darfur and northern parts of Kassala, two regions that host large displaced populations. Besides, the majority of settlements have been established in locations that were already occupied, and where the existing burden on forest resources may or may not have been sustainable.

In eastern Sudan, camp-related deforestation has been occurring for at least twenty years. Corrective measures (prohibitions) were put into place by UNHCR and the Forests National Corporation (FNC) to prevent refugees from cutting down trees for fuel, but as their ongoing energy needs were not addressed, these were not effective.

Figure 5.3   Deforestation at Nimule

The boundaries and names shown and the designations used on this map do not imply official endorsement or acceptance by the United Nations.

A brick kiln at Abu Shouk camp in Northern Darfur. One large tree is needed to fire approximately 3,000 bricks

In this mango orchard near Kalma IDP camp in Nyala, Southern Darfur, large amounts of clay have been extracted for use in brick-making. This has exposed the trees' root systems and will eventually lead to their death

CS 5.2   IDP brick-making, water use and deforestation in Darfur

Brick-making has become an important source of income for IDPs in Darfur, but has also caused considerable environmental damage around the camps. The impacts of the process include increased water consumption, damaged farmland and deforestation.

The clay for the bricks is dug from borrow pits by hand, in areas that were often previously farmed. In the wet season, these pits fill with stagnant water and contribute to environmental health problems such as malaria. The water necessary for the manufacturing process is obtained either from watercourses or from deep boreholes with submersible pumps installed by the aid community. The rate of extraction from such boreholes is not monitored, and may in some cases not be sustainable. Finally, trees are needed to fire the bricks in temporary kilns – local studies have found that one large tree is needed to fire approximately 3,000 bricks.

Simply banning such activities is not an appropriate or feasible option. A practical solution that still provides a livelihood for brick workers is urgently needed for Darfur as well as other parts of Sudan. One such option could be to use compressed earth technology rather than bricks. This would require a comprehensive introduction programme addressing both the demand and supply issues.

It should be noted that the international relief community is a major customer for the bricks, particularly to build the two-metre high compound walls required by international security standards. In Darfur especially, the relief economy has become a significant factor in the deforestation process.

In Darfur, fuelwood collection is effectively uncontrolled. Camp residents reported journeying up to 15 km to find timber, and UNEP fieldwork inspections revealed extensive deforestation extending as far as 10 km from the camps. This has contributed to a major security issue, as displaced women and girls are often at risk of rape, harassment and other forms of violence when they leave the camps to collect wood. This risk, however, is one they often have no choice but to take, since there are few other sources of cooking fuel or income available to them [5.9].

The fuelwood outlook for the major camps in Northern and Western Darfur is unpromising. Substantial deforestation has taken place over the last three years and the camps are likely to remain occupied for a number of years to come. In addition, renewed fighting since late 2006 has created a new wave of displacement and new camps.

It is possible that some camps in Darfur will exhaust virtually all viable fuelwood supplies within walking distance, resulting in major fuel shortages and/or high fuel prices. Without fuel for cooking, aid food such as cereals, legumes and flour cannot be eaten. This would add an additional facet to the ecological and human rights issues already troubling Darfur.

Some fuel conservation measures were noted by UNEP and reported by others. Though it is not universal, the use of fuel-efficient stoves, for instance, was found to be well established in Darfur. However, a detailed 2006 study by the Women's Commission on fuelwood and associated gender-based violence in Darfur showed that fuel conservation measures alone would not suffice, as the wood saved through the use of efficient stoves would continue to be gathered to be sold on local markets [5.9].

Finally, a number of very small tree plantations and nursery projects have been set up in Darfur, Khartoum state and Kassala (principally in the form of 'food for work' programmes for camp residents), but these are much too limited to meet current needs.

Land degradation in camp areas

Land degradation in camp areas is caused by over-harvesting of seasonal fodder and shrubs by camp residents and their livestock (commonly goats). Aside from its environmental impact, this activity places camp residents in direct competition and potential conflict with local residents (see Case Study 5.3).

The zone outside Abu Shouk camp in El Fasher, Northern Darfur, is completely devegetated

Camp residents seeking out livelihoods comb the drylands surrounding the camps. In this case, women are gathering fodder 13 km from the camp to sell on the local market

The prime agricultural land adjacent to the wadi in El Geneina has been cultivated by townspeople for many years, and is hence not available for camp residents

The IDP camps are all located on the fringes of town, facing waterless plains. Over the last four years, they have gradually become extensions of the town, which is benefiting economically from the associated influx of aid and labour

CS 5.3   Krinding IDP camp, Western Darfur

The Krinding IDP camp on the outskirts of El Geneina, Western Darfur, provides an example of the emerging urban environmental issues associated with the Darfur crisis. In economic terms, El Geneina in 2007 is a thriving town driven in part by the relief economy associated with the concentration of IDP camps and related activities in the region. Krinding (one of several camps in the El Geneina area) is located approximately four kilometres south-east of the town centre. Satellite images (see Figure 5.4) and photographs clearly show that the camp is effectively becoming an extension of the town, a fact confirmed by ground inspections. The environmental implication of this situation is that town and camp residents must now share or compete for the natural resources of this relatively dry and infertile region.

Most of the IDP camp residents were originally farmers, but the circumstances here and in most camps in Darfur severely restrict the potential for agricultural self-sufficiency and rural livelihoods. In El Geneina, the prime agricultural land next to the *wadi* was already being intensely utilized (principally for orchards and market gardens) prior to the creation of the camps. Unable to obtain a share in this prime land, the camp residents are left with very limited opportunities for agricultural livelihoods, as other available lands (to the east of the camps) are essentially waterless and suitable only for low-intensity grazing, fodder and fuel collection.

Thirteen kilometres outside the camp, UNEP interviewed a group of women from Krinding harvesting fodder for sale in Geneina markets. This provided a small insight into the practical links between the environment, natural resource competition, camp life and human rights. The women had walked from the camp without escort in a region UN security specialists considered so violent that the UNEP site visit required a dedicated armed escort from the African Union peacekeeping troops. The rape of female camp residents on such gathering missions is unfortunately routine in this region.

Money from the sale of gathered fodder and fuelwood is a small but vital supplement for camp residents who are otherwise completely dependent upon aid. Such efforts, however, bring camp residents in direct competition with locals (both pastoralists and agriculturists) for scarce natural resources, and undermine the sustainability of rural livelihoods in the area.

Figure 5.4a   El Geneina (12.02.2002)

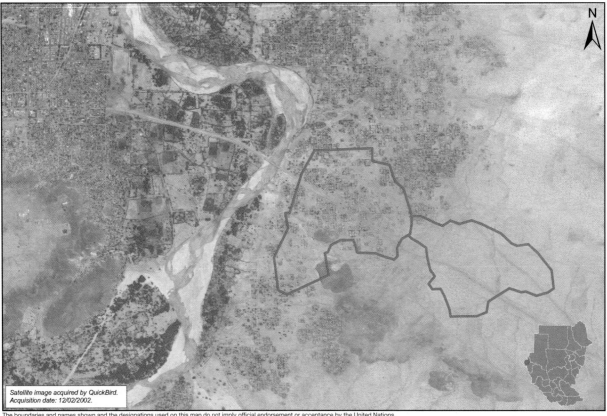

Satellite image acquired by QuickBird.
Acquisition date: 12/02/2002.

The boundaries and names shown and the designations used on this map do not imply official endorsement or acceptance by the United Nations.

Figure 5.4b   El Geneina (15.06.2006)

Krinding I camp

Krinding II camp

Satellite image acquired by QuickBird.
Acquisition date: 15/06/2006.

The boundaries and names shown and the designations used on this map do not imply official endorsement or acceptance by the United Nations.

The water container queue at a wellpoint in Abu Shouk camp. Each water point services over a thousand people

Unsustainable groundwater extraction in Darfur camps

The provision of clean water is a standard component of the aid supplied to the major camps and settlements in Sudan by the international community. This can be difficult to achieve for a combined camp population of several million, particularly in arid regions. The acknowledged standard for water supply is 15 litres per person per day, and wherever possible aid agencies aim for this as a minimum.

In Darfur, the larger camps are commonly supplied with water via a network of groundwater boreholes fitted with either hand pumps or electric submersible pumps. For the larger camps, supplying to standards all year round is proving to be possible but difficult, requiring numerous deep boreholes (between 30 and 40 m in most cases), and there is a major uncertainty as to whether this rate of supply is sustainable in the drier regions and areas with low-yield aquifers.

In some cases, there are signs that it is not: Abu Shouk camp in Southern Darfur has a population of 80,000 and rising, requiring more than 1,000 m³ of water per day. In 2006, five of twelve boreholes ran dry, indicating a substantial drop in the water table. Unfortunately, as of March 2007, groundwater level monitoring is not being conducted for any camp in Darfur, making it impossible to determine whether incidents such as the dry wells at Abu Shouk are isolated or rather the foretaste of a much larger problem looming in the future.

Short- to medium-term localized groundwater shortages are unlikely to have a major or permanent environmental impact. However, camps without a viable water supply may need to be moved, with all of the attendant issues, costs and risks that this would entail.

Water pollution

The concentration of a large number of people in temporary dwellings raises concerns for sanitation and bacteriological contamination of surface and groundwater. The standard solution is the construction of pit latrines, though these are not in place everywhere (this is particularly the case for IDP camps).

The most severe pollution problems were observed in IDP camps in the more humid regions of Sudan. UNEP field teams found major water pollution issues surrounding all informal camps visited in Southern Sudan. These same areas were epicentres for the cholera epidemic of 2006 (see Chapter 6). As detailed in Chapters 6 and 10, a lack of field data constrains more detailed analysis of this topic.

5.6 Other environmental impacts of displacement

Uncontrolled urban and slum growth

The majority of displaced people in Sudan are located in or close to towns and cities; there are over two million in the Khartoum region alone. Large-scale migration from the countryside to urban centres has been largely uncontrolled, with the result that a large number of urban slums or informal squatter settlements have been established. Urban slums are associated with a series of environmental and social problems, and are covered in detail in Chapter 6.

The urban issues associated with the north-south conflict have been ongoing for over twenty years. In contrast, the Darfur crisis is now creating new urban problems, as the majority of the displaced person camps are tightly linked to the regional towns and cities and are fast becoming a permanent part of those settlements.

Fallow area regeneration

One minor positive impact of historical displacement has been the natural regeneration of vacated lands. Large areas in the conflict zones have been partly or completely depopulated for a number of years, and this has eased the pressure on the land from farming, grazing, burning and timber-cutting. The fallow period for the vacated areas ranges from five to twenty-five years. In the moderate to high rainfall regions, the result has been the re-growth of forests. UNEP field teams saw 'new' forests of this type throughout the Nuba mountains and north of Bor in Jonglei state. The distinguishing characteristics of 'new' forests are heavier undergrowth, the lack of fallen and older trees and fairly uniform maximum tree sizes.

The 'new' forests represent a return to a wild habitat that is expected to be reversed if the displaced populations return in equal or greater numbers than were originally present. As such, they represent both a livelihood burden (as trees need to be cleared to grow the first crops) and a windfall asset that could in theory be sustainably managed.

5.7 Environmental implications of the return process

The population return process for Sudan has very significant environmental consequences, which are presently not being addressed. Two major return processes are currently underway:

1. The ongoing return process for the approximately four million people displaced by the north-south conflict. Due to a range of practical, economic and political constraints, this is expected to take several years;

The population return and recovery process in Southern Kordofan has led to a surge in deforestation

Wau township, Western Bahr el Ghazal. One environmentally significant consequence of the return process for Southern Sudan will be the rapid growth of urban centres

2. The future return process for the approximately two million displaced in Darfur. The current instability is preventing this process from even being planned.

The key environmental question is whether the return areas will be able to support the new populations. Unfortunately, in some specific cases, it is clear that the return process will not be environmentally sustainable in the medium to long term. In the worst cases, environmental issues will make the process unsuccessful and lead to renewed displacement.

The strongest evidence for this unwelcome prediction is the current condition of many of the proposed return areas, where long-term land degradation is visible even with a reduced population. This is particularly clear in the drier Sahel belt and the area immediately southwards.

Badly degraded drylands cannot support high-density rural populations: crop yields are low and livestock-rearing is problematic due to a lack of fodder. Populations living on badly degraded land are frequently forced to move; this is already a common occurrence in the drier states such as Northern Kordofan.

UNEP has conducted a preliminary analysis of the environmental sustainability of the return process for each of the twenty-five states, based on the following factors:

1. current population density;

2. future return population and net impact on population density;

3. current land quality/extent of degradation, estimated by using a combination of desk studies, field reconnaissance and satellite data; and

4. rainfall, as the strongest indicator of resilience to environmental stress, particularly from overgrazing.

While virtually every state has environmental issues associated with displacement and returns, the most vulnerable states are considered to be Darfur (all three states, but especially Northern Darfur), Southern Kordofan, eastern Kassala, northern Blue Nile, northern Upper Nile, and northern Unity state.

The situation in Darfur is particularly clear. Many regions of Northern and Western Darfur are undergoing desertification and land degradation at a significant rate. The rural areas of these regions are now partially depopulated due to the conflict, though some tribes (principally pastoralists) are still present. Given the current condition of the land and the increasingly dry climate, traditional rural livelihoods are no longer viable, so large-scale returns to these areas cannot be recommended.

For most of Southern Sudan, the situation is relatively positive in that the higher rainfall provides for greater agricultural productivity, and hence a greater capacity to absorb returnees. Nonetheless, certain areas – particularly those surrounding major towns – are expected to come under significant stress from the predicted large-scale returns.

Food distribution at the Bor way station. UN-supported returnee families are supplied with two months worth of foodstuffs, seeds, tools and other items to assist their re-establishment

Part of a group of 75 orphans delivered to Padak county under the care of three women and an elder. The urgent priority for returnees such as these is to establish a livelihood, usually crop-raising

The dominant livelihood for most rural Dinka people is a combination of cattle-rearing and slash-and-burn agriculture. While the planned return areas were found to be generally in good to moderate condition, it is doubtful that they could provide sustainable livelihoods for the projected 47 percent increase in population

CS 5.4   The environmental impact of the return of the Dinka to Bor county, Jonglei state

The return of the Dinka people to Bor county in Jonglei state provides a case study in the likely impact of returnees on the rural environment of the south. The Dinka people are agro-pastoralists, combining cattle-rearing with wet season agriculture, and migrating seasonally according to the rains and the inundation of the *toic* (seasonal floodplains). A large proportion of the Dinka in Jonglei state were displaced from their home rangelands by the north-south conflict, and fled to the far south of the country or to refugee camps in Uganda. Localized displacement also took place as people left conflict hotspots and fled to the towns for safety. The conflict and displacement were accompanied by major cattle losses due to theft and abandonment, though some stock was retained and transported south.

In 2006, the UN and a range of NGOs commenced a managed return programme for the Dinka. Able-bodied men drove the cattle up from the far south of the country to the rangelands, while women and children were transported by barge and truck. This organized process of preparation, transportation and provision of supplies resulted (by the second half of 2006) in approximately 7,000 people returning to Jonglei state over a period of six months. This was accompanied by a substantial number of spontaneous and unassisted returns. Each assisted family was supplied with approximately two months worth of food, shelter items, seeds and agricultural tools. Livestock was not supplied.

For the people arriving at the start of the rainy season, the immediate priority was to establish shelter, clear a smallholding and plant a range of crops. This resulted in an upsurge of slash-and-burn clearance and tree felling in the return areas. The geographic extent of this clearance was focused on areas with permanent water supplies and access to community services (roads, schools, and clinics).

The UNEP team inspected a wet season agricultural area located 5 km east of the Nile and 10 km north of the township of Padak. Residents within the local *payam* (district administrative unit) provided relatively detailed statistics on what the returnee process meant for them: for a 180 km² area, the *payam* had a population of 19,000, giving a density of approximately 100 per square kilometre or one person per hectare. The Jonglei state government had provided the local administrator with an estimate of 9,000 returnees to the *payam* over a few years. Several hundred had already arrived as of April 2006, but when or whether the figure of 9,000 would be reached (particularly when contrasted with the rate of return monitored by the UN) was unclear.

The region still had good tree cover and large patches of fallow land. There was no sign of major overgrazing, soil erosion or soil fertility problems. As such, it was determined that the agricultural livelihood of the current population was probably sustainable. However, whether the area could sustain the projected 47 percent population increase was far from clear, and a significant risk of environmental degradation and food insecurity remained for the longer term, as well as the possibility that some of the population might have to migrate further.

5.8   International aspects of environment and displacement in Sudan

The export of environmental problems to neighbouring countries

The countries neighbouring Sudan host some 700,000 Sudanese refugees. In addition to a range of chronic environmental problems, these countries suffer from the impact of numerous large camps.

Refugees from Darfur in north-eastern Chad, for example, are a considerable burden to their host communities due to their sheer number (400,000 people). Since their arrival in 2003, pressure has mounted significantly on scarce natural resources such as water, fuelwood and fodder for livestock, access to which has often been a source of conflict in the region.

Uduk refugees from the Upper Nile province now living in Gambella refugee camp in western Ethiopia have, in the thirteen years since the camp was established, seriously degraded an area of almost 400 km² by clearing it for agriculture. Rehabilitating this and other areas will require considerable time and resources if the welfare of hosting communities is not to be further degraded.

5.9   Conclusions and recommendations

Conclusion

The links between displacement and the environment in Sudan are clear and significant. Environmental degradation is one of the underlying causes of displacement in dryland Sudan. Unless the process of widespread desertification and other forms of land degradation are halted, large-scale displacement is expected to continue, whether or not major conflict goes on.

The displacement of over five million Sudanese into slums, camps and informal settlements has been accompanied by major environmental damage to the often fragile environments where these settlements have developed. The larger camps, particularly in Darfur, have been epicentres

of severe degradation, and the lack of controls and solutions has led to human rights abuses, conflicts over resources and food insecurity.

The population return process is expected to result in a further wave of environmental degradation in some of the more fragile and drier return areas. In the worst cases, such as Northern Darfur, large-scale rural returns may be simply untenable as the remaining natural resources are so limited and degraded that rural livelihoods can no longer be supported.

Background to the recommendations

Because international humanitarian aid organizations are by far the strongest actors in the area of IDP and refugee camp management, recommendations linked to camps and returns are generally addressed to this community. Nonetheless, close government involvement (by both GONU and GOSS) is necessary and assumed. All recommendations are short-term (0.5 - 2 years).

Two key policy issues must be addressed by the relief community. First, the current approach to the environmental impact of camps in Sudan, particularly regarding deforestation, is largely to ignore it (with some creditable exceptions). This is not due to local attitudes or a lack of standards or other guidance on this topic – what is missing is sufficient investment in this area. This needs to be addressed at the highest level to improve the current imbalance between daily humanitarian needs and long-term sustainability.

Second, a fundamental principle of displaced population assistance is the 'right to return' to the original site of displacement. For the drier parts of Darfur, however, this issue needs to be critically examined in the context of desertification and intense competition for natural resources. Assisting people to return to areas which can no longer sustain them is not a viable solution for camp closure.

In the detailed recommendations set out below, it should be noted that while UNHCR is designated as the primary beneficiary for its role in the oversight of the displaced population issue, the actual beneficiaries are the displaced populations themselves.

Elders from Shaggarab camp and hosting communities inspect a year-old acacia tree plantation

Together with officials from the Forests National Corporation and the Commissioner for Refugees, representatives from both the hosting and the refugee communities inspect progress on preparations for a community tree nursery in Fau 5 camp

CS 5.5 Community-based rehabilitation of refugee-impacted areas in eastern Sudan

Some of the largest and longest-lasting refugee caseloads in Africa have been those of Ethiopian and Eritrean refugees settled in eastern Sudan (principally in Gedaref and Kassala states). The impact of such a large number of people – some 1.1 million refugees at its climax in 1985 – has been significant in environmental as well as social and economic terms.

In October 2002, a multidisciplinary assessment mission developed a comprehensive proposal to address the issue of camp closure and rehabilitation needs in the affected area. Initiated by UNHCR and the Government of Sudan, the Sustainable Options for Livelihood Security in Eastern Sudan (SOLSES) Programme was conceived as a scaling-down exercise of UNHCR's presence in the region, with simultaneous preparation for the hand-over of assets to local communities and authorities, as well as some environmental rehabilitation. Needs assessments were carried out to evaluate peoples' actual and anticipated needs from a range of environmental resources, as well as for health and education facilities, and water and sanitation.

The environmental component of the SOLSES Programme is managed by IUCN - The World Conservation Union. Its point of departure is the engagement of beneficiaries (both refugees and local communities) with clear links to the state's development plan and processes, through community environmental management planning.

By November 2006, Community Environment Management Plans had been established for nine refugee-impacted areas in the central states (Sennar and El Gezira), as well as for the Setit region in Kassala state. The development of such plans has been an important part of the overall needs assessment of affected communities, some of which include refugees who are not able to return to Eritrea. Many of these people have lived in camps for more than thirty years, and are essentially already integrated into the local community (in some instances it is no longer possible to determine between a camp and a local village).

Support through SOLSES is intended to build peoples' capacities and expertise so that they might become self-sufficient and, at the same time, less reliant or better able to manage the natural resources they still depend on.

Agroforestry and community/compound tree-planting have become an important component of the work to support sustainable development and income generation. In its first year alone, the sale of products from a two-hectare irrigated agroforestry plot in the Mafaza former refugee camp generated USD 1,200 in revenue. Developing management plans for forests that were established in the past fifteen to twenty years, and ensuring that these resources are cared for in the future are also part of the overall strategy. In 2005, for example, more than 14,000 ha of forest were handed over to local communities or state authorities for future management. In addition, the programme is working with local communities and forestry authorities to reafforest important areas as community forests.

As community members become more familiar and convinced of the approaches promoted through SOLSES, the programme is also helping to respond to other pressing needs, far beyond the original concept of environmental rehabilitation, such as the provision of clean water and waste disposal, the use of agricultural chemicals, and the diseases caused by dirty water or mosquitoes. It is important that environmental concerns, issues and opportunities be pro-actively built into all SOLSES and related activities in the future.

Recommendations for the international community

R5.1 Implement an IDP and refugee camp environmental and technical assistance project for Darfur. This project should include the provision of training, technical advice and guidelines for camp planners and management staff, and a number of small demonstration projects at the larger camps.

CA: TA; PB: UNHCR; UNP: UNEP; CE: 1.5M; DU: 3 years

R5.2 Develop and implement a plan to resolve the Darfur camp fuelwood energy crisis. There are numerous options available and many studies have been conducted, so any major programme should be preceded by a rapid options analysis and feasibility assessment. Major investment is needed to address this large-scale problem.

CA: PA; PB: UNHCR; UNP: UNEP; CE: 3M; DU: 3 years

R5.3 Conduct an environmental impact assessment for the return process for Southern Sudan and the Three Areas, and develop plans for impact mitigation. The assessment should also provide guidelines for state, county and *payam* (district administrative unit) officials. Area plans should be developed for identified hotspots.

CA: AS; PB: UNHCR; UNP: UNEP; CE: 0.5M; DU: 1 year

R5.4 Conduct an environmental impact and feasibility assessment for the return process in Darfur. The assessment should be a multi-agency effort and focus on the potential for the projected return areas to adequately sustain rural livelihoods in the event of peace.

CA: AS; PB: UNHCR; UNP: UNEP; CE: 0.3M; DU: 1 year

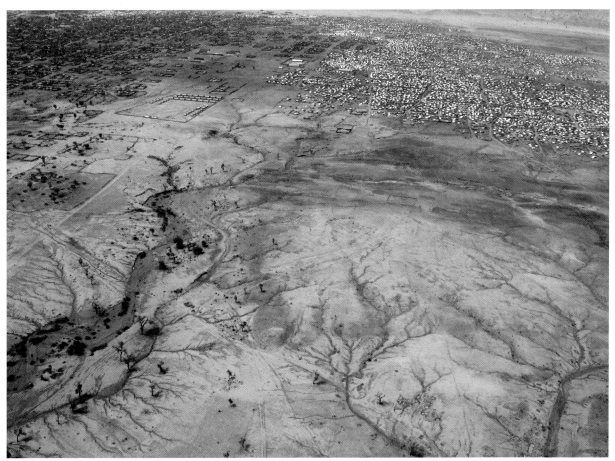

The devegetated outskirts of this IDP camp near Zalingei in Western Darfur clearly illustrate the impact of the concentrated exploitation of natural resources that were scarce to begin with

Urban Environment and Environmental Health

Urban environment and environmental health

6.1 Introduction and assessment activities

Introduction

Urban environment and environmental health issues are some of the most visible symptoms of the challenges facing Sudan. Sprawling slums, litter and polluted waterways are prevalent in most urban centres, and health and development statistics quantify in some detail the massive impact of this situation on the quality of life of the Sudanese population.

Shelter, potable water, sanitation and waste management are cross-cutting issues, and deficiencies in any of these areas can be categorized as development, health or environmental problems. This chapter focuses on the environmental aspects of these issues and the associated challenges in development and governance.

Assessment activities

Detailed desk study information was available on urban and environmental health issues, though statistical data on Southern Sudan was relatively scarce. UNEP's fieldwork included visits to urban centres of all sizes in twenty states. Particular attention was paid to the investigation of unplanned settlements, camps, waste management and sanitation. Three cities – Khartoum, Port Sudan and Juba – were selected for a closer assessment of urban services and housing.

Available statistics on environmental health and services, which are a combination of government and UN data, tell a sombre story of poverty and underdevelopment. On the national scale, even these numbers are overly optimistic, as much of the detailed data has historically been collected in the more developed areas of the northern states. On a more positive note, however, the economic development resulting from the oil boom is completely absent from older statistics, so that some areas such as Khartoum state are expected to show significant improvement from 2000 onwards.

Introductory field training in Juba for the newly recruited staff of the GOSS Ministry of Environment, Wildlife Conservation and Tourism

The capital Khartoum is by far the largest city in Sudan

In Southern Sudan, the major towns, such as Wau, consist of a small centre built in colonial times and a large fringe of informal settlements

The scope of the assessment was considered adequate to address but not fully quantify the issues at the national level. In addition, the statistical evidence collected and presented here should be treated with caution; it is considered sufficient to present trends but not to form the basis for detailed planning.

6.2   Overview of demographics and major urban centres

Demographics

The majority of Sudan's population (estimated to be between 35 and 40 million) lives in villages and hamlets in rural areas. Exact figures on the rural and urban populations are not available, but UNEP estimates, from a compendium of incomplete and obsolete sources, that approximately 70 percent live in villages, hamlets or lead a semi-nomadic existence, and 30 percent are town and city dwellers, or live in displaced persons settlements [6.1, 6.2].

Major urban centres

The urban population is concentrated in only a few cities. Greater Khartoum is by far the largest: its population was 2,918,000 in 1993, but it is estimated to have grown to more than five million in 2006. A study using 1993 census data for the northern cities showed that 64 percent of the total population of the nine largest urban centres lived in Khartoum.

Table 7.   Populations of the major cities in northern Sudan in 1993 [6.1]

City	Population	Percentage of total
Khartoum	2,918,000	64
Port Sudan	308,616	7
El Obeid	228,139	5
Nyala	220,386	5
Wad Medani	212,501	5
Gedaref	185,317	4
Kosti	172,832	4
El Fasher	141,600	3
El Geneina	127,187	3
Total	**4,427,578**	**100**

The busy port of Malakal, on the White Nile. Virtually all of the major urban centres in Sudan are located on rivers

Data on the size of the urban centres in Southern Sudan is extremely scarce. The largest towns are the state capitals of Juba, Wau, and Malakal, and the town of Yei. A 2005 urban planning study of Juba estimated the town population at 250,000 [6.3].

6.3 Overview of urban environment and environmental health issues

The UNEP assessment identified a long list of urban and environmental health issues in Sudan, but focused on those with the strongest link to the environment. In this sector, most issues are closely linked, so while the assessment could focus on individual problems, the solutions will need to be integrated. The issues investigated by UNEP were:

- rapid urbanization;
- urban planning;
- drinking water, sanitation and waterborne diseases;
- solid waste management;
- air pollution and urban transport;
- urban energy; and
- sustainable construction.

6.4 Urbanization and urban planning

Rapid urbanization

The two dominant demographic trends in Sudan are rapid population growth (estimated to be over 2.6 percent) and even faster urbanization, fuelled by population growth and a range of compounding factors including:

- drought and desertification eliminating rural livelihoods;

- mechanized agriculture schemes taking rural land from traditional farming communities;

- conflict-related insecurity forcing abandonment of rural livelihoods; and

- general flight from rural poverty in search of better livelihoods and services, such as hospitals and schools in the cities.

Moderately good data is only available for Khartoum (see Case Study 6.1). It shows growth estimates of over five percent per year from 1973 to 1993. Anecdotal evidence and data from studies conducted between 1993 and 2006 indicate that the explosive growth of Khartoum has not ceased [6.4, 6.5, 6.6]. Given the Khartoum-centred economic boom, the Darfur crisis, and the rural environmental problems of the north, UNEP's forecast for the capital is continued growth, with rapid inflows from northern states somewhat countered by outflow to Southern Sudan.

Following the signing of the Comprehensive Peace Agreement (CPA) in January 2005, displaced persons from the north and outside of Sudan have started to return to their homelands in the south. Only very approximate numbers of returns are available as of the end of 2006, but these are thought to be in the order of 300,000.

The exact percentage of these returnees relocating to southern towns is unknown, but the larger urban centres, such as Juba, Yei, Malakal, Wau and Rumbek, are clearly experiencing very rapid growth. Available data and estimates for Juba, for example, show a population increase from 56,000 in 1973 to 250,000 in 2006, which converts to a growth of 450 percent, or 14 percent (linear) per year [6.3, 6.7]. Growth rates since 2005 are expected to be much higher than this thirty-three year average.

This explosive urbanization is a severe challenge which has not been – and still is not – managed or adequately controlled by regional or local authorities. The result is chaotic urban sprawl and widespread slums, which are in turn associated with a number of health, environmental and social problems. UNEP teams observed informal settlements or slums on the outskirts of virtually every town visited in Sudan.

Urban planning

To date, not only has urban planning mostly been focused on metropolitan Khartoum, but the plans that have been developed have not been fully implemented due to under-investment in infrastructure and utilities, and underlying deficiencies in land tenure and the rule of law. While the capital has recently seen considerable investment, its size, high growth rate and historical lack of planning still constitute major challenges (see Case Study 6.1).

Large-scale informal settlements have multiplied in the Khartoum area since the 1980s. Most of these settlements have very limited access to water, and no sewage or waste management

CS 6.1 Urban planning and informal settlements in Khartoum

Metropolitan Khartoum, which comprises Khartoum, Khartoum North and Omdurman, has an area of 802.5 km². It is located at the point where the White Nile, flowing north from Uganda, meets the Blue Nile, flowing west from Ethiopia.

Founded as a military outpost in 1821, Khartoum soon became established as an important trading centre. It was chosen as the seat of government in 1823. Within the past century, the city has expanded 250 times in area and 114 times in population. The population of metropolitan Khartoum is now estimated to be more than five million, and it has a current estimated annual average growth rate of four percent, making it by far the largest and most rapidly increasing concentration of people in the country [6.6]. Some 40 percent of Khartoum residents are internally displaced persons (or children of IDPs) [6.17].

The capital is sprawling rather than dense: population density in metropolitan Khartoum was estimated at approximately 163 persons/km² in 2004 [6.4]. This low figure is due to the fact that 92 percent of Khartoum's dwelling plots contain one-level developments of 300-500 m² per plot. There are few multi-story residential buildings.

Key statistics for Khartoum are all obsolete and incomplete, but nevertheless illustrate the challenges in urban planning, transportation and provision of utilities and services.

Four master plans have been established for the development of Khartoum since independence. Most were only partially implemented, and a new plan is currently in process.

The most significant environmental health problems can be observed on the outskirts of the city, where the majority of unauthorized settlements are located. These settlements cover vast areas, contain no paved roads and offer negligible facilities for water, sanitation and solid waste management. The result is very poor sanitation, high disease rates, and difficulties in accessing basic services.

Khartoum authorities have attempted to address the issue of unauthorized settlements and squatters through a range of plans, initiatives and new settlement deals. Almost all of these have failed, and over the last ten years, authorities have turned to removing squatters by force, by bulldozing slum areas with little warning or compensation. Displaced persons settlements have been particularly vulnerable to this campaign.

At the same time, a sixty-five hectare central business district is currently being developed at the junction between the Blue and While Nile. The Almogran business district development, which is probably the largest such development in the region, includes plans for a six-hundred hectare residential estate and an eighteen-hole golf course built partly over the Sunut Forest Nature Reserve.

In sum, Khartoum's urban planning and utility provision challenges are considerable. In the absence of major investment and fundamental reforms in areas such as land tenure, the situation is likely to get significantly worse as the capital's population continues to grow.

Table 8. Key statistics for Khartoum [6.5]

Indicator	Statistic
Annual growth rate	4 %
Number of shanty towns surrounding metropolitan Khartoum (1986)	96
Estimated population of unauthorized settlements	2-3 million
Percentage ot central Khartoum covered by water network	71 %
Percentage of Khartoum connected to sewage system	28 %
Percentage of Khartoum using pit latrines or other basic systems	68 %

In Darfur, the cities of El Fasher, Nyala and El Geneina, as well as other urban centres, are severely impacted by the massive influx of displaced persons since the start of the conflict in 2003. The majority of the two million displaced are found on the fringes of urban centres which, in some cases, have increased in population by over 200 percent in three years [6.8, 6.9]. The experience of Southern Sudan indicates that a significant percentage of these 'temporary' settlements in Darfur will become permanent additions to the towns.

In Southern Sudan, urban planning challenges are twofold. First, urban populations are swelling due to the return of displaced people, and second,

Figure 6.1   Growth of Khartoum 1972-2000

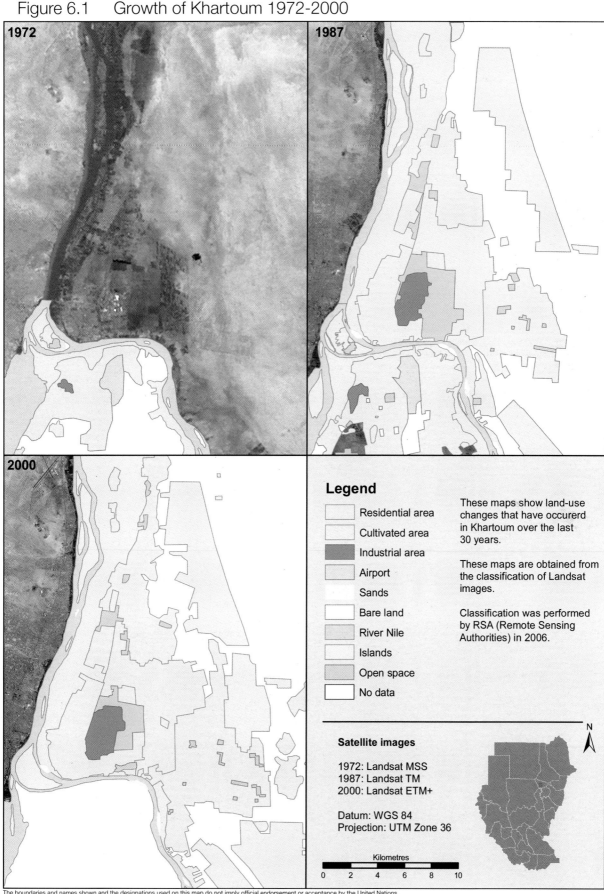

The boundaries and names shown and the designations used on this map do not imply official endorsement or acceptance by the United Nations.

some of the towns are inherently badly located: the Nile swamps and floodplains are home to several million people, but are very difficult places to develop urban centres in, due to high water tables, annual flooding and a lack of building materials such as sand, gravel, rock and suitable clay. Malakal is a classic example of the constraints imposed by location (see Case Study 6.2).

The Government of Southern Sudan launched a major urban development initiative for the ten state capitals in 2005. Planned infrastructure works include water and sanitation, roads and drainage, power supply and government buildings. The Juba civil works contracts, funded partly through the Multi-donor Trust Fund, were awarded in 2006 and on-site work is in progress. Discussions are currently being held to explore the financing of works in the other nine state capitals.

In parallel, UNDP has set up an Urban Management Programme for 2006-2009 to provide broad policy and technical support. UN Habitat has also commenced operations, and an international aid programme funded by USAID has started to conduct assessments and capacity-building in urban planning for Southern Sudan.

With limited soil absorption capacity and no gradient to allow for drainage, sewage remains stagnant in Malakal's town centre, increasing the risk of waterborne diseases

CS 6.2   Malakal: the environmental health challenges of urban development in the southern clay plains

Malakal (population approximately 200,000) is the capital of Upper Nile state. It is located on a flood plain near the junction of the White Nile and Sobat river. The town's location and local geology exacerbate the usual water and sanitation problems that afflict all of the major towns in Southern Sudan.

Indeed, the town is located on very flat ground consisting of heavy clay soil, and the water table is only 0.5 to 1.5 m below the surface. As a result, drainage is difficult. In the wet season, the town is frequently flooded for long periods of time. Because there are no significant rock or gravel deposits in the region, straightforward corrective measures like surface paving, minor relocations and raising settlements above the flood level are all extremely complex and costly, due to the need to import bulk materials.

Malakal's population is rising rapidly as people return from the north and from Ethiopia, and the limited public services are completely overstretched. There is no effective sewage system, and the open rainfall drains that serve as sewers in most of the town's streets commonly overflow in the wet season. Unsurprisingly, Malakal was one of the towns affected by the cholera epidemic of 2005-2006. Unless the problem of town sewage is addressed through a combination of investment and urban planning, preventing further outbreaks of waterborne diseases will be problematic.

6.5 Drinking water, sanitation and waterborne diseases

Access to safe and adequate drinking water

Sudan is one of the few countries in the world where the percentage of people with access to safe and adequate drinking water has declined over the last decade. Water access rates are comparable to poorer countries in sub-Saharan Africa.

Sudan actually has sufficient natural water resources in the form of rivers, lakes, seasonal streams and groundwater to supply drinking water for the population in virtually all areas, except for some parts of the northern desert (see Chapter 10).

The constraint in supplying adequate and safe drinking water is principally due to a lack of extraction and purification infrastructure. Under-investment and poverty are core obstacles for the supply of water throughout Sudan, and historical and current conflicts have exacerbated the problem.

Water availability for agriculture and industry (which can use over twenty times the amount required per capita for potable purposes) is much more limited, and constrained by the scale and reliability of the resources rather than just under-investment.

Table 9.   Overview of potable water statistics in Sudan [6.10, 6.11, 6.12]

Indicator	Statistic
Northern and national figures	
Urban populations without access to 20 litres per day (North, 2005)	40 %
Rural populations without access to 20 litres per day (North, 2005)	60 %
Khartoum population with improved water access (2005)	93 %
Blue Nile state population with improved water access (2005)	24 %
Primary schools without access to safe water	65 %
Percentage of daily income spent on water purchase by the urban poor	Up to 40%
Average water consumption per person per day from rural water points	< 6 litres
Darfur	
Average water consumption per person per day	< 7 litres
Southern Sudan	
Rural population without access to safe water supplies (2005)	75 %
Percentage of the estimated 6,500 water points currently not functioning properly	65 %

Water carts in Kassala state. Reliable water points are few and far between in the drier parts of Sudan. Many people rely on water purchased from vendors

A major aid-funded water drilling programme in Darfur has provided over a million people with access to clean water since 2003

Hand-operated well pumps provide a reliable water supply to millions of Sudanese people

The majority of the urban population of Sudan relies on basic latrines or septic tanks that are emptied by truck. In this case, the load is transferred to the Khartoum sewage works

Sanitation and sewage

Problems with sanitation are evident throughout Sudan, and inadequate facilities are the norm rather than the exception outside metropolitan Khartoum. Village fringes, disused lots and seasonal watercourses are commonly used as open toilets, with predictable health consequences.

Sanitation issues are most apparent in displaced persons settlements that have not been reached by international aid efforts. Such settlements are typically found on the outskirts of towns, and are generally very crowded and unsanitary. Large-scale aid-organized camps are usually in better condition but often face major challenges due to crowding and poor location.

Sewage systems have been installed in Khartoum, but these facilities, which cover only a quarter of the population [6.5], are now massively overstretched and not functioning properly. As a result, a large amount of untreated sewage is pumped back into the Nile, with obvious health implications for downstream communities. Most other cities have some form of sewage drainage system but no treatment, so that effluent is discharged directly into the nearest watercourse.

In the very dry areas and in towns without a sewage network, the standard solution for the more affluent communities (including the international aid community) is to use a septic tank. When tanks are full, they are emptied by a suction tanker and the contents are dumped, usually in the dry bed of a local seasonal watercourse. This process is particularly inequitable as it essentially transfers the waterborne disease risk from the affluent to the poor, who take their water from such watercourses.

Table 10. Overview of sanitation and sewage statistics in Sudan [6.10, 6.11, 6.12, 6.13]

Indicator	Statistic
Northern and national figures	
Urban population using improved sanitation facilities	80 %
Rural population using improved sanitation facilities	46 %
Primary schools with improved sanitation facilities	50 %
Percentage of Khartoum connected to sewage system	28 %
Darfur	
Population using improved sanitation facilities	< 20 %
Southern Sudan	
Population using improved sanitation facilities	< 30 %

In towns without sewage plants, septic waste tankers empty their loads on the city outskirts, in this case into the main wadi supplying drinking water to Port Sudan

Raw sewage flowing to the White Nile. Though there is a sewage network in Khartoum, it does not cover the entire city and no longer works properly, as it is stretched well beyond capacity

Waterborne diseases are a particularly severe problem in towns in Southern Sudan, due to the lack of water supply and sewage infrastructure in crowded informal housing areas like here in Juba

Waterborne diseases

The shortcomings in water quality and sanitation in Sudan are directly reflected in the incidence of waterborne diseases, which make up 80 percent of reported diseases in the country. The incidence of disease is highly seasonal: the greatest problems usually occur at the start of the wet season as the rains and run-off mobilize the faecal matter and pollution that have accumulated during the dry season.

The very limited water monitoring that has been carried out has confirmed bacteriological contamination of the Nile in Khartoum state and elsewhere in northern Sudan [6.12]. Limited groundwater monitoring in metropolitan Khartoum also confirmed bacteriological contamination [6.5]. There is practically no data for Southern Sudan.

Apart from the routine waterborne illnesses such as cholera, dysentery, hepatitis A and a range of parasitic infections like schistosomiasis, a number of tropical diseases including malaria, sleeping sickness, river blindness, guinea worm and visceral leishmaniasis are still prevalent. Southern Sudan is particularly afflicted, with an estimated 70 percent of the world's cases of guinea worm occuring there [6.13].

In 2005 and 2006, Southern Sudan experienced a major cholera outbreak in several cities including Yei, Juba, Bor and Malakal. The total number of victims recorded by WHO was over 16,000, with over 470 deaths [6.14]. Cholera is a waterborne disease linked to faecal pollution of drinking water. A UNEP team visited one of the epicentres of an outbreak in Juba in February 2006 (see Case Study 6.3) and found that water and sanitation problems were so severe and endemic that it would have been very difficult to pinpoint a single source, though according to WHO, untreated water from the White Nile and shallow open wells were the most likely suspects [6.15].

6.6 Solid waste management: consistent problems on a national scale

Solid waste management practices throughout Sudan are uniformly poor. Management is limited to organized collection from the more affluent urban areas and dumping in open landfills or open ground. In the majority of cases, garbage of all types accumulates close to its point of origin and is periodically burnt.

Carefully designed water points, such as this one that is connected to a deep well in Western Darfur, can help control the spread of waterborne diseases

Litter – plastic bags in particular – is a pervasive problem across the country, with Khartoum state being worst affected due to its population density and relative wealth.

UNEP field teams visited a number of municipal dumpsites in Port Sudan, Khartoum, El Obeid, El Geneina, Wau, Juba, Malakal and Bor, as well as in smaller towns and villages. Of all of the sites visited, only Khartoum and Juba were found to have organized systems of dumping waste into pre-defined moderately suitable locations. In all other cases, dumping took place on the outskirts of urban centres (see Case Study 6.4). Moreover, there was no waste separation at source, and slaughterhouse offal, medical wastes, sewage and chemicals were seen within the normal waste stream. Waste was also commonly dumped directly into seasonal watercourses or rivers, thereby contributing to water pollution and waterborne diseases.

Open air burning is the most common method of waste disposal in IDP settlements such as this one on the southern fringe of Khartoum

Wind-blown litter is an endemic problem in the countryside around major towns in northern and central Sudan

Offal and effluent from the slaughter yard flow past the well towards the White Nile

UNEP found that this hand-pump supplied both the slaughter yard and the nearby local settlement. Waterborne diseases such as cholera occurred in this area in 2006

CS6.3 Juba slaughter yard and community well

The slaughter yard on the eastern edge of Juba is the largest of several relatively small and primitive facilities used for slaughtering cattle, sheep and goats in the town. The site is surrounded by IDP settlements, and is approximately 200 m from the Nile and 400 m upstream of the town's municipal water extraction point.

The facility consists of an open concrete yard with a number of drains and open washbasins. On the day of UNEP's inspection, the facility was covered in blood and offal. Most of the non-commercial offal was washed into an open drain leading towards the river. The edges of the facility were used for dumping non-usable solid animal waste, and as an open latrine.

A community water point in constant use was located on the premises, within five metres of the offal drain and communal latrine. The surface of the water point was surrounded by stagnant noxious water and waste. The depth of the water table was estimated by the team to be in the order of two to three metres. Interviews of water point users revealed that many people in IDP settlements nearby had been struck with cholera.

This particular case of apparent contamination of community water supplies illustrates the problem of locating shallow groundwater wells in an urban setting in the absence of any real form of water and sanitation infrastructure or protection measures.

Since UNEP's visit, however, it has fortunately been reported that the replacement of the slaughter yard is being carried out as part of current infrastructure works in Juba. A new abattoir with modern facilities will be constructed on a new site to the north of the city.

A waste picker burns tires in order to retrieve wire to sell as scrap metal (left)

Abattoir waste was left in the open air for scavenging dogs and birds (top right)

Medical waste was found across the site and along the main road (bottom right)

CS6.4 The Port Sudan landfill

The case of Port Sudan (population approximately 500,000) illustrates the solid waste management problems that exist throughout Sudan. The city has several uncontrolled waste disposal sites on its fringes. The largest by far is located along the banks of a broad *wadi*, approximately six kilometres from the city centre.

The boundaries of the site are difficult to determine, as open dumping takes place along the access routes and in vacant or common land throughout the district. In total, it is estimated that no less than 5 km² are covered with a layer of mixed waste ranging from 0.1 to 1 m in thickness.

The site is virtually uncontrolled and presents obvious health and environmental hazards. Waste is burned and recycled by a resident group of waste pickers who live in terrible conditions on site. Animals observed feeding on the waste include dogs, goats, cattle and camels, as well as crows, kites and vultures.

The types of waste dumped on site include clinical wastes (syringes, catheters, blood packs, drugs and bandages), plastics and paper, drums and other metal scraps, small-scale chemical wastes, abattoir and food wastes, and septic tank solids and liquids.

The root cause of problems such as those seen at Port Sudan is inadequate investment in public services, including in all aspects of sanitation and waste management.

6.7   Air pollution and urban transport: a complete data vacuum

UNEP found no evidence of systematic air quality monitoring in Sudan. UNEP itself did not conduct any quantitative analysis, and thus cannot present any solid findings on the topic.

With respect to health, the most significant air pollutant in most of Sudan is dust generated by wind moving over dry and exposed soil. Indeed, large parts of northern Sudan are routinely enveloped in sand and dust storms, with high levels of atmospheric dust persisting for days at a time. This extent of exposure undoubtedly takes a toll on the population's respiratory health, although UNEP was not able to find solid statistics on this issue.

According to local authorities, the last significant air pollution and associated environmental health survey was conducted in Khartoum in 1990. This study reportedly focused on health impacts to traffic police, but the results were not available for interpretation. In 1979 and 1981, limited studies investigated particulate (dust) and sulphur dioxide (SO_2) levels in Khartoum; again, the results were not available.

On an anecdotal basis, industrial- and vehicle-based air pollution do not appear to be regional-scale problems in Sudan, though localized issues with factory and traffic emissions are evident in central Khartoum.

The current Environmental Framework Act of 2000 does include some general prohibitions on air pollution, but no numerical quality standards. As a result, there are no criteria against which the performance of individual facilities can be judged. There is also no measurement capacity within the regulatory authorities. Nonetheless, at least one state government has taken action on air pollution issues, forcing a cement factory to treat its emissions (see Case Study 7.3).

These and other positive steps at the local level should be supported via technical and legal development work, including data collection and the establishment of air quality and plant performance standards.

Carting firewood back to Juba: towns in Southern Sudan rely on a combination of firewood and charcoal for most energy needs

6.8   Urban energy: a declining dependence on wood

Sudanese cities are unusual even in the developing world in that the level of electrification is overall extremely low, and that the majority of the urban population still relies on wood for energy: a 1998 survey reported that 90 percent of urban households still depended on charcoal and wood for fuel. It is the energy needs of these ten million urban dwellers of northern and central Sudan that drive the large-scale and very unsustainable commercial charcoal industry (see Chapter 9).

There is some cause for optimism, however. Liquefied petroleum gas (LPG) is being introduced into northern Sudan – and Khartoum in particular (see Chapter 7). In addition, the electricity supplied by the Merowe dam project is expected to double the national electrical output in 2007-2008, ushering in a major switch to electricity (see Chapter 10). This move from one energy source to others with different environmental impacts is a typical example of the environmental trade-offs that occur with development.

6.9 Sustainable construction opportunities: alternatives needed to reduce deforestation rates

Sudan is currently experiencing a construction boom, which is greatly increasing the demand for construction materials, and particularly for bricks. All bricks in Sudan are baked using a low efficiency kiln system fuelled by firewood. The demand for wood has intensified the pressure on forests in most parts of the country, and especially in central Sudan and Darfur.

The cost of 'modern' construction remains extremely high, especially in Southern Sudan and Darfur, where transportation costs can be punitive. For example, the cement used for UN compounds built in 2006 in remote parts of Southern Sudan was generally airlifted – an extremely expensive approach for bulk commodity transport.

This building boom represents an opportunity to introduce sustainable and cost-effective construction techniques into the country. Techniques such as stabilized earth technology are already used on a small scale in Sudan and simply need promotion. Other practices, such as solar-aided hot water systems, have been introduced but have yet to be widely adopted.

6.10 Urban and health sector environmental governance: local management and funding issues

Under the terms of the 2005 Interim Constitution, practical management of the urban and health sectors in Sudan is largely the responsibility of state governments, which in turn delegate down to county and city governments. Cross-cutting this structure are federal ministries for physical development, health, water and irrigation, and transport.

Traditional buildings such as this barn under construction near Mabior in Jonglei state require a large number of young trees

Stabilized earth bricks are obtained by placing a mixture of clay, silt, sand and a stabilizing agent into a mechanical or hand-powered press, which crushes the mix into a hard, dense block that is then dried naturally

Stabilized earth construction techniques combine the advantages of traditional earth and modern brick construction. Compressed earth blocks have been used in the construction of several buildings in Khartoum

CS 6.5   Sustainable construction using stabilized earth blocks: an opportunity for the UN and others to do less harm to the environment

Traditional soil construction techniques are used in 80 percent of buildings in Sudan, and this figure rises to over 90 percent in rural areas (2000 data). The advantages of soil are its very low cost, its local availability and the simplicity of construction. Its disadvantages are its low strength and durability, particularly in high rainfall areas. The more affluent Sudanese therefore rely on brick construction instead, and the demand for fuel to fire bricks is one of the causes of the deforestation occurring in Sudan.

Compressed and stabilized earth construction techniques combine the advantages of both traditional earth and modern brick construction. The method can be summarized as follows: suitable moist soil consisting of a mixture of clay, silt and sometimes sand, is blended for uniformity before a stabilizing agent such as cement, lime, gypsum or bitumen is added. The material is then placed in a mechanical or hand-powered press, which crushes the soil-stabilizer mix into a hard, dense brick that is dried naturally, gaining strength in the process. The bricks obtained can be used just like fired clay or concrete bricks.

Modern compressed earth technology has proven effective in many parts of the world, and several buildings, such as the Haj Yousif experimental school in Khartoum North, have already been constructed in Sudan as demonstration projects [6.18, 6.19].

The environmental savings are significant, as studies have shown that compressed earth construction uses approximately only one to two percent of the energy for material development per cubic metre that cement and fired bricks use [6.18]. For Sudan, this translates into potentially major savings in fuelwood.

The economics of compressed earth indicate that – if introduced correctly – the technology can be commercially self-sustaining, as it can compete with brick and cement on cost grounds. The main obstacle to market entry is its novelty and a lack of local knowledge.

UN agencies in Sudan and elsewhere in developing countries use considerable amounts of fired bricks to build their offices and residential compounds. In fact, the MOSS (Minimum Operating Security Standard) requirement for a two-metre high solid wall surrounding compounds is the direct cause of the felling of thousands of trees in Sudan and elsewhere. Compressed earth technology offers the opportunity for the UN and other international aid organizations to reduce the negative impact of their presence and extend the 'do no harm' principle to include the environment.

The main issue for state governments in Sudan (outside of Khartoum) in areas such as urban planning and environmental health is insufficient funding: local officials are generally quite aware of the problems but cannot act in the absence of funds.

The second major obstacle to tackling urban and environmental health issues is the pace of urban growth and slum development: it is difficult to enforce basic planning and environmental health standards when uncontrolled settlements are set up on land that is either unsuitable for inhabitation or needed for the provision of adequate infrastructure. A particular problem arises where illegal settlements are established in flood plains and partly block existing drainage basins and corridors, resulting in increased flooding and the spread of waterborne diseases.

6.11 Conclusions and recommendations

Conclusion

While urban environment and environmental health issues are clearly apparent to all living in Sudan, attempts to change this situation have met with little success to date. The main obstacle for improvement in these areas is a lack of investment, but other problems, such as the widespread lack of adequate urban planning, also play a role.

Background to the recommendations

Water and environmental sanitation are major areas for international humanitarian funding; in the UN, work in these sectors is led by UNICEF. Solid waste management and urban planning are traditionally not well supported, though this is now changing.

It is extremely clear that neither humanitarian nor development aid efforts in these sectors will be fully successful or sustainable without greater government support, principally increased government funding. Issues such as land tenure, unauthorized settlements and chronic solid waste management problems can also only be resolved by national and local authorities.

On this basis, UNEP's recommendations are focused on increasing government capacity and support for these sectors rather than implementing site-specific projects. The exceptions are the need for practical solid waste management and sustainable construction projects in one or more locations to demonstrate the way ahead. It should be noted that a substantial humanitarian water and sanitation programme is separately promoted and managed by UNICEF and others on an annual basis, and is hence not repeated here.

Recommendations for the Government of National Unity

R6.1 Invest in urban planning capacity-building for all northern and central states, and for Darfur. This will entail a process of importing expertise and 'learning by doing' through improved master planning for each state capital. Particular attention should be given to Darfur state capitals, where the need is greatest due to the influx of people displaced by the conflict. To improve political support, assistance should be channeled in part by the Governor's office in each state.

CA: CB; PB: GONU state governments; UNP: UN Habitat; CE: 2M; DU: 3 years

R6.2 Increase investment in environmental health-related infrastructure and services in all northern and central states, and in Darfur. There is no substitute for significant investment in solving issues such as sanitation and solid waste management. Any major investment programme should proceed in stages, attempt to introduce self-sustaining financing and involve the private sector. A proportion of the total cost should be directed toward human resource capacity-building and awareness-raising. Note that this recommendation is not costed, but that the investment required to attain even a basic level of service is anticipated to be in excess of USD 1 billion over a period of more than a decade.

CA: GI; PB: GONU state governments; UNP: UN Habitat; CE: NC; DU: 10 yrs+

R6.3 Promote the growth of the LPG market in major urban centres. This measure will directly reduce the pressure on remaining forests in dryland Sudan by substituting for charcoal as an urban fuel source. Promotion may entail some form of initial subsidization of the LPG cylinders. Fuel should not be subsidized, as this would create a distorted market in the long term. Costs and duration of the programme are flexible and scalable.

CA: GI; PB: Public via MoF; UNP: UNEP; CE: 1M; DU: 2 years

R6.4 Complete a stabilized earth technology demonstration project for Khartoum and three other states including Northern Darfur. This should entail the construction of a UN and government-used building in a prominent position to maximize exposure, and should include extensive capacity-building components. The technology and capacity already exist within the Ministry of Environment and Physical Development.

CA: CB; PB: MEPD; UNP: UNOPS; CE: 1M; DU: 2 years

R6.5 Complete a stabilized earth technology demonstration project for Juba and three other states. The technology and capacity already exist within the GONU Ministry of Environment and Physical Development, and GONU assistance to GOSS on this topic would be a positive example of north-south cooperation.

CA: CB; PB: MEPD; UNP: UNOPS; CE: 1M; DU: 2 years

Recommendations for the Government of Southern Sudan

R6.6 Invest in urban planning capacity-building for all southern states. This will entail a process of importing expertise and 'learning by doing' through improved master planning for each state capital. To improve political support, assistance should be channeled in part by the Governor's office in each state.

CA:CB; PB: GOSS state governments; UNP: UN Habitat; CE: 2M; DU: 3 years

R6.7 Increase investment in environmental health-related infrastructure and services in all southern states. This recommendation matches R6.2 above with similar anticipated costs and time scales.

CA: GI; PB: GOSS state governments; UNP: UN Habitat; CE: NC; DU: 10 yrs+

Recommendations for the United Nations in Sudan

R6.8 Construct a MOSS-compliant compound perimeter for at least one base in Southern Sudan using stabilized earth technology. Such a demonstration project potentially has very high added value if explicitly endorsed by the UN.

CA: PA; PB: GONU MEPD; UNP: UNMIS and UNOPS; CE: 1M; DU: 2 years

Industry and the Environment

Industry and the environment

7.1 Introduction and assessment activities

Introduction

Sudan's industrial sector is currently undergoing rapid change and expansion. Historically limited to utilities and small-scale food processing, the sector is now booming thanks to oil production, which began in 1999.

Environmental governance of industry was virtually non-existent until 2000, and the effects of this are clearly visible today. While the situation has improved significantly over the last few years, major challenges remain in the areas of project development and impact assessment, improving the operation of older and government-managed facilities, and most importantly changing attitudes at the higher levels of government.

Industries covered in this chapter include oil production, power generation, food-processing, transportation, chemicals and construction.

Assessment activities

UNEP teams visited a range of industrial facilities across the country. In some cases, a full tour of the facility was possible; in others only brief inspections were carried out due to limited time or access. The sites visited include:

Port Sudan region, Red Sea state:

* harbour operations and warehousing (site meetings and full tour);
* several very light industry sites (site inspections);
* saltworks (full site tour);
* desalination plant (full site tour);
* power station (external viewing only); and
* refinery (site meeting only).

Khartoum state:

* Comfort soap and toothpaste factory (brief site visit); and
* GIAD car assembling complex (brief site visit).

Gezira state:

* Baggier industrial complex (brief site visit);
* Aqsa cooking oil factory (brief site visit); and
* Hibatan tannery and leather factory (closed).

Chlorine storage cylinders outside a chemical plant in Barri, Metropolitan Khartoum. UNEP's assessment of the industrial sector included visits to many factories. Access was normally granted without restriction

A small sesame seed oil pressing plant in Port Sudan. Food processing represents a significant part of the light industry sector in Sudan

Sennar state:

- Kenana sugar factory (full site tour); and
- Asalaya sugar factory (full site tour).

Southern Kordofan:

- Heglig crude oil production complex (site meetings and full tour).

Jonglei state:

- oil exploration seismic survey base and line sites (site meeting and tour).

Northern state:

- Merowe dam site (Khartoum meetings, no access to the site, visited the downstream region, see Chapter 10 for details); and
- Atbara cement factory (brief site visit).

The number of sites visited was considered sufficient to evaluate the environmental governance of industry in Sudan; the assessment was supported by an analysis of both general and site-specific legislation and enforcement practices.

Oil-related sites were visited, but not in sufficient depth and number to gain a comprehensive picture of the industry. The implications of this data gap are addressed further in this chapter.

7.2 Overview of the industrial sector in Sudan

General industrial structure

Sudan is experiencing rapid industrialization due to the growth of the oil industry and associated service industries and imports. For the purposes of this environmental assessment, industry is divided into five sectors, as follows:

1. the upstream oil industry;
2. the downstream oil products industry;
3. utilities (power generation and water supply);
4. food processing (sugar, sesame oil, cereals); and
5. miscellaneous (including mining, textile manufacturers, tanneries and workshops).

Oil, utilities and food processing dominate the industrial sector. Until recently, virtually all of the major industries in Sudan were state-owned or controlled. This has now changed, as many of the main manufacturers have been privatized. Apart from the newer oil facilities, the industrial sector has suffered from a lack of investment which is reflected in the condition of the plants and their environmental performance.

Oil industry structure

The oil industry is conventionally divided into three sectors:

- The **upstream** sector, which covers exploration for crude oil and gas, extraction, and transport via pipelines and tankers to markets;

- The **downstream oil products** sector, in which the supplied oil and gas are refined and converted into usable products (petrol, diesel, lubricants) and sold to customers; and

- The **petrochemicals** sector, in which oil and gas are converted into chemicals and materials such as solvents and plastics.

Sudan's upstream oil industry is set to dominate industrial activity in the country for the next generation. UNEP interviews indicated a nationwide concern about the environmental impacts of exploration and extraction of oil, and this topic is addressed in some detail below. In contrast, the downstream sector in Sudan is relatively small and set for moderate growth only. There is no petrochemical industry in Sudan yet.

Oil industry exploration and production history

Oil exploration in Sudan started in 1959, but the first major find was only made in 1980 by the US company Chevron (now Chevron-Texaco), north of Bentiu in Western Upper Nile state (now renamed and boundaries changed to Unity state). Further finds were made in 1982, 70 km north of Bentiu in the Heglig district, in Southern Kordofan [7.1, 7.2].

Oil production in Heglig and Bentiu was delayed until 1996 by the north-south civil war, which was itself partly caused and sustained by the competition for control of the oilfields. The conflict and political changes during this period were accompanied by a shift in international oil development partners. Most western companies gradually withdrew, due in part to pressure in their home countries. They were replaced by Chinese, Malaysian and Indian national oil companies, which now manage the oilfields in Sudan together with representatives from the Government of National Unity.

Well casings lined up beneath the Heglig drilling rig. Oil production is rapidly increasing in Sudan, as new fields are developed and transport infrastructure such as trunk pipelines and marine terminals is constructed

A crowned crane on 'toic' grassland near Padak. Much of the planned oil exploration is set to take place in the Nile flood plain, an environmentally very sensitive area

Current oil industry activities

Sudan started exporting oil in 1999. According to official figures, oil production in Sudan was approximately 400,000 barrels per day as of mid-2006, and was expected to rise to 500,000 barrels per day within a short period of time [7.3, 7.4]. Based on an oil price of USD 67 per barrel [7.5], the latter production level equates to a theoretical revenue stream of USD 33.5 million per day or USD 12.2 billion per year, which represents 14 percent of the 2005 estimated gross domestic product for Sudan (USD 85.5 billion) [7.6].

Sudan also has significant gas reserves (some 3 trillion standard cubic feet) [7.7, 7.8] and currently produces gas as a by-product of oil production in central Sudan. Unfortunately, no large market has yet been developed for this gas in Sudan. As a result, most of it is burned off by flaring. Efforts are ongoing to tap this supply by increasing the existing liquefied petroleum gas (LPG) market.

As of mid-2006, the principal oil and gas production facilities in Sudan are:

- production wells and initial treatment complexes in the fields of Heglig (Southern Kordofan), Bentiu (Unity state), Thar Jath, Muglad and Adar (Upper Nile state); some of these facilities are still under development but expected to start or increase production within the next two years;

- four crude oil export pipelines connecting the fields to Port Sudan, with a combined length of 3,900 km; and

- a marine oil export terminal at Port Sudan.

Oil exploration and production plans

Sudan's commercially recoverable oil reserves are currently in the approximate range of 500 to 800 million barrels, and total oil reserves are estimated to be up to eight billion barrels [7.8, 7.9, 7.10]. At present and projected extraction rates, these reserves will last for approximately a decade, though it is expected that further reserves will be discovered and exploited over time. Current plans are to expand production to 1.5 million barrels per day by 2008 [7.3, 7.11].

Only a small portion of central Sudan has been explored thoroughly, and only a fraction of that small area is in production. Before the Comprehensive Peace Agreement (CPA) was signed in 2005, exploration was limited to military-controlled areas in the north-south border regions. The establishment of peace and security is now allowing exploration to expand into the rest of Southern Sudan, as well as Southern Darfur.

There are nine exploration concessions in Sudan (see Figure 7.1), totalling approximately 250,000 km^2 or ten percent of the country's land area. Most of the important unexplored areas are in Southern Sudan. Accordingly, large-scale oil exploration and perhaps development are expected to come to Southern Sudan within the next ten years. Some activity has already started: the White Nile Petroleum Company has been conducting seismic surveys in Padak County, Jonglei state since 2006 (see Case Study 7.2) and plans to commence drilling in the second quarter of 2007.

7.3 Overview of industry-related environmental issues

Industry-related environmental issues can be divided into those applicable to all industries and those applicable to the upstream oil industry only.

General issues are:

- absence of environmental considerations in the development of new projects; and
- poor environmental performance at operating sites.

Upstream oil industry issues are:

- isolation from governance and scrutiny;
- existing impacts and future risks of oil exploration;
- produced water;
- produced gas flaring and utilization; and
- oil spill risks from sea transportation.

7.4 General industry-related environmental issues

An absence of environmental considerations in the development of new projects

Environmental issues have rarely been considered in the development of major industrial projects in Sudan over the last forty years. This has been the case throughout Sudan for all aspects of project implementation: design, feasibility, site selection, and facility construction and operation.

As a result, a number of large projects have had very negative impacts on the environment. Unfortunately, new projects are still being implemented without environmental consideration today (see Chapter 10 for section on dams). What's more, development in Sudan has historically been driven by a series of national-level plans and mega-projects, such as the Gezira agricultural scheme and the Jonglei canal. These schemes tend to have high-level political backing and progress rapidly from conception to construction, without opportunity for assessment or public consultation.

The construction of this major new harbour facility in Port Sudan proceeded without an environmental impact assessment or mitigation of its impacts

Figure 7.1 Sudan oil industry

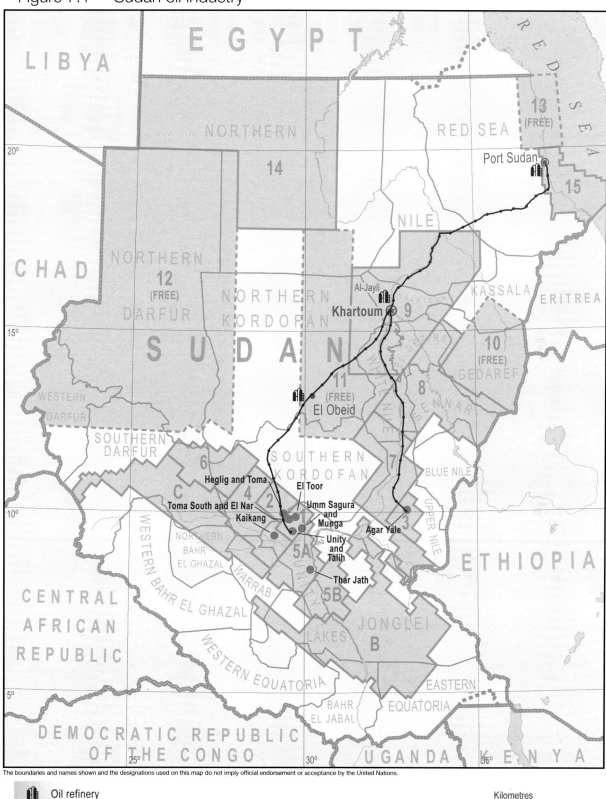

The boundaries and names shown and the designations used on this map do not imply official endorsement or acceptance by the United Nations.

Oil refinery
Oilfield
Oil pipeline
Concession block and id

UNEP/DEWA/GRID~Europe 2006

Kilometres
0 100 200 300 400 500
Lambert Azimuthal Equal-Area Projection

Sources:
ECOS; SIM (Sudan Interagency Mapping); USGS; vmaplv0, NIMA;
various reports, maps and atlases; UN Cartographic Section.

Stand of mangroves located some 500 m from the power station

The waste oil that is regularly dumped outside this Port Sudan power station migrates towards a lagoon and mangrove forest on the outskirts of the city

CS 7.1   Port Sudan power station waste oil dumping

The lack of environmental governance in the industrial sector is readily apparent throughout Sudan. In Port Sudan, for example, electricity is supplied by several government-operated oil-fired power generation stations. Power Station C is located 5 km south of the city on the Port Sudan-Suakin road. It is built on what were previously salt marshes and located approximately 200 m from a shallow lagoon, 500 m from one of the only remaining mangrove forests and at an equal distance from the principal coastal recreation site outside the city.

The diesel generators require regular oil changes, generating large quantities of waste oil. The UNEP inspection team witnessed this oil being simply poured onto the ground in vacant land next to the station, whence it gradually flowed into the lagoon; open channels had been cut in the sediment to aid its flow.

Poor environmental performance at operating sites

UNEP site inspections revealed chronic serious environmental problems at the majority of industrial facilities visited. The issues noted ranged from air emissions and water pollution to hazardous and solid waste disposal. There was no correlation with scale: large facilities had the same performance as smaller ones, if not worse. Air and liquid discharges were found to be mostly uncontrolled, and untreated effluent was seen to be discharged directly into watercourses at several sites.

Used asphalt drums dumped on the outskirts of Port Sudan

Fuel oil spillage at the Rabak cement factory, in White Nile state

The environmental performance of the two utilities visited by UNEP in Port Sudan – the water desalination plant and Power Station C – was very poor (see Case Study 7.1). Utilities are still generally owned by the state and suffer from a lack of investment. They are also effectively immune from legal sanctions because they provide vital services that cannot be interrupted.

At the country's five main sugar estates, the key problem was the release of effluent. All sugar factories were found to be releasing factory wastewater directly into the Blue and White Nile without pre-treatment. This wastewater contains an elevated biological oxygen demand (BOD), which can reach 800-3,000 ppm. The resulting pollution of river water is suspected to be the leading cause of frequent fish kills, particularly in the Blue Nile. It should be noted that the Kenana factory is in the process of constructing a wastewater treatment plant to address this problem. Others have yet to follow suit.

Waste oil discharged onto the ground from a lubricant factory in Khartoum state

Untreated effluent flows from the Assalaya sugar factory to the White Nile

7.5 Environmental issues specific to the upstream oil industry

Generic issues

The generic environmental impacts and risks associated with the oil industry are well known and include:

- oil spills during any part of the process with a particular risk related to sea transport;

- very large-scale intrusion into previously undeveloped or inaccessible areas via access roads for exploration, production plants and pipelines;

- generation of water pollutants (produced water from well fields is a particular problem);

- generation of general and chemical solid wastes;

- air emissions, particularly from gas flaring; and

- secondary development impacts as the oil facilities attract populations seeking employment and other benefits.

The significance of these impacts can vary dramatically from one oilfield or plant to another, depending on the scale of the facility, the sensitivity of the location and the standards of operation.

As noted in the introduction, UNEP's assessment did not cover the full extent of the industry. Detailed comments are hence restricted to what was physically viewed and verified by the UNEP team, and to what was reported by oil industry personnel. Unverified statements with significant implications are recorded as such.

UNEP also received numerous and generally extremely negative anecdotal reports from southern Sudanese, which focused on the following:

- discharge of untreated produced water;

- damage to pastoral land and dwellings from road building; and

- oilfield chemical dumping.

Figure 7.2   Um Sagura seismic survey grid

The boundaries and names shown and the designations used on this map do not imply official endorsement or acceptance by the United Nations.

The seismic lines and access roads in the Abyei region were cleared by bulldozer. They are visible as a grid at least ten years after completion of the survey, indicating significant damage to the vegetation and drainage patterns

Airboats used for seismic surveying access in the swamps and flood plains of Jonglei state, reducing the need for access roads in the first stages of oil exploration

A UNEP inspection of a portion of the seismic line through wooded savannah in Padak county revealed minimal long-term impact due to the limited clearance methods used

CS 7.2   Seismic surveys for oil exploration

The first stage of oil exploration that has any significant impact in the field is the construction of access roads and seismic surveying. Seismic surveys entail the capture of subsurface data in a grid pattern over thousands of square kilometres with line spacing of anywhere between 500 m and 5 km. Each line requires access by truck, and it is common practice to use a bulldozer to cut a track of four to twelve metres in width. This process can be very destructive in wooded regions and in wetlands, though the extent of the damage depends on the habitat, survey method and behaviour of the clearance teams.

Seismic lines in the Bentiu and Abyei districts, which were placed in the 1990s on behalf of the Greater Nile Petroleum Company, cross relatively open terrain and soft ground. These lines are still clearly visible in 2003 satellite images, indicating a deep cut method of clearance with significant impact on the vegetation and drainage patterns (see Figure 7.2).

In contrast, UNEP inspected a one month-old seismic line placed by Terra Seis on behalf of White Nile Petroleum in sparsely wooded and settled terrain in the Padak region. The method of clearance used was scrub clearance, avoiding trees and dwellings by offsetting the line by a few metres. The UNEP team walked one line for two kilometres and found negligible impact, apart from the stated scrub clearance.

These two examples indicate that while oil exploration will inevitably impact the environment of Southern Sudan, the impact can be greatly reduced with appropriate controls.

Additional accounts of environmental problems have been documented in some detail by a number of NGOs and international observers over the last ten years [7.12, 7.13, 7.14, 7.15]. These accounts are not reproduced here due to lack of verification by UNEP on these critical and sensitive issues.

Upstream oil industry isolation from governance and scrutiny

The upstream oil industry in Sudan is essentially self-regulated and has never been subject to independent technical scrutiny. Due to the limited scope of the assessment, UNEP cannot comment in detail on the actual performance of the upstream oil industry in Sudan. Elsewhere in the world however, the general experience is that the industry's level of environmental performance is closely linked to the level of external scrutiny – secrecy is bad for performance.

Existing impacts and future risks of oil exploration

If it is not well managed, the exploration process can have the greatest impact on the environment of all the phases of oil production, due to the large areas affected and the temporary nature of the work. Exploration is unsuccessful in over 90 percent of cases, and when the results are negative, oil companies abandon the areas surveyed. Unless it is remediated, the environmental legacy of exploration can last for generations.

The most significant of these impacts are access roads for very heavy equipment, seismic survey lines and drilling sites. The damage is mainly physical, comprising deforestation and devegetation, erosion and watercourse siltation, and disrupted drainage patterns. Extensive damage of this type was observed by the UNEP team north of the Heglig facility in Southern Kordofan. Inspections of seismic lines in Jonglei state, however, revealed a much lower level of impact (see Case Study 7.2).

The areas targeted for oil exploration in Southern Sudan are particularly vulnerable to exploration-related damage, as they do not have many existing roads, are relatively well forested, have very soft soils, and flood for several months a year. Control of such impacts should therefore be a top priority for the industry. While appropriate control measures would increase the cost of exploration, exploration itself would not be undermined, as it would be prohibited only in the most sensitive areas, and then only at certain times of the year.

Produced water

The single most significant environmental issue for crude oil production facilities in Sudan is the disposal of produced water. Produced water is the water extracted from the reservoir along with crude oil, and separated from it before the oil is transported via pipeline. The volume of water can be very large, particularly in the later years of production, when the wells tend to produce more water and less oil as reservoirs become depleted. The Heglig facility alone currently generates over ten million cubic metres of produced water annually. Full production of the central Sudan fields in ten years time may yield five to twenty times that amount.

Appropriate treatment and disposal options exist for produced water, but they can be costly. In the absence of regulations, it is unfortunately common practice around the world to simply discharge it to the nearest watercourse. Legislation and investment in treatment facilities are required to protect the environment from this type of pollution.

UNEP's inspection of the Heglig facility in March 2006 noted an operational produced water treatment facility based on reed bed technology. However, the GONU State Minister for Energy and Mining, as well as oil industry personnel, reported to UNEP in November 2006 that produced water was now being discharged untreated from the complex; volumes were not specified. The reasons given for the lack of treatment were a recent major increase in produced water flow rates and under-sizing of the treatment plant.

Produced water flowing into a holding pond at Heglig. Produced water can be difficult and expensive to treat, but has serious impacts on the environment if released untreated

Experimental reed bed for the treatment of produced water at Heglig. Like all treatment facilities, it needs to be properly designed, sized and maintained to be efficient

Produced gas flaring and utilization

The gas produced as a by-product of crude oil in Sudan is presently not all used. Some of it is flared (burned off) at the production site. Precise figures for gas flaring were not available to UNEP at the time of the assessment, but irrespective of scale, this practice has three negative impacts:

- needless emission of large volumes of greenhouse gases;

- waste of an energy resource that could feasibly replace much of the charcoal that is the cause for extensive deforestation in central Sudan; and

- local air quality issues (generally a minor problem).

The petroleum gas that is being flared could potentially be converted to bottled LPG. Though there is still ample room for growth (present market penetration is approximately 18 percent [7.7]), the market for LPG is currently developing in Sudan. In 2005, the domestic consumption – mainly in

cities in the northern states – was 102,000 tonnes, but the potential domestic demand for LPG has been estimated by government sources at 554,000 tonnes per year. Sudan also exports LPG through a terminal at Port Sudan, and this market could be expanded as well.

The development of the domestic LPG market and other uses for co-produced gas, such as electricity generation, would reduce the demand for fuelwood dramatically. In the long term, this could be the single most important factor in reversing the deforestation observed in the central and northern states.

Sea transport oil spill risks

There are two main sources of risk for oil spills arising from export operations in the Red Sea. The first is the process of loading the ships from the shore; the second is the navigation of the loaded tankers through the Red Sea.

Spills associated with loading have occurred, but have apparently been very minor. One such incident reported by the Government in 2004 was a spill of approximately 10 m^3 at the loading point of the marine oil terminal (details not verified). Given that the marine oil terminal facilities are very modern, the risk of a major spill occurring during the loading process is considered moderate to low, provided operations are well managed.

Oil tanker transport presents a larger risk. The Red Sea is a busy shipping corridor connecting Europe to the Arab Gulf states and Asia. The traffic at the Port Sudan oil terminal is a new and growing load, with over 200 tankers anticipated per year as the industry develops.

The Red Sea generally has relatively calm weather but it is littered with navigational hazards in the form of over 1,000 very small islands, sandbars and shallow submerged coral reefs. Much of the coastline is fringed by reefs and there are few safe havens able to take large vessels. In addition, the presence of coral reefs and seagrass beds makes the Red Sea highly sensitive to pollution.

Oil-spill response resources in Sudan and elsewhere are structured according to a recognized international scale:

Tier 1 Small spills that can be managed using the resources available to the facility (or to a local government unit in the case of small ship or coastal spills);

Tier 2 Small- to intermediate-scale spills that require a coordinated response using local and national resources; and

Tier 3 Large spills requiring both national-level mobilization and the importation of international specialized spill response resources. There are many centres worldwide capable of providing such equipment, but only three major centres (Southampton, Singapore and Dubai) are designed for rapid and large-scale international responses.

The marine oil terminal and Port Sudan both have Tier 1 facilities (not verified). The oil terminal management has conducted several training exercises to build capacity, including spill containment boom deployment. However, there is reportedly no oil dispersant (surfactant) capacity in country, and UNEP interviews indicated that Tier 2 planning was not well advanced due to

difficulties in communication between different ministries and government bodies. The Ministry of Energy and Mining reported that the marine oil terminal had a Tier 3 agreement with Oil Spill Response Limited in Southampton (not verified).

Interviews also revealed that small oil slicks (1-10 m³) caused by passing ships clearing bilges in international shipping lanes were very common in Sudanese territorial waters. This is an endemic international problem, and is not linked to Sudan's oil industry.

To summarize, while it is impossible to eliminate the threat of a major oil spill, the risks observed and the safeguards reported to be in place for Sudan's oil export industry appear to be generally in line or only slightly below those for oil export facilities worldwide. The most important areas for improvement would be the ability to mobilize surfactant-based responses, and better coordination at the Tier 2 level. Notwithstanding the response capacity, the risk of an oil tanker incident is still considered relatively high due to the abundance of navigational hazards.

Waves breaking on a coral reef just off the marine terminal in Port Sudan

Industrial waste burning on vacant land in Khartoum state. Waste management and water pollution are two areas in need of improved governance

7.6 Industrial sector environmental governance

General industrial facilities

Industry is subject to national- and state-level environmental legislation, but the enforcement of existing laws is limited and difficult.

At the national level, Sudanese industry is governed by the Environmental Framework Act of 2001. In some cases, it is also regulated by the need to obtain and renew operating licences issued by state governments. While there is no specific national-level statute addressing the environmental impacts of industry, individual operating permits may have provisions regarding air emissions or effluents.

The most direct form of environmental governance observed by UNEP during the assessment was at the state level, where local complaints of large-scale air and water pollution had led to action by the State Governor and a form of state-level environmental council. In two cases reviewed (a cement factory and a tannery), the action was successful: the cement factory was upgraded and the tannery was shut down (see Case Study 7.3). In one other case, the facility (a lubricant plant) was resisting control.

Settlement pond under construction at the Kenana Sugar Company, located near Kosti, which has recently invested heavily in the construction of water treatment facilities

The Atbara cement factory is now privately owned

The newly installed bag house filter treats emissions from the main furnace

CS 7.3   Upgrade of the Atbara cement factory

The Atbara cement factory in Northern state is a positive example of the potential benefits of local governance and foreign investment in improving environmental performance.

The factory is one of only two major cement production facilities in Sudan. It was established in 1947 as a private sector shareholder company and began production in 1949, with second-hand equipment. It was nationalized in 1970, before being privatized and purchased by a foreign company in 1994. One of the conditions for privatization was that the existing plant emissions be significantly reduced. An eighteen-month window was given for the installation of the necessary equipment.

When this had not occurred by the deadline, the Governor of Nile state closed the plant by decree. Within three months, the company had completed installation of a filtration system and the plant was permitted to re-open. Emissions are now reported to be significantly lower and the plant is undergoing a number of other improvements.

Oil industry

The oil industry in Sudan is managed by the Ministry of Energy and Mining, and governed by directives from the highest levels of the Government of National Unity (GONU). Oil industry staff report that, in terms of environmental performance, companies are regulated by clauses of the 1998 Petroleum Wealth Act.

The White Nile Petroleum Company is an exception, as it is not controlled by GONU. Rather, the Government of Southern Sudan (GOSS) is a minor shareholder in the venture, and the company's government counterpart is the GOSS Ministry of Industry and Mining. However, UNEP's assessment of the company's operations and the Ministry's capacity has made clear that the company is effectively self-regulated.

In theory, the Environmental Framework Act of 2001 applies to the oil industry, but discussions with the GONU Ministry of Environment and Physical Development revealed that MEPD personnel could generally not gain access to oil industry sites and had never applied any form of sanction for violation of any legislation.

In addition, UNEP enquiries did not uncover any form of publicly available environmental or social impact assessment for the oil industry, although interviews with industry personnel indicated that some environment-related studies had been conducted. One management document, the (now obsolete) Marine Oil Spill Response Plan, was publicly available [7.16].

Project development and environmental impact assessments

As detailed above, environmental impact assessment (EIA) processes exist on paper in Sudan but are not followed in practice. The Environmental Framework Act of 2001 includes a basic EIA and approval process, which is not applied effectively to the majority of projects, and not applied at all to upstream oil projects.

7.7 Conclusions and recommendations

Conclusion

Environmental governance in the industrial sector of Sudan is problematic and in need of major improvement and reform. Due to the relatively limited level of industrial development to date, environmental damage has so far been moderate, but the situation is expected to worsen rapidly as Sudan embarks on an oil-financed development boom.

The main problems include:

- absence of sector-specific legislation and statutory guidance;

Oil well drilling pits such as these at Heglig are normally remediated after use. At present, however, there is no oversight of the oil industry's performance or detailed environmental standards for such work

- lack of performance standards and enforcement capacity; and

- immunity of the oil industry, state-owned firms and major new projects to public scrutiny.

The upstream oil industry and water pollution from industrial sites are sources of particular concern. There are, however, some positive examples of governance at the state level for individual facilities.

Background to the recommendations

Two key issues strongly influence the recommendations for Sudan's industrial sector. First, unlike many other sectors of the economy, industry generally has the capacity to invest its own funds in improving environmental performance, and site-specific solutions are usually straightforward. If required, capacity-building can also be purchased in the commercial market. For GONU and GOSS, industrial environmental performance is considered first and foremost to be a governance issue.

Second, the environmental impact of the oil industry in central and Southern Sudan clearly has the potential to catalyse conflict between the industry and local interests. Accordingly, resolving this issue is considered to be of the highest priority.

Recommendations for the Government of National Unity

R7.1 UNEP or another fully independent body should undertake an environmental assessment of the upstream oil industry. The scope of this assessment should encompass the impacts of past exploration, current operational practices and proposed exploration. The agreed final results should be made public, so as to eliminate the atmosphere of suspicion caused by the current information vacuum.

CA: AS; PB: MOEM; UNP: UNEP; CE: 0.4M; DU: 6 months

R7.2 Develop a national oil industry environment act with accompanying statutory guidelines and standards. This would be a major venture requiring a cooperative approach with the oil industry and GOSS. Due to the complexity, and political and

financial implications of this recommendation, the highest levels of political will and cooperation as well as international assistance are required. The cost estimate is for legislation development. The cost of legislation implementation is expected to be tens of millions of US dollars over five years to be adopted by industry into existing projects and then implemented as standard.

CA: GROL; PB: MOEM; UNP: UNEP; CE: 0.5M; DU: 2 years

R7.3 Develop a national-level, independent environmental enforcement unit for the industrial sector, including the oil industry. This would entail greatly strengthening the capacity of the Ministry of Environment and Physical Development (or a similar body) to enable it to review EIAs, issue environmental permits, conduct inspections, support prosecutions and carry out similar governance tasks.

CA: GROL; PB: MEPD; UNP: UNEP; CE: 2M; DU: per annum

Recommendations for the Government of Southern Sudan

R7.4 Establish an interim environmental screening and industrial permitting process for all new projects on GOSS territory. This would be designed to cover the urgent requirements for project assessment before adequate longer-term controls can be established. A multi-ministry committee could be appointed to review all significant project proposals and issue construction and interim operating permits (up to five years).

CA: GROL; PB: MEWCT; UNP: UNEP; CE: 0.3M; DU: 2 years

R7.5 Monitor GONU progress on R7.2 and R7.3; if not implemented within one year, commence a regional governance programme similar to that described above. Development of the oil and general industry sector will go ahead in Southern Sudan, and governance is definitely and urgently needed. A uniform approach at the national level is the preferred approach, and GOSS should lobby for this.

CA: GROL; PB: MIM; UNP: UNEP; CE: 0.7M; DU: 2 years

It is completely feasible to reduce the environmental impact of oil exploration and production to acceptable levels in all but the most ecologically sensitive areas. That, however, requires both commitment and substantial investment

Agriculture and the Environment

In this view of the Jebel Berkel archeological site in Northern state, a thin irrigated strip of date palms bordering the Nile is visible in the background. The Nile has supported agriculture in the Sahara desert for over 5,000 years, but upstream dam construction is threatening the existence of this ancient and previously sustainable form of cultivation.

Agriculture and the environment

8.1 Introduction and assessment activities

Introduction

Agriculture, which is the largest economic sector in Sudan, is at the heart of some of the country's most serious environmental problems: land degradation in its various forms, riverbank erosion, invasive species, pesticide mismanagement, water pollution, and canal sedimentation.

The significance of land degradation in Sudan cannot be underestimated: not only are 15 percent of the population partly or wholly dependent on imported food aid, but the population is growing by more than 2.6 percent per annum and per hectare crop yields are declining. In addition, conflict linked to competition over scarce agricultural resources continues in Darfur.

Without major action to stop the wave of degradation and restore land productivity, the natural resource base will simply continue to shrink, even as demand grows. Resolving this issue is thus central to achieving lasting peace and food security.

Assessment activities

UNEP first conducted a thorough desk study based on a large body of national and local knowledge on the subject of agriculture in Sudan. In the field assessment phase, UNEP teams were able to cover all principle farming systems and regions in the country. Agricultural sites were visited in twenty-one states (excluding Unity, Warrab, Eastern Equatoria and Upper Nile) and particular attention was paid to thirteen of these: Blue Nile, Gedaref, El Gezira, Jonglei, Kassala, Khartoum, Northern Kordofan, Nile, Northern, Red Sea, Sennar, Southern Kordofan, and White Nile.

Early morning at a Dinka cattle camp, Jonglei state

Large areas of woodland are being cleared for crop-planting by the returning population in Southern Kordofan

In addition to these core team efforts, UNEP – in cooperation with the Food and Agriculture Organization of the United Nations (FAO) – commissioned the World Agroforestry Institute (ICRAF) to lead a consortium of local NGOs and institutes in a detailed study of rural land use changes and degradation in fourteen locations across Sudan. The ICRAF team first performed remote sensing analyses – each covering approximately 2,500 km² – of the fourteen target areas. Field teams then visited nine of these sites to conduct ground truthing.

8.2   Overview of agriculture in Sudan

The largest economic sector in Sudan

Estimates of Sudan's cultivable area range from 84 to 105 million hectares, or 34 to 42 percent of the country. Of this cultivable area, between 12.6 and 16.65 million hectares or 15-16 percent (1980-2002 data) are actually farmed in a given year, depending largely on rainfall levels [8.1, 8.2, 8.3]. Hence the frequent claim that Sudan is the potential 'breadbasket' of Africa and the Middle East.

The FAO country report for 2004 indicates that the agricultural sector is the main source of sustained growth and the backbone of Sudan's economy in terms of contribution to the gross domestic product (GDP). Although the sector's economic stake is declining with the emergence of the oil industry, Sudan continues to depend heavily on agriculture, whose share currently fluctuates around 40 percent of the GDP [8.1]. The value of the crop and livestock sub-sectors, which together contribute 80 to 90 percent of non-oil export earnings, is almost equal at 47 and 46 percent respectively [8.4].

Five main types of farming are practised in Sudan, and each has a specific set of environmental impacts:

- mechanized rain-fed agricultural schemes;
- traditional rain-fed agriculture;
- mechanized irrigation schemes;
- traditional irrigation; and
- livestock husbandry/pastoralism.

Fifty-eight percent of the active workforce is employed in agriculture, while 83 percent of the population depends on farming for its livelihood: 70 percent depends on traditional rain-fed farming, 12 percent on irrigated agriculture and only 0.7 percent on mechanized agriculture [8.4]. Sorghum, millet and maize are the main food crops. Other important produce for the domestic market includes sugarcane, dates, wheat, sunflower, pulses and forage. The principle export crops are cotton, gum arabic, sesame, groundnuts, fruits and vegetables.

Commercial agricultural activities are mostly concentrated in a belt at the centre of the country, which extends approximately 1,100 km from east to west between latitudes 10° and 14° north, in the semi-arid dry savannah zone. Small-scale subsistence agriculture is found throughout Sudan, and is dominant in Southern Sudan and Darfur. On average, traditional and mechanized agriculture account for 55 and 45 percent respectively of the rain-fed cultivated area [8.3, 8.4]. Due to the vagaries of rainfall, however, and to the fact that significant swathes of mechanized agriculture have been abandoned because of land degradation, economic collapse and conflict, these estimates are only indicative.

Figure 8.1   Major agricultural schemes

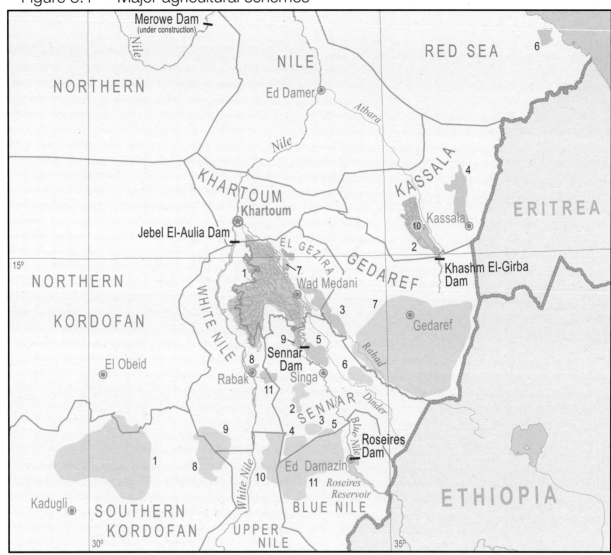

The boundaries and names shown and the designations used on this map do not imply official endorsement or acceptance by the United Nations.

Irrigated Agricultural Schemes

1. Gezira and Managil	870'750 ha	7. Guneid Sugar	15'795 ha
2. New Halfa	152'280 ha	8. Assalaya Sugar	14'175 ha
3. Rahad	121'500 ha	9. Sennar Sugar	12'960 ha
4. Gash Delta	101'250 ha	10. Khashm El-Girba	18'225 ha
5. Suki	35'235 ha	11. Kenana Sugar	45'000 ha
6. Tokar Delta	30'780 ha		

Mechanized Agricultural Schemes
(planned and unplanned)

1. Habila	7. Gedaref
2. El-Dali	8. Southern Kordofan
3. El-Mazmum	9. White Nile
4. El-Raheed	10. Upper Nile
5. El-Sharkia	11. Blue Nile
6. Dinder	

Agricultural schemes boundaries are approximate.

Kilometres

0   50   100   150   200   250
Lambert Azimuthal Equal-Area Projection

Sources:
SIM (Sudan Interagency Mapping); FAO; vmaplv0, gns, NIMA;
The Gateway to Astronaut Photography of Earth, NASA;
various reports, maps and atlases; UN Cartographic Section.

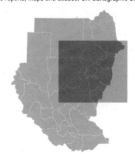

UNEP/DEWA/GRID~Europe 2006

The largest irrigated area in sub-Saharan Africa

Sudan boasts the largest irrigated area in sub-Saharan Africa and ranks second only to Egypt on the continent. Given that only two-thirds of the estimated potentially irrigable area of 2.8 million hectares are utilized and that this figure does not include Southern Sudan's virtually unused vast potential, there is significant opportunity for further expansion.

Irrigated agriculture in Sudan falls into two broad categories: traditional irrigation and modern schemes. Approximately 90 percent of the irrigated area is managed under the latter [8.1, 8.2]. Sorghum is the main cultivated crop, followed by cotton, fodder, wheat, vegetables, groundnuts and sugarcane.

The importance of the irrigated sub-sector is reflected in the fact that while it makes up only 7 percent of the cultivated area, it accounts for more than half of the crop yields. Although large-scale irrigation schemes have been Sudan's leading economic investment in the past century, various studies indicate that their performance has been considerably below potential. Of the 1.9 million hectares prepared for irrigation, only half was actually cultivated in 2005, owing largely to dilapidated irrigation and drainage infrastructure [8.1]. Environmental factors such as canal sedimentation have also contributed to low irrigation returns.

A livestock herd of over 130 million

Estimates of grazing land vary between 97 and 117 million hectares, or 39 and 47 percent of the country. Rangeland is found in almost all of Sudan's ecological zones, with the exception of montane and real desert areas. As is the case with arable land, however, an overwhelming proportion (80 percent) is found in semi-desert and low rainfall savannah zones characterized by unpredictable rainfall and frequent droughts [8.1, 8.5]. The rangeland's vulnerability to overgrazing is thus high, and its overlap with cultivation is a major source of potential conflict.

The livestock population consists mainly of camels, sheep and goats in the desert and semi-desert areas, and of cattle in the low to high rainfall savannah and Upper Nile floodplains. The estimated 134 million livestock in Sudan are almost entirely reared under nomadic and semi-pastoral systems [8.5].

8.3 Cross-cutting environmental issues and impacts

A broad array of issues and impacts were observed in the course of the assessment. The majority related to one or two of the agricultural sub-sectors only, but four cross-cutting issues were noted:

- population pressure, conflict and displacement linkages;
- climate and climate change;
- desertification and land degradation; and
- invasive species, namely the mesquite tree in northern and eastern Sudan.

Population pressure, conflict and displacement linkages

As discussed in Chapters 4 and 5, the issues of conflict and displacement, environmental degradation and Sudan's rising population are considered to be intrinsically linked. The situation in many of the drier parts of rural Sudan today can only be described as an intense and unremitting competition amongst an impoverished population for scarce and diminishing natural resources. Episodic events such as droughts, conflicts and waves of displacement are important, but considered to be part of a larger trend of rural landscapes stretched beyond their limit and declining in long-term capacity as a result.

Climate and climate change

This issue is addressed in detail in Chapter 3. In sum, the agricultural sector in Sudan is highly vulnerable to shortages in rainfall. There has been a substantial decline in precipitation in the dryland parts of the country, and global warming models predict that this trend will continue.

Desertification and other forms of land degradation

Land degradation is a critical issue throughout the country, including in areas with the highest rainfall. Its various forms are deforestation, devegetation and species changes, loss of soil fertility and seed bank, and the physical loss of soil through erosion. In the drier regions, degradation is usually referred to as desertification. In Sudan, its principal causes are crop cultivation, overgrazing, cutting trees for firewood and charcoal, and climate change.

Invasive species: the mesquite tree in northern and eastern Sudan

The invasive tree species known as mesquite (*Prosopis juliflora*) has taken over large areas of land in both pastoral regions and irrigation schemes. While it is a particular problem for spate irrigation schemes, it has proven highly useful for dune stabilization in other areas (see Case Study 8.1). Because of its negative impacts, the government of Sudan passed a law in 1995 to eradicate the tree. This has proven very difficult, however, as the species has very deep-seated root systems and can regenerate even if cut down below ground level.

Mesquite is currently still spreading, and complete eradication of the tree in Sudan is considered by UNEP and others in the forestry and environmental management field to be physically impossible, economically unviable and more importantly, not warranted. The recommended alternative is control, with elimination in high-value irrigated land only. Because mesquite seed pods are distributed in the droppings of animals, any control measure

will need to address the issue of the uncontrolled communal grazing of existing tree stands.

At the same time, efforts need to be made to maximize the benefits of mesquite. If managed from seedlings, mesquite can grow in a manner that allows it to be used for shade, fruit, fuelwood and construction timber. Given the dire deforestation situation in northern and central Sudan, the opportunity of this renewable resource should not be underestimated.

Though there are potentially viable native alternatives to mesquite, their use in new dune stabilization projects has been limited to date. It is therefore recommended that greater investment be made in researching the potential of native plants and trees, and capitalizing on indigenous knowledge in environmental rehabilitation and desertification control. Some of the promising native plant species include *Tamarix aphylla* (Tarfa), *Leptadenia pyrotechnica* (Markh), *Salvadora persica* (Arak), *Imperata cylindrica* (Halfa) and *Capparis decidua* (Tundub).

Figure 8.2   The spread of mesquite in the Tokar delta

The boundaries and names shown and the designations used on this map do not imply official endorsement or acceptance by the United Nations.

Clearing mesquite in the Tokar delta, Red Sea state   *A mesquite thicket in Red Sea state*

CS 8.1   Positive and negative aspects of mesquite

The mesquite tree (*Prosopis juliflora*) is the most important invasive species in Sudan. It is a fast growing and highly drought-resistant small tree that is spread by the distribution of its seed pods in the droppings of grazing animals. The tree is characterized by a high density of long, sharp and hard thorns, and very tangled dense growth. Mesquite out-competes a range of native species in arid areas. Where conditions are most suitable, it can become the dominant form of vegetation, forming monoculture thickets and forests.

Mesquite was reportedly first brought to Sudan from Egypt and South Africa in 1917 by a British government botanist. It was then deliberately introduced on a large scale into northern and eastern parts of Sudan in the 1970s and 1980s, for the purposes of dune stabilization. It has since spread in an uncontrolled manner.

The species has proven to be well suited for dune stabilization, but overall problematic for Sudan. For pastoralist societies, its principle disadvantage is that its foliage is essentially inedible by all herd animals, so that it provides negligible fodder compared to the native species it replaces. For farmers, mesquite is a major menace in the wetter *wadi* regions most prized for crop-raising, where it crowds out native and edible plants, blocks drains and irrigation canals and forms dense impenetrable thickets. These same features, however, make mesquite trees ideal for use as dune stabilizers and windbreakers. Besides, the plant also yields fruit, timber for construction, and fuelwood.

The contrasting views on mesquite are best illustrated in two case study locations: the Tokar delta and the Gandato irrigation scheme. The Tokar delta in Red Sea state is a water-rich and fertile oasis in an otherwise very arid and barren coastal desert environment. Water and sediment from the neighbouring mountains converge onto the delta and replenish it on an annual basis, providing perfect conditions for high-yield agriculture without irrigation. The area was used for cereal production for centuries, before being developed as a major cotton production centre during the 20th century.

In 1993, the border conflict between Sudan and Eritrea engulfed the delta region, forcing the local population off the land, which then lay effectively untouched until early 2005. Within this twelve-year period, the approximately 50,000 hectares were completely covered by a dense thicket of mesquite. Early efforts at hand clearance proved ineffective, but a major mechanical clearance project (funded by the European Commission) commenced in 2004. By February 2006, approximately 3,000 hectares had been cleared and converted back to agriculture. While this type of mechanical clearance may be economically viable for recovering high-value agricultural land, it is unlikely to be viable for low-value pastoral land, where other solutions such as land abandonment or reduction in grazing intensity may be required.

In the Gandato irrigation scheme, in White Nile state, traditional farmers have used mesquite to stabilize dunes which would otherwise overrun prime farming land. Thanks to its bushy habitus with branches down to the ground, *Prosopis* is one of the best tree species to use in shelterbelts against sand and wind encroachment. Shelterbelts or buffer zones of mesquite trees can reduce the speed of wind to half of what it is in bare landscapes, and trap the sand carried by the wind so that villages and cultivated fields inside the shelterbelt are almost entirely protected. Physical protection against sand invasion is a highly important positive environmental service provided by *Prosopis*.

Given the impossibility of eradication and the continuing need for dune stabilization, the recommended strategy for mesquite is a combination of control and better utilization in areas where it is already established, and replacement by native species as a preferred option for new stabilization projects.

8.4 Mechanized rain-fed agriculture sector impacts and issues

A history of rapid and uncontrolled development

Generally speaking, the development of mechanized agriculture in Sudan has been accompanied by large-scale destruction of the environment. Not only does the sector have major environmental problems of its own, but its uncontrolled expansion and replacement of other forms of agriculture have triggered a wide range of negative impacts in other sectors as well.

The core of the issues related to mechanized agriculture can be found in the lack of control and planning that accompanied the rapid development of the sector during the last half of the 20th century. The mechanization of rain-fed agriculture was initiated by the British in Gedaref in 1944 to meet the food needs of their army in East Africa. Following independence in 1956, the government adopted a policy to expand mechanized farming and encouraged the private sector to invest in new schemes [8.2].

Today, mechanized agriculture occupies a swathe of the clay plains in the high rainfall savannah belt estimated to be 6.5 million hectares, extending from the Butana plains in the east to Southern Kordofan in central Sudan. This area covers parts of the states of Gedaref, Kassala, Blue Nile, Sennar, White Nile, Upper Nile and Southern Kordofan. The principle crops cultivated are sorghum, sesame, groundnuts and, to a lesser extent, cotton and sunflower. UNEP visited three mechanized farming areas: Habila in Southern Kordofan; Dali-Mazmum in Sennar state; and the region bordering Dinder National Park in Gedaref.

Original plans called for the government to set aside large blocks of land (up to several hundred thousand hectares) and divide them into plots of 420 or 630 hectares. Half of the parcels were to be leased to private tenants, while the other half was left as grass fallow. After four years, farmers were to exchange the formerly leased land with adjacent fallow plots to allow the soil to recover [8.2].

A typical mechanized agriculture landscape in Dali, Sennar state, with Mount Moya providing some relief to an otherwise flat topography

exhaustion. The resultant suite of environmental, social and economic consequences, which has been highly damaging, includes the destruction of forests and pre-existing agricultural and social systems, soil erosion and increased flash floods, soil depletion and a collapse in yields.

To counter this accelerating environmental degradation, the federal Ministry of Agriculture and Forestry has required of new leases since the mid-1990s that 10 percent of the proposed scheme area be allocated to shelterbelts. UNEP observed, however, that this requirement was by and large ignored; a fact that was also widely corroborated in discussions with the responsible authorities. Reasons for this failure include limited outreach to farmers and lack of incentive, as shelterbelts are the property of the forest authorities. Moreover, farmers' interest in planting *A. senegal* shelterbelts fluctuate with gum market prices.

Even if it were implemented, the 10 percent quota would be insufficient. In addition to shelterbelts, which should be implemented at more frequent intervals (i.e. every 250 m rather than the current 500 m), forest reserves equivalent to no less than 25 percent of the farmed area should be created within and around the overall scheme. This would contribute to enhancing soil fertility and mitigating the impacts of flash floods.

These problems have been well documented by national and international researchers, but no significant or proactive corrective measures have been introduced to date. In contrast, the GONU Ministry of Agriculture and Forestry's 2006 plans (the 2006 'Green Programme') call for further investment in the large-scale expansion of mechanized agriculture.

Although authorities require that at least ten percent of all new mechanized agricultural schemes be protected by shelter belts, implementation is irregular and problematic

This model, however, has almost never been followed in practice. As demand outstripped the capacity of government to demarcate land, not only were fallow periods increasingly not observed, but private farmers illegally seized large areas outside the designated blocks. In Gedaref, for example, almost 66 percent of the 2.6 million hectares under mechanized agriculture in 1997 were unauthorized holdings, referred to as non-planned schemes [8.6]. In the Habila region, some 45 percent of mechanized farms in 1985 were unsanctioned [8.7]. In Sennar state, officials from the State Ministry of Agriculture confirmed that mechanized schemes were introduced in the 1950s with virtually no planning, and that pastoral routes were adversely affected as a result. The Ministry's reports reveal that 60 percent of Sennar's two million hectares under rain-fed agriculture are occupied by non-authorized mechanized schemes, while 30 percent are under planned mechanization and 10 percent under traditional agriculture. These changes in land use continue to lead to violent clashes between farmers and nomads, as in Dali and Mazmum.

Mechanized farming in Sudan has in effect degenerated into a crude form of extensive shifting cultivation with a tractor, exploiting land to

Destruction of forests and pre-existing agricultural and social systems

Land taken by mechanized schemes was generally not vacant. Instead it supported either pastoralism, traditional shifting rain-fed agriculture or wild habitats, principally open woodlands and treed plains. This was all appropriated without compensation and is now permanently lost. Important wildlife habitats and sources of wood products have vanished, and mechanized farming is now even encroaching on legally protected areas like Dinder National Park. The clearing has been so disorderly that forest authorities believe that in some cases the real intent was charcoal and firewood production rather than agriculture. Forest officials in Southern Kordofan reported that they had at times been obliged to issue permits for forest clearance even where trees covered more than 50 percent of the land.

Soil depletion, yield collapse, desertification and abandonment

Mechanized agriculture schemes have traditionally used neither fertilizers, nor organized crop rotation or fallow systems. The inevitable and well documented result has been a collapse in per hectare yields. In Gedaref state, for example, sorghum and sesame yields in 2002 had reportedly dropped by about 70 and 64 percent respectively from 1980 levels in established areas [8.8]. Given the region's wide climatic variations and patchy agricultural data, more detailed analysis is required, but a general trend of diminishing harvests is evident. As a direct result of this decline, sponsors of mechanized schemes have been forced to expand the total area under cultivation just to maintain output.

The final stage of mechanized agriculture as it is practised in Sudan is the abandonment of land due to yields dropping below economic limits. The total area abandoned to date is unknown, but estimated by GONU Ministry of Agriculture and Forestry officials to be in the order of millions of hectares. Abandoned land is generally found in the northern part of the mechanized scheme belt. Desertification is clearly apparent in such regions, particularly in Khartoum state, Kassala and Northern Kordofan. In a country with massive food insecurity and ongoing conflicts over land, such waste of natural resources is tragic and raises the spectre of the intensification of existing problems.

A new and serious development with both environmental and conflict-related implications is that there is now little available land left for expansion of the schemes in northern and central states. Major new schemes can only be developed in two areas, with serious environmental, social and political consequences in either case:

The tractor has enabled a massive expansion of mechanized agriculture, fundamentally altering the landscape of central Sudan, as here in Gedaref state

- Southern Darfur and southern parts of Northern Darfur, on the sandy *goz* soils, which are well recognized as very fragile, thin and prone to wind and water erosion; and

- territory within the Three Areas and ten states of Southern Sudan, which may be more suitable for agriculture but are currently occupied (mainly by pastoralists) and extremely sensitive politically and socially. The introduction of such schemes into Southern Kordofan and Blue Nile state was a catalyst for conflict in the past and would in all likelihood be in this case as well.

Given this track record of problems and the ongoing loss of fertile land, GONU plans for further expansion of the sector are a source of deep concern.

8.5   Traditional rain-fed agriculture sector impacts and issues

Population pressure and lack of development

The principle problem facing the traditional rain-fed sector is population pressure driving unsustainable rates of exploitation. This is also a main cause of deforestation in Sudan (see Chapter 9). This issue is actually a missed opportunity as well as a symptom of under-development: in the bid for immediate food security, traditional farmers are burning and clearing forests that would have a much higher return as agroforestry plantations than as short-term crops. In Southern Sudan, high-value timber trees are being burnt simply to clear land for a few years of low-intensity maize production.

The core of food security for Sudan

A majority of Sudanese farmers (70 percent) rely on rain-fed farming for their sustenance. This is generally a low input/low yield production system characterized by small farms ranging from two to thirty hectares in size and relying on labour-intensive cultivation with hand tools. Available estimates (virtually all from northern and central Sudan) show that the traditional rain-fed sector contributes the entire production of millet, 11 percent of sorghum, 48 percent of groundnuts and 28 percent of sesame in the country [8.1]. Despite its importance, this sub-sector has suffered from low social and economic investment, resulting in negligible technical development. Given the heavy dependence on food crops produced by traditional rain-fed agriculture, however, its critical role in upholding food security cannot be overemphasized.

Unsustainable land clearing and crop-raising observed in all areas

Across Sudan, UNEP noted a general trend of intensification of traditional rain-fed agriculture and associated land degradation. In the drier areas, repeated monoculture without crop rotation and adequate fallow periods has led to a decline in soil fertility. This has, in turn, increased run-off and topsoil erosion, further degrading the soil and inhibiting re-establishment of non-pioneer vegetation and potential restoration of wildlife habitats.

In the very dry regions of Northern Kordofan and Darfur, farmers have long relied on a relatively sophisticated system of rotation and inter-cropping, producing both cereal crops and gum arabic from *Acacia senegal* trees. This system is now breaking down due to pressures from drought, desertification, population increase and mechanized agriculture (see Case Study 8.2).

Farmers outside of Mornei, Western Darfur. Traditional rain-fed agriculture is very labour-intensive

A gum arabic farmer from the Jawama'a tribe in El Darota in the heartland of Northern Kordofan's gum belt

This badly degraded land near El Azaza maya, now dominated by Calotropis procera, used to be vegetated by Acacia senegal

A freshly exuded 'gum tear'. Sudan is the world's largest exporter of gum arabic, though its stake is reportedly declining

CS 8.2   Gum arabic production: an age-old system under extreme pressure

Acacia senegal (hashab) – the tree that produces gum arabic – grows naturally in the low rainfall savannah zone, an area extending from eastern Darfur to the Blue Nile and covering one fifth of the country. A 1989 survey estimated the number of mature *A. senegal* trees to be 400 million, approximately one tenth of which was found in gum gardens [8.9].

A. senegal has effectively been 'domesticated' through the development of an indigenous bush-fallow system, whereby agricultural cropping and forest regeneration are practiced in sequence. With the completion of the forest rotation (the bush period), the land is cleared for crop farming. At the same time, important trees such as *Balanites aegyptiaca* (heglig) are left intact. Fertilized by the nitrogen-fixing acacia, yields are typically high and cultivation can continue for five to seven years before the land is forsaken for another bush rotation.

Traditionally, farmers would organize their land into five blocks under a system managed on a twenty-five year rotation. This was successful as long as the farm functioned as a single unit. With the growing population and fragmentation of holdings, however, farmers can no longer afford the space to pursue twenty-five year gum garden rotations. In many cases, rotations have been shortened to only ten or twelve years, which is far too short to restore soil fertility [8.6]. Moreover, the *goz* sands (arenosols) on which *A. senegal* flourishes are highly susceptible to wind and water erosion. As a result, extensive land degradation, particularly along the belt's upper extent, has ensued.

In the sandy plains of Bara province, the removal of acacia trees has led to dune mobilization and sand encroachment on agricultural lands. The situation has been further exacerbated by recurrent droughts. The 1989 drought alone is reported to have killed up to half the gum trees – an event from which the gum belt has not yet fully recovered.

The general trend is of a southward decline of the gum belt: the Gum Arabic Research Station in El Obeid has reported that *A. senegal* is no longer found north of 13° 45' and that it is sparse north of 13°. This represents a contraction of 28 to 110 km compared to the Harrison and Jackson baseline of 1958. This decline also correlates with a southward shift of isohyets. These changes, however, are not fully substantiated and more detailed scientific evidence is needed to document fluxes in the gum belt. Similar problems have beset other traditional bush-fallow systems reliant on indigenous tree species, such as *Acacia seyal*, from which gum is also extracted.

Population increases and displacement are also forcing the size of individual plots down, with the average size falling to around four hectares in some northern states. This is too small a land base to practice bush-fallow shifting cultivation. As farmers become locked into shorter rotations, the pressure on the land increases, inhibiting the restoration of soil fertility.

Gum farmers are trying to cope with these pressures by switching from sequential rotation to simultaneous inter-cropping of *A. senegal* with food crops such as millet, sorghum, faba beans, sesame and groundnuts. The Gum Arabic Research Station is also promoting the adoption of such agroforestry practices, but limited resources to conduct research and a poor agricultural extension service are curtailing its efforts. In addition, the profitability of gum cultivation has been affected by changes in real producer prices, making it less attractive to farmers.

In the wetter regions of Sudan, the stress on the land is evidenced by the gradual replacement of *harig* (slash-and-burn) patterns of vegetation with large areas that remain permanently cleared of forest. The UNEP-ICRAF analysis and fieldwork indicated a similar pattern of deforestation and growth in rain-fed agriculture in Yambio, Yei, Wau, Aweil and Bor. In certain areas of Southern Sudan such as Yei and Yambio counties, population pressure has reduced the fallow period from an estimated average of twenty years to five years or less. Such short turnover periods are insufficient for forest regeneration or restoration of soil fertility (see Figure 8.3).

The Nuba mountains are in a comparable but more severe situation. During the conflict, Nuba people lost access to some of their best land and were constrained to continuously farm the same holdings, causing serious soil impoverishment. Peace has unfortunately not significantly improved the situation, as much of the land remains unavailable, having been taken over by mechanized agricultural schemes [8.10, 8.11].

Difficult choices facing the sector

Traditional rain-fed agriculture has been practised in Sudan for millennia and has proven to be stable and self-sustaining when population density is low. Demographic, political, and technical challenges are now upsetting this balance, and Sudan is experiencing a breakdown in long-held patterns and an unsustainable intensification of farming.

There are only two viable options available to reverse this trend and both are difficult. Firstly, the introduction of modern hybrid methods of sustenance agriculture, such as agroforestry, will benefit areas where it is not already practised (gum gardens are an example of agroforestry that existed well before the term was developed). Secondly, large-scale out-migration from rural areas could act to ease the pressure before major and permanent damage is done. Without these measures, large-scale out-migration will occur regardless, as a result of food insecurity.

Figure 8.3   Expansion of slash-and-burn agriculture in Yambio

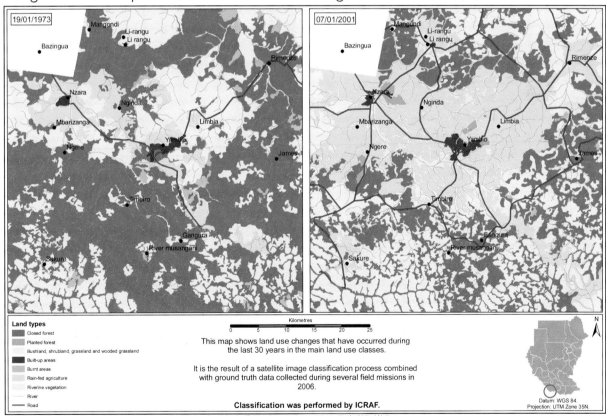

Land class analysis of satellite images from Yambio district in Western Equatoria, Southern Sudan, illustrates the pace and scale of the expansion of slash-and-burn agriculture in the region. Between 1973 and 2006, cleared agricultural land increased from 6.8 percent of the study area to 27.7 percent, mainly at the expense of closed forest and wooded grasslands

8.6 Mechanized irrigation sector environmental impacts and issues

The mechanized irrigation sector is associated with a range of environmental issues, including:

- ongoing use of pesticides and a legacy of obsolete pesticide stocks;
- water pollution from sugar factories;
- potentially unsustainable expansion plans into desert regions; and
- canal siltation, soil salinization and yield reduction.

These issues are considered to be significant, but potentially more manageable than those related to mechanized rain-fed schemes.

The major irrigation schemes

The Gezira irrigation scheme (including its Managil extension) between the Blue and White Nile covers nearly half of Sudan's total irrigated area and is reportedly the largest contiguous irrigation scheme under single administration in the world. Alone, it consumes 35 percent of Sudan's share of Nile waters [8.12]. The other two major schemes are the Rahad on the bank opposite Gezira, and the New Halfa on the Atbara river. The latter was until very recently severely affected by an infestation of mesquite, but the scheme administration reported that 60-70 percent had been cleared as of mid-2006.

In addition, there are five major sugar schemes of which four are government-run. The fifth and largest sugar plantation is the Kenana Sugar Company, which is an international public-private joint venture.

The few irrigation schemes in Southern Sudan (the Aweil rice scheme, and Mongalla and Melut sugar companies) ceased operations during the conflict, but there are plans to revive them as well as initiate new projects.

Ongoing pesticide management problems

The use, storage and disposal of pesticides are some of the most serious environmental issues related to the agricultural sector, which is by far the leading user of chemicals in Sudan. The application of pesticides in large-scale irrigation schemes and the treatment of obsolete pesticides are particular causes for concern.

The Gezira scheme main canal and the Managil extension are used by farmers for drinking water and fishing

In addition to the lack of protective gear, derelict and leaky equipment exposes workers of the Crop Protection Department in El Kajara, Gedaref, to serious occupational health hazards

The bulk of pesticide application in irrigated schemes is carried out by aerial spraying under the command of the respective scheme administrations. The Gezira Board has reported that an estimated 125,000 to 205,000 hectares of cotton and 62,000 hectares of wheat fields are sprayed annually. Past studies have revealed widespread pollution of surface waters and irrigation canals due to extensive aerial spraying, and it is likely that this remains a problem today [8.13, 8.14].

Aerial spraying of pesticides is a particular issue in the Managil extension, where the irrigation supply canal is also the main source of drinking water. There is no pesticide monitoring programme or any regular surveillance system to analyse the environmental fate of pesticides in water, soil or food. Most studies date back to the early 1980s and there is a major information gap regarding the current situation. Previous analysis has shown that DDT and its derivatives were the most widespread contaminants. Moreover, residue testing on food products, such as goat milk in the

Over 250,000 ha of cultivated land are sprayed annually in the Gezira scheme

The head of the Technical Centre for Pesticide Spraying at the Kenana Sugar Company explains the use of modern application techniques and selective pesticides

Gezira region, has indicated that organochlorine pesticide levels including the POPs heptachlor, aldrin and dieldrin, as well as endosulfan and HCH significantly exceeded standards set by the FAO/WHO Codex Alimentarius [8.1, 8.13].

Most workers queried had not received training in pesticide handling and application, and lacked protective equipment or refused to use it due to its unsuitability in a tropical climate. Surveys conducted in 1989 showed that pesticide applicators were largely ignorant of the hazardous nature of the chemicals handled and did not observe safety measures [8.13]. The same was evident during UNEP visits. Moreover, protective gear examined was often of sub-standard quality, and replacements were reportedly not provided if damaged. Mixing and spraying equipment was derelict, corroded and often leaking. As a result, the risk of occupational exposure and soil and water contamination from spills was considered to be very high.

In Gezira, there has been a positive policy shift to reduce pesticide application by discontinuing routine calendar spraying and linking application to field checks of pest infestation levels. This has

reportedly resulted in a reduction of pesticide spraying on cotton from a previous average of nine to eleven times a year to an average of two to three times a year. Other positive measures include the application of selective rather than broad-spectrum pesticides that can harm beneficial insects and lead to pest resistance. To reduce contamination from spillage, greater use is intended of closed mixing/loading systems, as well as GPS technology to limit the risk of aerial spray drift into sensitive areas such as irrigation canals. Use of this advanced equipment, however, remains the exception and not the norm. The adoption of integrated pest management practices is reportedly intended, but has not been implemented in a systematic manner due to lack of resources.

Pesticide management appears to be considerably better in the sugar companies, particularly in Kenana, where there are well-defined procedures for the use of chemicals. The company's recent adoption of a corporate environmental strategy – one of the few of its kind in Sudan – should help reinforce responsible pesticide stewardship [8.15]. This could provide a model for other agricultural corporations.

Obsolete pesticide stockpiles: a major hazard

Sudan has very large stockpiles of obsolete pesticides that are stored in very hazardous conditions across the country.

A preliminary inventory by the Plant Protection Directorate (PPD) in the early 1990s estimated the expired stock at 760 tonnes and 548 m³ of contaminated soil [8.16]. A survey completed in 2006 under a GEF-POPs project found this stock to have increased to 1,200 tonnes of obsolete pesticides and 16,000 m³ of contaminated soil [8.17]. These figures do not include several hundred tonnes of expired dressed seeds and containers. Moreover, the survey only covered some of the provincial capitals in Darfur and Southern Sudan and is therefore incomplete for those regions.

UNEP visited four stores where large stocks of expired chemicals were kept, including Hasahesa and Barakat (Gezira scheme), El Fao (Rahad scheme) and the Gedaref PPD store. In addition, a visit to the Port Sudan commercial harbour revealed a large stock of expired pesticides and other chemicals. While storage conditions were overall very poor, three sites in close proximity to inhabitations (Hasahesa, El Fao and Gedaref) were considered dangerous toxic 'hotspots' (see Case Study 8.3).

Obsolete pesticides constitute a severe environmental and public health threat and must be treated as hazardous waste. Now that an inventory of the stockpile has been completed (except for Southern Sudan and Darfur), the first step should be to collect all the materials – with a special emphasis on persistent organic pollutants (POPs) and contaminated soil – for storage in one central location.

Elsewhere in the world, safe disposal or destruction by incineration of unwanted organic pesticides and highly contaminated soil costs in the order of USD 500 to 2,000 per tonne (not including any international transportation costs). UNEP estimates that the total cost of safely resolving the pesticide legacy problem in Sudan would exceed USD 50 million. Given this amount, a permanent solution is expected to take some time and interim measures to reduce the risks are clearly needed.

Corroding drums of obsolete pesticides are stored in unsuitable conditions at Port Sudan, 30 m from the water

This cement-lined pit in Hasahesa – where an obsolete pesticide stockpile has been buried – has cracked, releasing a strong stench and exposing groundwater to a high risk of contamination. Highly hazardous and persistent heptachlor was buried in Hasahesa (inset)

An estimated 110,000 litres of very hazardous endosulfan have leaked into the ground at the main Rahad Irrigation Scheme warehouse in El Fao

CS 8.3   Obsolete pesticide storage: three extremely hazardous sites

UNEP visited three expired pesticide storage sites in central Sudan that were considered to represent a significant risk to human health and the environment.

In Hasahesa – a controversial site commonly known as the 'pesticide graveyard' – a misguided decision was made in the mid-1990s to bury a large stockpile of pesticides in a cement-sealed pit in the ground. UNEP observed that the cement casing had cracked, releasing a strong stench and exposing the groundwater to a high risk of contamination. The site was unguarded and people and livestock were seen to be trespassing. Moreover, the powder contents of torn bags, cardboard boxes and empty drums littered the site, which was adjacent to a residential community.

In El Fao, obsolete pesticides were kept in an open shed with a dirt floor. The shed was clearly not designed for long-term storage. The drums were all damaged and had leaked an estimated 110,000 litres of liquid endosulfan (a persistent organochlorine) into the soil. The gravity of the situation was amplified by the fact that an irrigation canal was located some 12 m behind the shed. Although at the time of its construction in 1977, the Fao facility was situated far from any inhabitation, migrant labourers soon settled around it. By 1993, it was decided to transform the informal settlement into a planned residential area, event though the pesticide warehouse was in its midst. The airstrip used by the pesticide spraying aircraft was also divided into residential plots within this housing scheme, clearly reflecting a poor level of land use planning [8.18].

At the Gedaref PPD store, pesticide containers were scattered haphazardly all over the site and large piles of exposed treated seed were decaying. None of the site guards had protective or first-aid equipment, or basic services such as water and electricity.

In the three aforementioned sites, complaints of ailments and allergies by neighbouring inhabitants were attributed to the noxious smell and polluted run-off, particularly during the rainy season.

Potentially unsustainable expansion plans into desert regions

Major plans for irrigation schemes downstream of Khartoum in Nile and Northern states are likely to give rise to significant environmental concerns in the next fifteen to twenty years. In Northern state, for instance, ambitious estimates by official planning place the potentially irrigable area at 800,000 to 2 million hectares. This represents a two and a half to sixfold increase of the presently cultivated area. The planned expansion is almost entirely in the upper terraces of the Nile, and a substantial proportion (around 300,000 hectares) is to be irrigated by the Merowe dam reservoir once it is completed [8.12, 8.19, 8.20]. The long-term sustainability of these reclamation projects is questionable, and they should proceed with care based on prior environmental impact assessment studies.

Water pollution from sugar factories

The main environmental problem associated with the country's five major sugar estates is the release of effluent from the sugar factories. Industrial water pollution issues are discussed in Chapters 7 and 10.

Canal siltation, soil salinization and yield reduction

Most of the major schemes have been seriously affected by heavy siltation in canals, a process that is accentuated by upstream watershed degradation. For example, the average sediment load entering the main canal in Gezira increased more than fivefold between 1933 and 1989, from 700 ppm to 3,800 ppm. It is estimated that 15 percent of the Gezira scheme is now out of production due to siltation [8.17]. Sedimentation of canals also leads to water stagnation and the emergence of weeds that provide an ideal habitat for the proliferation of water- and vector-borne diseases, in particular schistosomiasis and malaria. Chronic incidence of these diseases has been exceptionally high in the irrigation schemes.

Due to the nature of the heavy clay cracking soils, the two major problems of soil salinization and water logging typically associated with irrigated agriculture are not prevalent in Sudan's schemes. Nevertheless, there is reportedly significant salinization at local levels in the drier north-western reaches of the Gezira near Khartoum, as well as in the Guneid sugar scheme. Monoculture farming and poor implementation of crop rotation has also led to deterioration in soil fertility and a significant decline in yields.

Sugarcane is one of the major crops of the mechanized irrigated agriculture sector

8.7 Traditional irrigation sector impacts and issues: a highly productive system under threat

Traditional irrigation is concentrated on the floodplains of the main Nile downstream of Khartoum, but is also practised over substantial areas along the White and Blue Nile and the Atbara river, as well as on the Gash and Tokar deltas. Crops are irrigated in three ways. The method most widely used is based on cultivation of quick maturing crops on the highly fertile lands (*gerf*) that are exposed following the withdrawal of annual floods. This technique capitalizes on the residual moisture in the soil profile that is available when the floodwaters recede. The second type of traditional irrigation, which is based on the *shaduf* (hand-operated water lever) and the animal-driven water-wheel (*saqia*), has been almost entirely replaced by small-scale irrigation pumps. The third type, known as spate irrigation, relies on the capture and redirection of seasonal run-off to flood wide areas of arable land.

Traditional irrigation is not considered to have significant environmental impacts: in contrast, it is a relatively sustainable sector that is actually under threat from external factors including environmental problems. UNEP identified three such environmental

Cultivation of the highly fertile 'gerf' lands in Khartoum state

threats, which in combination are anticipated to significantly reduce this sector's output:

• sand dune encroachment (see Chapter 3);
• riverbank erosion, including downstream erosion from the new Merowe dam (see Chapters 3 and 10); and
• mesquite invasion.

All of these factors lead to the loss of arable land, which in turn increases poverty levels and threatens the food security of local communities. Riverbank erosion and sand dune encroachment have both had major socio-economic consequences resulting in the abandonment of entire villages.

Encroaching sand dunes, seen here in Arji in Northern state, threaten to smother the narrow strip of arable land along the Nile's flood plain, which sustains thousands of communities

Encroaching sands have displaced entire communities, such as the people of the village of Jadallah in Nile state

8.8 Livestock husbandry impacts and issues

Rangeland degradation and shrinkage

Rangeland degradation due to the overuse of shrinking resources is the most prominent environmental problem associated with livestock husbandry in Sudan. Although there is no systematic and quantitative inventory of rangeland conditions or rangeland carrying capacity on a national scale, discussions with national experts and various studies point to three negative trends:

- explosive growth in livestock numbers, particularly in central Sudan;

- major reduction in the total area of available rangelands; and

- widespread deterioration of the remaining rangelands, caused largely by drought, climate change and overstocking.

Extensive annual rangeland burning in south and central Sudan is another important environmental issue, as this practice degrades and alters the natural environment in low rainfall savannah regions.

The evidence for rangeland degradation

Though the degradation of rangelands has not been quantified, it has been extensively documented and was again confirmed by UNEP and ICRAF fieldwork and satellite image analysis in 2006 (see Case Study 8.4).

At the ground level, the most visible indicator of overgrazing is simply less forage and more exposed earth, though it is difficult to quantify the rate of degradation using such anecdotal indicators without a baseline. The UNEP-ICRAF satellite image analysis found that it was also extremely difficult to distinguish between bare earth caused by overgrazing and bare earth associated with tilled and empty fields for crops. Only in one image – of Renk district in Upper Nile state – was it possible to confidently quantify land degradation within

areas that had remained rangelands (see Figure 8.4). In this case, the proportion of degraded land as marked by bare earth increased from 0.8 percent of the total area in 1973 to 15.4 percent in 2006.

The second indicator of overgrazing is the marked replacement of palatable perennial grasses by annuals of low environmental and nutritional value. This has been confirmed by technical studies in at least six states (Northern, Gedaref, Kassala, Northern Kordofan and Northern Darfur). In Gedaref, the Range and Pasture Administration estimates that 50 percent of the state's rangelands are in a degraded state, with a severe incidence of invasive species. There are reports of valuable range species vanishing, including *Blepharis edulis* in Butana, *Andropogon gayanus* in western Kordofan, *Blepharis lenarrifolia* in Northern Kordofan and *Aritida paposa* in Northern Darfur [8.5, 8.21].

Figure 8.4   Land degradation in Renk district

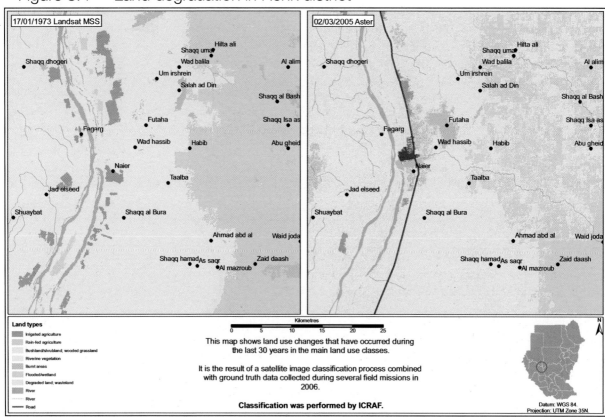

Land degradation in Renk district, Upper Nile state. In this 2,500 km² area, the rangeland is a mix of open grassland and bushland. In 1973, open rangeland made up 6.9 percent of the total land area, but had fallen to 2.8 percent by 2006, when fragmentation was very apparent. Bare and degraded land increased from 0.8 percent of the total area in 1973 to 15.4 percent in 2006. Some of the abandoned cultivated land has reverted to bushland and could potentially be used for grazing but it has major access constraints

Herders set fire to the Um Hureiza forest reserve in Sennar state before the onset of the rains

Some heavily grazed areas have undergone a notable shift from grassland to woody thickets. The encroachment of mesquite in rangelands in Kassala, Red Sea state and Gedaref, for instance, is linked to overgrazing not only because its seed is carried in droppings, but also because degraded landscapes favour the spread of such competitive pioneer species.

Bare earth in non-desert areas is an indication of both overgrazing and livestock trampling damage. Excessive trampling in dry conditions can lead to the break-up of soil, which accelerates wind erosion, and to compacting, which reduces water infiltration capacity. This is particularly noticeable around boreholes and rainwater storing dugouts known as *hafirs*, as well as along livestock migration routes throughout Sudan.

A host of factors have enabled uncontrolled overgrazing to develop, but there are two critical forces driving this process:

- explosive growth in livestock numbers over the last fifty years, resulting directly in overstocking and overgrazing; and

- a reduction in available grazing land due to desertification and unfavourable land use changes.

Agricultural encroachment onto pastoral migration routes, as evidenced here by the uprooted path markers in the region of Wad el Kabo in Gedaref state, is a major cause of conflict

When pasture is limited, pastoralists often resort to slashing trees trunks and branches to enable their livestock to feed on the otherwise unreachable parts of the tree, as seen here in the Al Ruwashida forest reserve in Gedaref state

A Mundari tribe cattle camp by the White Nile in Central Equatoria at the start of the wet season 2006

CS 8.4   Land degradation due to cattle-rearing in Southern Sudan

Pastoralist societies in Southern Sudan have developed a lifestyle closely tuned to the challenges presented by the climate and geography of the region. Each area has its own nuances, but the general pattern is of a semi-nomadic (transhumant) lifestyle dominated by cattle-rearing, with agriculture practised in the wet season only.

The possibilities for cattle-rearing in the great plains of Southern Sudan are largely constrained by the availability of water and by disease. Though the wet season generates extensive flood plains, the hot climate results in rapid evaporation and limited water supplies in the dry season.

In the wet season, the problems of mud and insect-borne diseases in the flooded areas drive pastoralists to drier ground, generally found to the north or further from the Nile tributaries. In the dry season, however, cattle camps concentrate along the fringes of swamps and watercourses.

In the far south-eastern corner of Sudan, near the Kenyan and Ethiopian borders, the climate is much drier but the soil is poorer, resulting in a lower yield of fodder and a different annual migration pattern.

UNEP has carried out a qualitative assessment of land degradation in Southern Sudan and the Three Areas using a combination of remote sensing and ground reconnaissance. Results indicate that the land is in overall moderate condition, with some clear negative trends and problem areas.

Within the southern clay plains, land degradation is generally limited to strips alongside watercourses, though topsoil losses can be critical at the local level. In the drier south-east however, land degradation is severe. Regional problems are also evident on the boundary between the large-scale agriculture schemes in the north and the southern pastures, and a band of degradation surrounds some of the larger towns.

The Imatong region south-east of Kapoeta in Eastern Equatoria consists of a number of mountain ranges separated by gently sloping valleys. The region is climatically linked to the drylands of the Kenyan Lake Turkana district, and the low valleys receive 25 to 50 percent less rainfall than the plains to the north. Nomadic pastoralism is the main rural livelihood in these dry valleys. Figure 8.6 clearly shows the soil erosion that is occurring: bare subsoil exposure is visible as ochre in contrast to the more vegetated uplands and riverine strips (in green). The primary cause of this degradation is overgrazing of pastures that are naturally vulnerable to erosion due to poor soil quality and low rainfall.

The Government of Southern Sudan hopes to develop the rural sector and improve cattle production through water projects and the provision of veterinary assistance. The warning signs of land degradation indicate that any increase in cattle numbers would constitute a risk of significant damage to pastures which are already worked close to or over their sustainable yield. Any such rural development project should accordingly include land condition and sustainability components to avoid creating new problems. In degraded regions, development projects should avoid increasing stock levels and look instead for options for rehabilitation and resource recovery.

Figure 8.5   Grazing impact in Bor county, Jonglei state

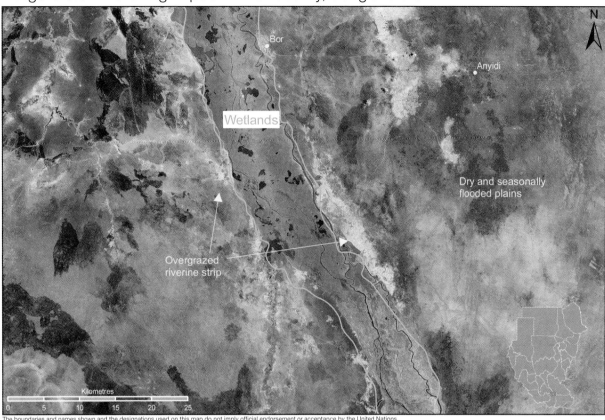

The boundaries and names shown and the designations used on this map do not imply official endorsement or acceptance by the United Nations.

Figure 8.6   Grazing impact in Kapoeta county, Eastern Equatoria

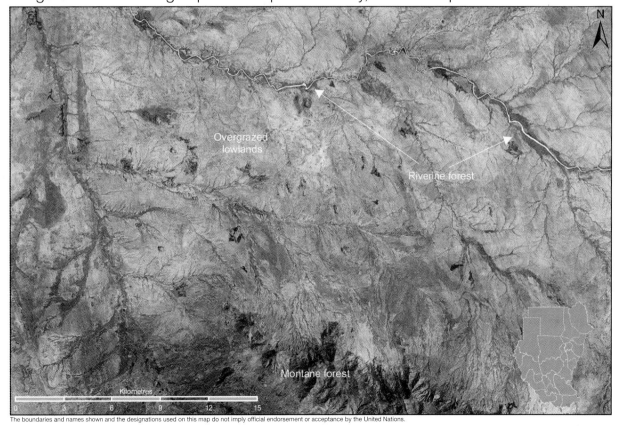

The boundaries and names shown and the designations used on this map do not imply official endorsement or acceptance by the United Nations.

The primary cause of overgrazing: overstocking

With the second largest herd on the continent (after Ethiopia), livestock is a central component of Sudan's agricultural sector. Livestock-rearing is typically categorized into three types: (i) pure nomadism, based largely on the herding of camels, sheep and goats by the Abbala in the semi-arid and arid north; (ii) semi-nomadic agropastoralism, combining the herding of cattle and some sheep with a form of cultivation by the Baggara and Dinka/Nuer in central and south Sudan as well as in the seasonal *wadis* of the north; and (iii) a sedentary system, where cattle and small livestock are reared in close proximity to villages, mainly in the central belt from Gedaref to Kordofan/Darfur [8.22].

Livestock husbandry in its various forms is practised by an estimated 40 percent of the population. This figure is even higher in Southern Sudan, where over 60 percent of the population depend on livestock [8.23]. Geographically, livestock-keeping is found virtually throughout the country, with the exception of the extreme arid north and the tsetse fly-infested areas in the far south.

The livestock population (cattle, sheep, goats and camels) is impressive, with a head count of approximately 135 million in 2004. Its rate of growth has been equally remarkable: the stocking rate has increased sixfold in less than fifty years, from a population size of 22 million in 1959. No livestock census has been carried out recently in Southern Sudan, where estimates of the population range from 12 to 22 million [8.5, 8.22].

Table 11.   Growth of the livestock sector

Livestock type	1961 (million)	Percentage of population	1973 (million)	Percentage of population	1986 (million)	Percentage of population	2004 (million)	Percentage of population	Times population has increased
Cattle	10.4	36	14.1	35	19.7	36	39.8	30	3.8
Sheep	8.7	30	13.4	33	18.8	34	48.9	36	5.6
Goats	7.2	25	10.5	26	13.9	25	42.2	31	5.9
Camels	2.3	8	2.7	7	2.7	5	3.7	3	1.6
Total	**28.6**	**100**	**40.7**	**100**	**55.1**	**100**	**134.6**	**100**	**4.7**

Cattle herders in Kosti, White Nile state. Livestock populations in central Sudan have increased sixfold in the last forty years

The second cause of overgrazing: a major reduction in rangelands in central and northern Sudan

Concurrent with the increase in livestock, a substantial reduction in rangeland areas has occurred over the past several decades due to three factors:

- uncontrolled expansion of mechanized and traditional rain-fed agriculture;
- desertification; and
- expansion of irrigation schemes (a lesser issue).

Rangeland reduction is most prevalent in northern and central Sudan. The UNEP-ICRAF rural land use study provides an indication of the overall trend.

Table 12. Changes in rangeland cover at UNEP-ICRAF study sites across Sudan

Study area and state	Original and current pasture land (% of total area)	Annual linear rate + (period loss)	Comments
North, east and central Sudan			
Ed Damazin, Blue Nile	18.5 to 0.6 from 1972 to 1999	- (96.7 %)	Loss due to the expansion of mechanized agriculture and increase in bush and shrubland
El Obeid, Northern Kordofan	50.4 to 33.5 from 1973 to 1999	- (33.5 %)	Loss due to the expansion of mechanized agriculture, increase in closed forests
Gedaref and Kassala states	13.0 to 8.2 from 1972 to 1999	- (37 %)	Decrease due to expansion of rain-fed agriculture and increase in closed forests
Kassala B	36.1 to 26.4 from 1972 to 2000	- (2.6 %)	Increase in wetland, loss of soil fertility due to wind erosion resulting in loss of pasture lands
Sunjukaya, Southern Kordofan	39.2 to 13.7 from 1972 to 2002	- (34 %)	Loss due to the expansion of mechanized agriculture, increase in bush and shrubland, riverine vegetation and wooded grassland
Tokar delta, Red Sea state	10.0 to 11.7 from 1972 to 2001	+ (1.7 %)	Increase in wooded grassland, and decrease in bush and shrubland, flooded/wetland and riverine vegetation
North-east and central Sudan		**- (50 %)**	**Highly variable but a major loss of rangeland overall due to agricultural expansion**
Darfur			
Jebel Marra, Western Darfur	5.9 to 23.0 from 1973 to 2001	+ (289 %)	Increase in open forest land, decrease in closed forest and bush and shrubland
Timbisquo, Southern Darfur	65.4 to 59.3 from 1973 to 2000	- (9.3 %)	Loss due to the expansion of mechanized agriculture, bush and shrubland, and flood and wetland
Um Chelluta, Southern Darfur	42.4 to 32.7 from 1973 to 2000	- (65 %)	Loss due to the expansion of mechanized agriculture, increase in degraded areas and flooded land, and decrease in grassland area
Darfur		**NA**	**No simple trend: Jebel Marra anomalous, Southern Darfur similar to Southern Sudan with agricultural expansion**
Southern Sudan			
Aweil, Northern Bahr el Ghazal	78.4 to 63.9 from 1972 to 2001	- (18 %)	Increase in rain-fed agriculture and riverine vegetation
Wau, Western Bahr el Ghazal	39.2 to 47.1 from 1973 to 2005	+ (20.1 %)	Decrease in closed forest, degraded land and riverine vegetation, and increase in burnt areas due to slash-and-burn agriculture
Renk, Upper Nile	6.9 to 2.8 from 1973 to 2006	- (59.4 %)	Pastureland lost due to increased land degradation and bush and shrubland
Yambio, Western Equatoria	26.0 to 27.7 from 1973 to 2006	+ (6.5 %)	Increase due to decrease in closed forests
Yei, Central Equatoria	30.9 to 17.5 from 1973 to 2006	- (42.7 %)	Loss due to increase in bush and shrubland, and decrease in wooded grassland
Southern Sudan		**- (18.5 %)**	**Highly variable but loss of rangeland overall due to agricultural and pastoral expansion**

Figure 8.7   Loss of rangeland in El Obeid district

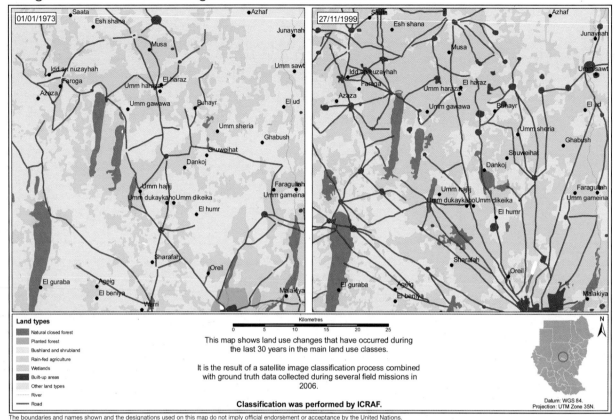

This time lapse satellite image of El Obeid shows a 57.6 percent increase in cultivated land over the period 1973 to 1999. This increase is achieved at the expense of pastoralism, as indicated by the 33.5 percent reduction in rangeland over the same period. In one generation, a third of the pastoralists' territory has been lost or converted to cultivation. Given that this region is considered to be extremely vulnerable to desertification, the sustainability of the intense land use noted here is highly questionable

In summary, the last generation of pastoralists has seen rangelands shrink by approximately 20 to 50 percent on a national scale, with total losses in some areas. It should be noted, however, that the UNEP-ICRAF study focused on the semi-desert and wetter regions. It did not include the losses due to desertification in historically important regions that are now desert or badly degraded semi-desert.

In addition to direct land loss, the reduction in rangelands has caused problems for the pastoralists' mobility. Pastoralists in Sudan have historically been very mobile, but have kept their annual herd migrations to relatively well-defined routes. Their general pattern is to move north and south to optimize grazing conditions and minimize pest problems. In the dry season, the movement is southwards towards the better pastures and later rainfall; in the wet season, it is generally northwards to follow new growth and avoid the flooding, mud, and insect-borne diseases prevalent in the more humid

regions. A similar pattern of migration, though over shorter distances, occurs in the hilly regions, where valleys are grazed mainly in the dry season and high rangeland mainly in the wet season.

In order to reach new pastures, pastoralists pass through agricultural regions. In a land without fences where agricultural and grazing zones are not clearly delimited, competition for land is at the heart of many local conflicts. Indicative pastoral routes for Sudan and Darfur are shown in Figures 8.8 and 8.9, respectively. The indicated routes are general and include only the largest scale movements. Numerous and often contrasting smaller scale movements occur on a local and seasonal level.

This major reduction in the amount, quality and accessibility of grazing land is considered to be a root cause of conflict between pastoralist and agriculturalist societies throughout the drier parts of Sudan, as discussed in Chapter 4.

Figure 8.8   Annual pastoral migration routes in Sudan

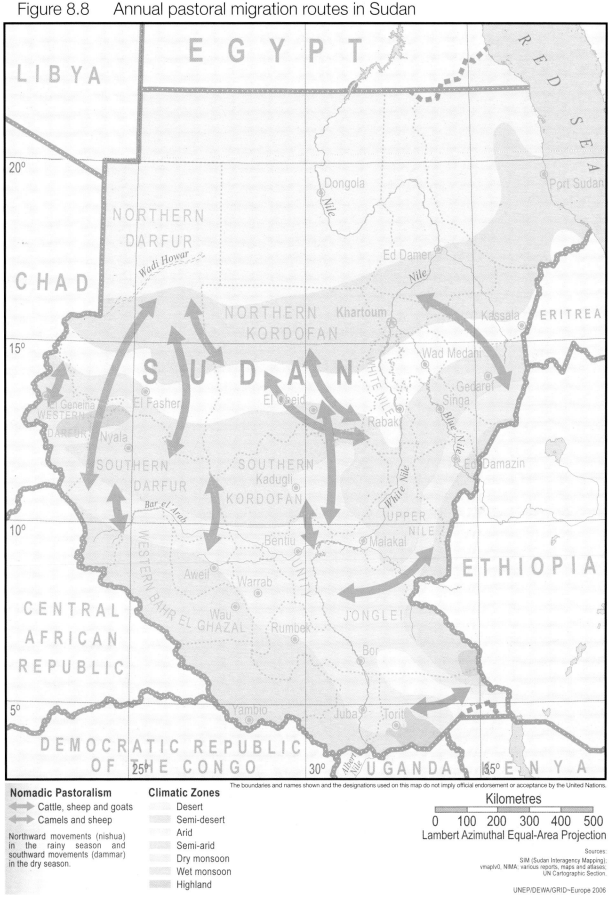

Figure 8.9   Annual pastoral migration routes in Darfur

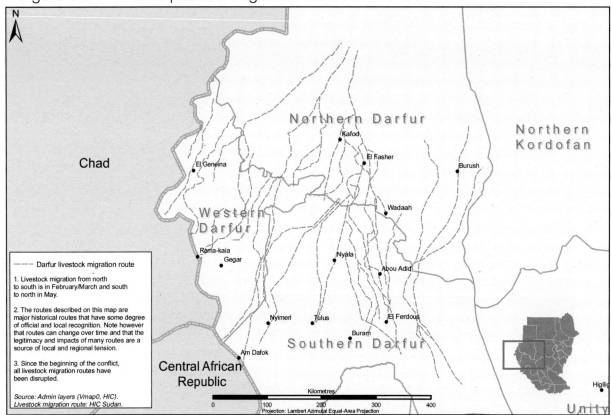

The boundaries and names shown and the designations used on this map do not imply official endorsement or acceptance by the United Nations.

Pastoral migration routes in Darfur. The very mapping or classification of pastoral routes in Darfur is a contentious issue, particularly as many routes have been blocked or changed by the recent conflict. These routes as indicated from government sources show the scale of seasonal migration and the multiplicity of potential routes but the actual lines of travel and the associated rights are not always confirmed or agreed, either in a legal sense or in the sense of having community-level acceptance

Rangeland burning in south and central Sudan

The dry season in Sudan is also the burning season. Grassfires are visible in pastoralist regions throughout the country, while slash-and-burn clearance can be observed in the southern half.

The great majority of pasture burning is deliberate. Herders set fire to the dry grass to remove old unpalatable growth, fertilize the soil with ash and promote new shoots that are more suitable as fodder. The scale of the pastoralist burning can be gauged by satellite and by aircraft (see Figure 8.10). The open clay plains of Jonglei and Upper Nile states, for example, are heavily burnt, and UNEP estimates that virtually the entire region is burnt on a two- to four-year cycle.

There is no doubt that annual burning succeeds in its purpose of short-term pasture regeneration, but it also has a number of negative impacts even when timed and executed with care. When done poorly or with hostile intent, it is highly destructive for the environment, the rural economy and society. Regular burning destroys young trees and shrubs, thus maintaining much of central and south Sudan as open plain, when its undisturbed natural state is open woodland savannah. The great plains of Southern Sudan may appear to be 'wild' but are in fact highly modified environments.

One of the long-term negative effects of very regular burning is the loss of nutrients and soil organic matter, which are lost to combustion, and water and wind erosion. For sloping terrain regions such as the Nuba mountains, such losses are clearly important. Pasture burning can also cause problems between different communities with intermingled land uses. In the extreme case of Darfur, pasture burning is used as a weapon to destroy competing livelihoods.

8.9 Agricultural sector environmental governance

Sector governance structure and issues

Governance of the agricultural sector is relatively straightforward and well structured: both GONU and GOSS have ministries of agriculture and ministries of animal resources. These ministries, however, are under strong pressure to provide policies and projects that will rapidly increase food security. This in turn results in a tendency to promote major agricultural development projects that are often environmentally unsustainable. Insufficient technical capacity and under-funding also constrain the ministries. Furthermore, the linkages between the agricultural and livestock ministries and the environment ministries are weak in both GONU and GOSS.

The most environmentally damaging aspect of government policy has been the promotion of rain-fed mechanized agriculture and the subsequent failure to address its negative consequences when these first became clearly apparent. Likewise, the lack of governance in the area of pesticides management has left Sudan with a difficult and expensive environmental legacy. Land tenure, as detailed below, is another important failure.

Land tenure

The land tenure situation in Sudan constitutes a major obstacle to sustainable land use. Prior to the 1970s, communal title to shared rural land was generally acknowledged at the local level but undocumented. The traditional community-based land management systems that were in place were reportedly reasonably effective. This situation was radically changed in the 1970s by a number of ill-planned initiatives, the consequences of which are still felt today.

The imposition of the 1971 Unregistered Lands Act effectively sequestered most of the untitled land (the majority of rural Sudan) as government property. In the same year, the People's Local Government Act took the authority away from the pre-existing traditional land management systems, which had until then provided vital checks and balances in the absence of a modern land tenure system [8.24].

Figure 8.10   Rangeland burning in Jonglei state

The boundaries and names shown and the designations used on this map do not imply official endorsement or acceptance by the United Nations.

As a result of this legislation and subsequent related acts, the majority of Sudanese now farm and rear livestock on government land, without any real supervision or form of title. As the pre-existing control measures are either weakened or completely destroyed, there is an effective governance vacuum on rural land use in much of the country.

This deficiency in rural land tenure is one of the root causes of many agricultural, environmental and social problems in Sudan. Without ownership, people have little incentive for investment in and protection of natural resources. Land owners, and smallholders in particular, are also vulnerable to more economically powerful or better armed groups, who may wish to dispossess them in order to use the land for their own purposes.

The Comprehensive Peace Agreement envisaged the immediate establishment of a new body, the Land Commission, to analyse land tenure issues and propose a way forward. As of end 2006, it has yet to be formed.

8.10 Conclusions and recommendations

Conclusion

Sudan's major investment in agricultural development over the past century has proceeded with little consideration of environmental sustainability. The resulting environmental issues are uniformly worsening and now represent a major threat to Sudan's food security. In the absence of significant action on these problems, large-scale ecological and social breakdown in the dryland regions of Sudan are considered to be a real risk in the medium to long term. It could be argued that this has already occurred to some extent in Darfur.

Agricultural authorities in the north and in Darfur face the most severe challenges, with an array of environmental problems closely tied to the social, political and economic issues affecting the region. The ongoing destruction resulting from the current system of rain-fed mechanized agriculture schemes in northern and central Sudan needs to be halted if food insecurity and conflicts are to be avoided in the future. This does not call for a reversion from mechanization back to traditional methods, but for a revision of current practices in order to combine the best of both approaches in a sustainable manner.

At present, Southern Sudan only faces severe agriculture-related environmental issues along its northern boundaries, but there are numerous warning signs that action is needed to forestall damaging overtaxing of the environment in the more populated regions in the far south. It is therefore extremely important that lessons from other regions be learnt, and that agricultural development in the south proceed with extreme care to ensure its environmental sustainability.

Background to the recommendations

GONU government reform and capacity-building in land use planning and environmental sustainability are the central themes of the recommendations for this sector. Specific environmental rehabilitation programmes are definitely needed, but in the absence of major reform in the approach to agricultural development in northern and central Sudan, further ad hoc investment in environmental initiatives is considered to be highly risky.

In Southern Sudan, the rapidly developing agricultural policies as seen by UNEP in late 2006 appear to be generally sound, with one major gap. A high priority should be given to conversion of traditional agricultural systems to more modern hybrid systems such as agroforestry, which preserves tree cover and boosts per hectare productivity while improving environmental sustainability.

Recommendations for the Government of National Unity

R8.1 Establish the proposed Land Commission. The proposed commission is a key part of the CPA and a good initiative that warrants support. The international community already has funds set aside for this initiative.

CA:GROL; PB: MAF; UNP: FAO; CE: nil; DU: 3 years

R8.2 Impose a moratorium on new mechanized rain-fed agriculture schemes and conduct a major review and study on the way forward. The objective is to understand the real impacts and control the unplanned expansion of mechanized agriculture, and improve sustainability. Priority states are Southern Kordofan, Blue Nile, Gedaref, White Nile and Sennar.

CA: GROL/AS; PB: MAF; UNP: FAO; CE: 0.2M; DU: 2 years+

R8.3 Invest in technical assistance, capacity-building and research in seven environment-agriculture subject areas. The overall objective is to embed the culture and capacity for the sustainable development of agriculture into the Ministry of Agriculture and Forestry, the Ministry of Animal Resources and a number of linked institutes. The investments need to be spread between the federal and state levels and various ministries. The target subjects are:

- meteorology services;
- sustainable rural land use planning;
- rangeland conservation;
- agroforestry;
- Water Use Associations (WUA) in irrigation schemes;
- integrated pest management and pesticide management; and
- rehabilitation of desert regions using native species.

CA: TA; PB: MAF; UNP: FAO; CE: 8M; DU: 3 years

R8.4 Develop policies and guidelines to prevent future accumulation of pesticide stockpiles. Policy development should be based on multi-stakeholder consultations involving relevant government authorities, industry, aid agencies and development banks, and farmers.

CA: GROL; PB: MAF; UNP: FAO; CE: 0.1M; DU: 1 year

R8.5 Collect all obsolete pesticide stocks for safer long-term storage, treatment and disposal, and conduct a feasibility assessment for safe final disposal. Prior to final disposal, the stocks disseminated across the country will need to be assessed, categorized, and made safe for transport and interim storage. A single well-sited, well-designed and maintained interim storage place would be a major improvement on the current situation. Any major investment in final disposal will require a cost and feasibility study to select the best option and assist financing.

CA:PA; PB: MAF; UNP: UNEP; CE: 3M; DU: 2 years

R8.6 Assess the full extent of riverbank erosion and invest in practical impact management

plan based on Integrated Water Resource Management (IWRM).** This should be considered an investment in the preservation of high-value agricultural land.

CA:PA; PB: MAF; UNP: FAO; CE: 3M; DU: 2 years

R8.7 Develop a national strategy and priority action plan for mesquite control in the agricultural sector. The Presidential Decree should be amended at the same time as the plan is developed to avoid a legislation-policy clash.

CA:GROL; PB: MAF; UNP: FAO; CE: 0.3M; DU: 1 year

Recommendations for the Government of Southern Sudan

R8.8 Impose a moratorium on new mechanized agriculture schemes in southern states, and a major review and study on the way forward. The objective is to understand the real impacts and control the unplanned expansion of mechanized agriculture, and improve sustainability. For GOSS, applicable to Upper Nile state.

CA: GROL/AS; PB: MAF; UNP: FAO; CE: 0.2M; DU: 1 year

R8.9 Invest in technical assistance, capacity-building and research in a range of environment-agriculture subject areas. The overall objective is to embed the culture and capacity for the sustainable development of agriculture into the Ministry of Agriculture and Forestry, Ministry of Animal Resources and a number of linked institutes. The investments need to be spread between the federal and state levels and various ministries.

CA: TA/CB; PB: MAF; UNP: FAO; CE: 4M; DU: 3 years

R8.10 Design and implement agroforestry demonstration projects in each of the ten southern states. The objective is to demonstrate the benefits of switching from shifting agriculture to more sustainable land use models.

CA: PA; PB: MAF; UNP: ICRAF; CE: 5M; DU: 5 years

Forest Resources

Plantations such as this teak stand in Kagelu, Central Equatoria, are a valuable asset and potential source of hard currency for Southern Sudan. Commercial exploitation of the forest resources of Southern Sudan is expected to expand with peace and road network improvements. The challenge will be to develop the industry in an environmentally sustainable manner.

Forest resources

9.1 Introduction and assessment activities

Introduction

The rural population of Sudan, as well as much of its urban population, depends on forests. Trees are the main source of energy and provide timber for roofing and building. In rural Sudan, the extensive benefits derived from forests include grazing, hunting, shade, forest foods in the form of tree leaves, wild fruits, nuts, tubers and herbs, tree bark for medicinal purposes, and non-wood products such as honey and gum arabic. In addition, the commercial lumber industry is a small but growing source of employment. According to FAO, the forestry sector contributes as much as 13 percent to the gross domestic product of Sudan [9.1].

This valuable resource is threatened, however, by deforestation driven principally by energy needs and agricultural clearance. Moreover, the unbalanced distribution of forests in Sudan – most of the remaining forests are found in the south, while the demand for forest products is highest in the north – presents a potential threat for north-south peace, but also a significant opportunity for sustainable north-south trade development.

Assessment activities

Forestry was a priority topic for the UNEP assessment, and was also included in the scope of the ICRAF study on rural land use changes commissioned by UNEP in cooperation with FAO. In addition, the forestry sector assessment was marked by strong and welcome support from the Forests National Corporation (FNC) in northern and central Sudan.

UNEP teams visited forests in over twenty states. Particular attention was paid to deforestation pressures in different regions. Satellite imagery analysis of fourteen sites included a quantitative assessment of deforestation, and satellite reconnaissance was widely used to search for deforestation 'hotspots'. In Southern Sudan, the Kagelu Forestry Training Centre worked with ICRAF to provide UNEP with detailed information on the Equatorian states timber reserves. However, security constraints prevented access to important forests in Darfur; the Jebel Marra plateau, for instance, was almost completely inaccessible at the time of the survey.

A commercial mahogany stand in the Nuba mountains, Southern Kordofan. Northern Sudan's major timber deficit is currently being met principally through unsustainable logging in central Sudan. Viable and sustainable alternatives include increased use of plantations

Figure 9.1   Sudan forest cover

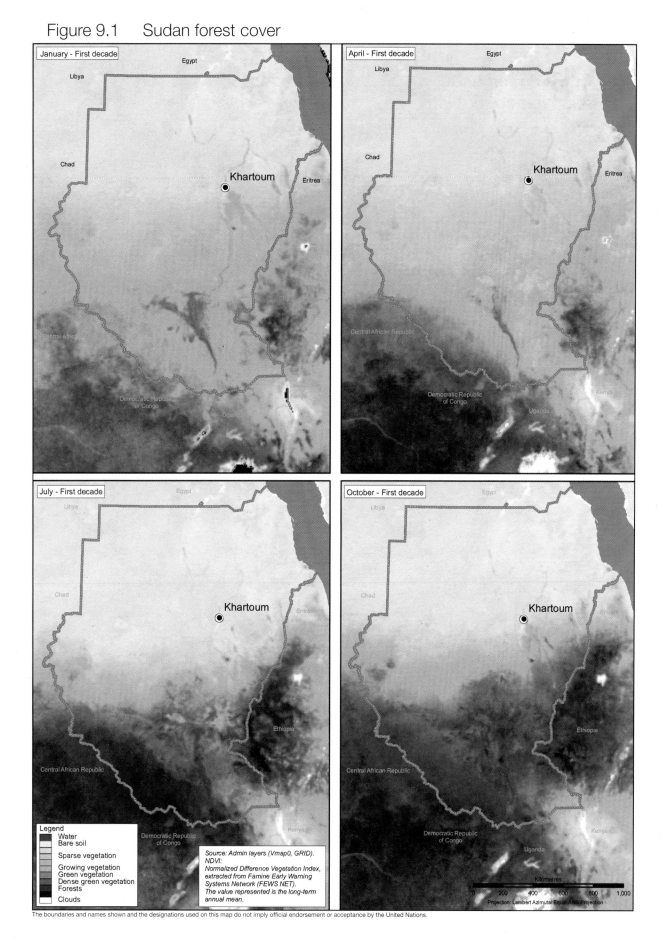

The boundaries and names shown and the designations used on this map do not imply official endorsement or acceptance by the United Nations.

195

With the exception of central Darfur, UNEP's forestry-related activities were considered comprehensive enough to develop an accurate picture of the status of Sudan's forests and prevailing trends across the country.

9.2 Overview of forest resources

A wide range of forests and related vegetation types is found in Sudan due to regional variations in soil and rainfall. The most important types are listed below, in rough order of distribution from the arid north to the tropical south:

- desert and semi-desert trees and shrubs;
- riverine forests;
- low rainfall woodland savannah;
- high rainfall woodland savannah;
- montane and gallery forests;
- tropical forests; and
- plantations.

Most trees in Sudan grow in open to semi-closed woodlands with numerous under-storeys of grasses and shrubs. Fully closed forests are only found in a few of the most humid areas in the south. This complicates attempts to quantify the extent of forests and deforestation in the drier regions, as there is rarely a clear deforestation or ecosystem boundary, but rather a gradual thinning out of trees over a large area.

The long-term Normalized Difference Vegetative Index (NDVI) is a measurement of the overall vegetation density, including trees, shrubs and grasses over different seasons. The images in Figure 9.1 (see previous page) clearly show the dominant impact of the Sahara desert and low rainfall zones on vegetation cover and the associated north-south difference in tree cover.

Desert and semi-desert trees and shrubs

Desert vegetation in the northern states (Northern, Northern Darfur, Northern Kordofan, Kassala and Red Sea) is limited to xerophytic (drought-resistant) shrubs, such as *Acacia ehrenbergiana, Capparis decidua, Fagonia cretica* and *Leptodemia pirotechnica*. Scrub formations occur in the semi-desert zone (the northern half of Kordofan and Blue Nile states, all of Khartoum state, most of Red Sea state, and some parts of Darfur), where the vegetation is a varying mixture of grasses and herbs with widely scattered shrubs.

Forest resources in the desert and semi-desert northern states are extremely limited and in continual decline

Acacia nilotica in Sennar state. The density and variety of tree cover increases further south, following rainfall patterns

Riverine forests

Riverine forests are a critical resource for the northern states. They occupy the lands that are flooded when rivers rise in the latter part of the wet season. *Acacia nilotica* – the dominant species – is found as pure dense stands over large areas from the Egyptian border in the north to as far south as Jebelein on the White Nile, and Roseires on the Blue Nile. The species also occurs along the Dinder and Rahad rivers. In less frequently flooded basins along the Atbara river and in some inland sites, *Acacia nilotica* is replaced by *Hyphaene thebaica* (Dom palm) forests.

Low rainfall (< 900 - 1,000 mm) woodland savannah

The low rainfall woodland savannah region lies in the centre and south of the country, with the exclusion of the flood region. Rainfall is confined to a few months of the year (March or April to July), and is followed by a long hot dry season. The vegetation is composed of mixed grass types with bushes and trees, but species distribution within the low rainfall savannah zone varies with rainfall and soil type. Sandy soils dominate in the west and central regions, and clay soils are prevalent in the east and south. In the drier parts, trees are nearly all thorny and low in stature, with a predominance of species of acacia. Broadleaved deciduous trees become prevalent in the wetter parts, but there is not as great a variety of species as in the high rainfall woodland savannah, and thorn trees are usually present. The gum arabic belt lies within this zone. The belt occupies an area of 520,000 km² between the latitudes of 10° and 14° N, accounting for one-fifth of the total area of the country. Its importance is reflected in the fact that it accommodates approximately one-fifth of the population of Sudan and two-thirds of its livestock, and that it acts as a natural barrier to protect more than 40 percent of the total area of Sudan from desert encroachment [9.2].

High rainfall (> 900 – 1,000 mm) woodland savannah

The high rainfall woodland savannah extends into most parts of Bahr el Ghazal and Equatoria states in the south. Trees in this region are generally tall and broadleaved. Coarse tall tussocks of perennial grasses predominate and fires are hence usually fiercer than in the low rainfall woodland savannah. The most important tree species are *Khaya senegalensis* and *Isoberlina doka*. Other species are *Parkia oliveri, Daniella oliveri, Afzelia africana, Terminalia mollis, Burkea africana* and *Vitellaria paradoxa*.

Tropical forests

Sudan's tropical forests are confined to a few small and scattered localities: the Talanga, Lotti and Laboni forests at the base of the Imatong mountains and the Azza forest in Maridi in Western Equatoria, and other small areas on the Aloma plateau and near Yambio. Species occurring in these tropical forests are similar to those found in the drier parts of the forests of West Africa. The most common are *Chrysophyllum albidum* and *Celtis zenkeri,* with *Holoptelea grandis* in the Azza forest. A number of valuable timber trees are also found, including *Khaya grandifolia* (mahogany), *Chlorophora excelsa,* and *Entrandrophragma angolense.*

Montane and gallery forests

Mountains in Sudan are characterized by higher rainfall, resulting in different and more robust woodlands than in the surrounding areas. The Jebel Marra plateau in Darfur is the most important ecosystem of this type in the drier parts of Sudan.

Coniferous forests occur in the montane vegetation of the Imatong and Dongotona ranges in Eastern Equatoria state, as well as in the Red Sea hills in the north-east. Important species include *Podocarpus milanjianus, Juniperus procera* and *Pinus radiata.* Planted exotics include *Eucalyptus microtheca* and *Cupressus spp.* In the more humid areas of the Imatong and Dongotona ranges, the vegetation is similar to that of low rainfall woodland savannah.

Gallery forests occur on the banks of streams. They are generally found in relatively deep U-shaped valleys, and benefit from both the extra water supply from the streams and the protection against fires afforded by the steeply sloping banks. Important species are *Cola cordifolia, Syzygium guineense* and *Mitragyna stipulosa* in swampy places.

Plantations

Plantations were first established in Sudan by the Anglo-Egyptian administration. The most significant of these were the teak (*Tectona grandis*) plantations of Southern Sudan, many of which are still standing (see Case Study 9.1). This process was continued by the government forestry administration, and by the mid-1970s, plantations totaled some 16,000 additional hectares of hardwoods and 500 to 600 hectares of softwoods [9.3].

Today, most of the remaining plantations are found in Central and Eastern Equatoria states, in Southern Sudan. They include stands of teak in the far southern regions and pine in the higher elevations of the Imatong mountains. Elsewhere in Sudan, plantations are comprised of riverine *Acacia nilotica* forests, *Acacia senegal* plantations in abandoned mechanized farms, inside forest reserves, in private gum orchards, and in isolated shelter belts planted in Northern Kordofan and other central states, pine and eucalyptus plantations in the Jebel Marra region in Darfur, and eucalyptus in the irrigated agricultural areas.

Southern Sudan still retains the majority of its forest cover, but deforestation is occurring at a steady rate

These teak trees have not been tended for 20 years, so the productivity of the plantation is well below potential. The plantations, however, are a valuable asset

CS 9.1   Yei county teak plantations: a valuable colonial legacy

Teak (*Tectona grandis*) plantations are spread all over Yei county. Prior to the conflict, the largest and best managed plantations were located in Kagelu, 8 km south-west of the town of Yei, between 04°03'34" N and 30°36'56" E.

The community living around the plantation, the Kakwa ethnic group, mainly practises subsistence agriculture, though some members also plant their own woodlots for cash income and construction materials. Before the war, the community benefited from the infrastructure provided by the government forest plantation project in terms of employment, education, health services and improved road access. Other benefits included extension services, fuelwood and other forest products from the reserve.

Between independence and the second civil war, the teak plantations in Yei county were managed by the Sudan German Forestry Team, funded by GTZ (German Technical Aid), but the project was shut down in 1987 due to the intensification of the conflict. During the war, all of the teak plantations were subject to uncontrolled felling and export to Uganda. The entire process was managed on the black market by foreign-owned logging companies, and royalties from the timber went to the SPLA.

With the end of the conflict and the establishment of the GOSS Ministry of Agriculture and Forestry, H.E. Martin Elia Lomoro ordered a review and evaluation of commercial logging activities. The committee that conducted the review found that all of the contracts that were issued were illegal and that they did not conform to best forestry practices. This prompted the Minister to issue a decree annulling all the contracts and banning logging in both the teak plantations and natural forests. This ban, while admirable, is not expected to hold much beyond 2006 due to the need for foreign currency and construction timber in Southern Sudan.

There is accordingly an urgent need for the GOSS to develop an appropriate governance regime, including a transparent licensing process, strict quotas and reforestation obligations.

Table 13.   Teak plantations in Yei county [9.8]

Name of forest reserve	Size in hectares
Loka	918
Kagelu	1,045
Kajiko North	750
Kajiko South	90
Korobe	50
Mumory	30
Yei Council	2
Total	**2,985**

9.3   Forest utilization

A range of ecosystem services

The forests of Sudan have economical, ecological, and recreational values, known collectively as ecosystem services.

Wood products from the forestry sector include fuelwood, sawn timber and round poles. The Forest Product Consumption Survey conducted by the FNC in Northern Sudan in 1995 found that the total annual consumption of wood was 15.77 million m³. FAO calculated that in 1987, Sudan produced 41,000 m³ of sawn timber, 1.9 million m³ of other industrial round wood, and more than 18 million m³ of firewood. Each of these categories showed a substantial increase from production levels in the 1970s [9.4].

The ecological benefits of forests include sand dune stabilization in fragile semi-desert environments, amelioration of soil through nitrogen fixation, and the provision of natural ecosystems for wildlife and the conservation of biodiversity.

Fuelwood and charcoal production

The felling of trees for fuelwood and charcoal production occurs throughout Sudan, but the pressure is generally greater on the more limited resources of the north and the areas surrounding the country's urban centres. An additional growing use for fuelwood in all parts of Sudan is for brick-making. In Darfur, for instance, brick-making provides a livelihood for many IDP camp residents, but also contributes to severe localized deforestation (see Case Study 5.2).

Fuelwood market in Nyala, Southern Darfur

Brick kilns on the banks of the Blue Nile, in El Gezira state. The brick-making industry is a major market for fuelwood

As is the case for many natural resource management issues in Sudan, the data on wood consumption is incomplete and often obsolete. What is available, however, provides a picture of substantial and increasing demand. The 1995 FNC survey indicated that fuelwood contributed 78 percent of the energy balance of Sudan, the rest being provided by oil (8 percent), generated electricity (8 percent) and agricultural residues (6 percent). With a per capita annual consumption of approximately 0.68 m³, the total fuelwood requirement for 1995 was estimated at 22 million m³ [9.4, 9.9]. These figures were extrapolated by UNEP to estimate the fuelwood requirement for 2006 at 27-30 million m³.

In theory, forest authorities in northern and central Sudan direct the commercial logging of *Acacia nilotica* and *Acacia seyal* for supply of firewood and charcoal to the cities. Wood is meant to be extracted mostly from the thinning of branches of *Acacia nilotica* in reserved riverine forests, and the clearing of *Acacia seyal* and other species from areas allocated for agriculture. In practice, however, the process is much less controlled and the felling less selective.

Rural inhabitants use most of the tree species in the low rainfall savannah for fuelwood. The removal of dead trees and branches is permitted for people living around forests in all parts of Sudan.

A charcoal market in Khartoum

Sawn timber

In the northern and central states, logging for the production of industrial timber is carried out by contractors under the supervision of sawmill and industry managers who are directly responsible to the State Director of Forests within their respective states. In the southern states, the industry is currently stagnant, but was managed by the military forces on both sides during the conflict.

The sawn timber in the north is mainly from *Acacia nilotica*; in the south, it is extracted from a range of high rainfall savannah woodland species including *Isoberlinia doka, Khaya grandifolia, Milicia excelsa, Khaya senegalensis, Olea hochstetteri, Afzelia africana, Daniellia oliveri, Sclerocarya birrea,* and *Podocarpus milanjianus.*

Traditional construction

There is no detailed data available on wood product usage in traditional construction. One figure much quoted to UNEP, from unknown sources, is that it takes approximately ten young trees to build one *tukul* (traditional round dwelling). With a rural population of over thirty million, the total demand is therefore significant, but anticipated to be much below the fuelwood demand from the same population.

Sawn teak in Wau, Western Bahr el Ghazal

Dried wild fruit for sale in the Tokar region, Red Sea state. Non-wood forest products such as fruit, nuts, and medicinal herbs are important but often under-valued components of the overall value of forests

Non-wood forest products

Gum arabic is Sudan's most important non-wood forest product, with an annual exported crop of approximately 45,000 tonnes. The grey-barked *Acacia senegal* produces *hashab* gum, while the usually red-barked *Acacia seyal* gives *talh* gum. The latter is inferior in quality. The dom nut, a vegetable ivory, is obtained from *Hyphaene thebaica*. Dom nuts are sliced and used as button blanks; an average of 1,500 tonnes is exported annually. Minor products include bee honey and bees wax, the latter being exported at a rate of 80 tonnes per year, palm oil (*Elaeis guineensis*), garad tanning pods obtained from *Acacia nilotica*, lulu (shea oil and butter) from *Vitellaria paradoxa* and the fruits of the shrub species *Capsicum frutescens*. Other vegetal non-wood forest products are fodder (e.g. *Ziziphus spp., Acacia spp.*), edible oils (e.g. *Balanites aegyptiaca*), medicines (e.g. *Tamarindus indica*), dyes (e.g. henna from *Lawsonia inermis, Prosopis africana*), fibres (e.g. *Borassus aethiopum*) and latex (e.g. *Landolfia ovariensis*).

9.4 Forestry sector environmental impacts and issues

There are three key environmental issues for the forestry sector in Sudan:

1. deforestation;
2. the charcoal industry, which constitutes a potential north-south conflict 'flashpoint'; and
3. the southern timber industry development opportunity.

Deforestation – an overall and effectively permanent reduction in the extent of tree cover – is the dominant environmental, social and economic issue affecting the forestry sector in Sudan. The removal of trees has a range of very negative impacts, including increased land and water resource degradation, and the loss of livelihoods from forest ecosystem services.

The second important issue is the risk of renewed conflict over the exploitation of timber resources for charcoal in the north-south border regions. Directly linked to this is the economic opportunity afforded by the forests of Southern Sudan and the challenge of developing a significant new industry while at the same time avoiding deforestation.

A further issue for the forestry sector is the management of invasive species, and specifically of mesquite (*Prosopis juliflora*), which was discussed in the previous chapter. It should be noted that the solutions to this problem are linked to improved management of this resource rather than its elimination.

9.5 Deforestation rates and causes

Measuring the rate of deforestation at the national scale

In the late 1970s, FAO estimated that the country's forests and woodlands totaled approximately 915,000 km², or 38.5 percent of the land area. This figure was based on a broad definition of forests and woodlands as 'any area of vegetation dominated by trees of any size'. It also included an unknown amount of cleared land that was expected to have forest cover again 'in the foreseeable future' [9.5].

An estimate by the forestry administration in the mid-1970s, however, established the total forest cover at some 584,360 km², or 24.6 percent of the country's land area. More than 129,000 km² (about one quarter) of this amount was located in the dry and semi-arid regions of northern Sudan [9.9].

Given this nearly 50 percent difference in baseline depending on definition, it is difficult to make a comprehensive quantitative comparison of deforestation on the national scale since the 1970s, and UNEP has not attempted to do so for this assessment. More exhaustive and rigorous information is available from 1990, when FAO Forest Resources Assessments (FRAs) started to cover Sudan in more detail. The latest assessment work, which was released in 2005, is set out in Tables 14 to 16.

Table 14. Extent of forest and other wooded land in Sudan [9.6]

Extent of forest and other wooded land			
FRA 2005 categories	**Area (1,000 hectares)**		
	1990	**2000**	**2005**
Forest	76,381	70,491	67,546
Other wooded land	–	54,153	–
Forest and other wooded land	**76,381**	**124,644**	**67,546**
Other land	161,219	112,956	170,054
...of which with tree cover	–	–	–
Total land area	**237,600**	**237,600**	**237,600**
Inland water bodies	12,981	12,981	12,981
Total area of country	**250,581**	**250,581**	**250,581**

Table 15.   Characteristics of forests and other wooded land in Sudan [9.6]

Characteristics of forest and other wooded land						
FRA 2005 categories	**Area (1,000 hectares)**					
	Forest			**Other wooded land**		
	1990	**2000**	**2005**	**1990**	**2000**	**2005**
Primary	15,276	14,098	13,509	–	–	–
Modified natural	53,467	49,344	47,282	–	54,153	–
Semi-natural	1,528	1,410	1,351	–	–	–
Productive plantation	5,347	4,934	4,728	–	–	–
Protective plantation	764	705	675	–	–	–
Total	**76,381**	**70,491**	**67,546**	**–**	**54,153**	**–**

Table 16.   Growing stock in forests and other wooded land in Sudan [9.6]

Growing stock in forests and other wooded land						
FRA 2005 categories	**Volume (million m³ over bark)**					
	Forests			**Other wooded land**		
	1990	**2000**	**2005**	**1990**	**2000**	**2005**
Growing stock in forests and other wooded land	1,062	980	939	–	–	–
Commercial growing stock	–	–	–	–	–	–

It should be noted that the above table is the result of various inventories and assessments over time, and that the calculation of the change rate is based on World Bank 1985 (reference year 1976) and Africover data (reference year 2000). Due to different classification systems, the change rate was calculated on the combined area of forest and other wooded land and allocated proportionally to the two classes according to the latest estimate (Africover 2000).

Though some agricultural land that was abandoned due to the conflict has regenerated naturally, the clear trend overall has been for significant and consistent deforestation across the country: according to FAO, Sudan lost an average of 589,000 hectares (5,890 km²) of forest per year between 1990 and 2000. This amounts to an average annual deforestation rate of 0.77 percent. Between 2000 and 2005, the rate of deforestation increased by 8.4 percent to 0.84 percent per annum. In total, between 1990 and 2005, Sudan lost 11.6 percent of its forest cover, or around 8,835,000 hectares.

Balanites trees provide vital shade for livestock in 40°C heat

Measuring the rate of deforestation at the district scale

The ICRAF study included detailed remote sensing analysis of fourteen regions over time periods of up to thirty-three years. Each study site covered an area of 2,500 km² and included a number of different land uses. The rate of deforestation was estimated for each site, and is set out in the table below. Note that 'deforestation' here refers to calculated changes in percentage of land use from forested land forms to others, including from closed forests to more open wooded grasslands.

Table 17. Summary of deforestation rates in Sudan from 1973 to 2006

Study area and state	Original and final forest and woodland cover	Annual linear deforestation rate + (period loss)	Comments
North, east and central Sudan			
Ed Damazin, Blue Nile	7.5 to 0.1 from 1972 to 1999	3.6 % (98.6 %)	Wooded grassland replaced by rain-fed agriculture. Some regrowth of closed forest (verification required).
El Obeid, Northern Kordofan	12.0 to 8.7 from 1973 to 1999	1.05 % (27.5 %)	Wooded grassland replaced by rain-fed agriculture. Shelter belts remain.
Shuwak, Kassala	–	–	Non-measurable arid zone, now with both irrigation and mesquite invasion.
New Halfa, Kassala	–	–	Non-measurable arid zone, now with both irrigation and mesquite invasion.
Sunjukaya, Southern Kordofan	29.2 to 8.4 from 1972 to 2002	2.37 % (71.2 %)	Wooded grassland replaced by traditional rain-fed agriculture. Some regrowth as scrubland.
Tokar delta, Red Sea state	15.8 to 26.8 from 1972 to 2001	Mesquite + 2.4 % (+ 170 %)	Reforestation. Non-precise arid zone with mesquite invasion replacing agriculture.
North, east and central Sudan case study averages	Natural forest only	2.37 % (65.7 %)	**Complete deforestation is two-thirds complete by 2001. Predicted to be over 70 % by 2006. Extrapolated near total loss within 30 years.**
	Including invasive species	1.15 % (31.8 %)	Reforestation by invasive species is compensating in total cover by 50 % but still a major net loss.
Darfur			
Jebel Marra, Western Darfur	50.7 to 35.8 from 1973 to 2001	1.04 % (29.4 %)	Closed forest changing to open forest land and burnt areas.
Timbisquo, Southern Darfur	72.0 to 51.0 from 1973 to 2005	1.33 % (29.1 %)	Closed forest and wooded grassland replaced by burnt areas and rain-fed agriculture.
Um Chelluta, Southern Darfur	23.8 to 16.1 from 1973 to 2000	1.20 % (32.4 %)	Closed forest replaced by burnt areas, pasture and rain-fed agriculture.
Darfur case study averages		1.19 % (30.3 %)	**Rapid and consistent deforestation approximately one-third complete by 2006.**
Southern Sudan			
Aweil, Northern Bahr el Ghazal	11.9 to 7.2 from 1972 to 2001	1.38 % (39.4 %)	Closed forest changing to wooded grassland and pasture.
Wau, Western Bahr el Ghazal	76.5 to 51.8 from 1973 to 2005	1.00 % (32.3 %)	Closed and riverine forest and wooded grassland replaced by traditional rain-fed agriculture.
Renk, Upper Nile	6.5 to 0 from 1973 to 2006	> 5 % (100 %)	Wooded grassland and riverine forest replaced by degraded land.
Yambio, Western Equatoria	80.2 to 51.5 from 1973 to 2006	1.12 % (35.8 %)	Closed forest and wooded grassland replaced by traditional rain-fed agriculture.
Yei, Central Equatoria	29.8 to 19.3 from 1973 to 2006	1.53 % (35.2 %)	Closed forest and wooded grassland replaced by open forest and traditional rain-fed agriculture.
Southern Sudan case study averages		> 2 % (40 %)	**Rapid and consistent deforestation approximately 40 % complete by 2006. Extrapolated near total loss within 50 years.**
National average based on FAO study	30.4 to 26.9 from 1990 to 2005	0.76 % (11.5 %)	**Remote sensing work only.**
National average based on UNEP case studies	Natural forest only	> 1.87 % (48.2 %)	**Rapid deforestation has resulted in the loss of the majority of forests in the north and the same pattern is visible elsewhere in Sudan.**

Figure 9.2   Jebel Marra deforestation

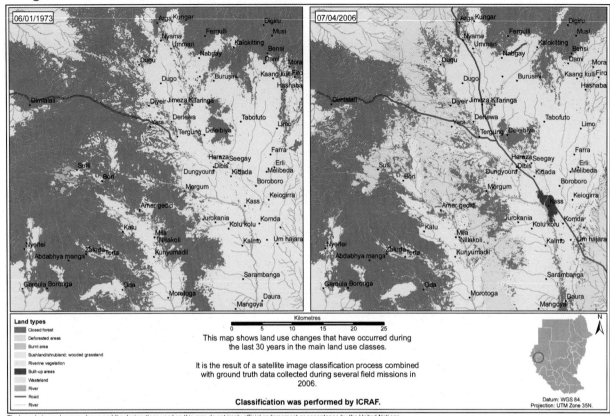

This time lapse satellite image of Jebel Marra shows a very destructive pattern of land use change. The closed forest has been extensively degraded to burnt areas and open woodland, with a deforestation rate of 1.04 percent per annum. This clearing has not been matched by an increase in agricultural areas. The only gain has been a marginal increase in grazing land on the steep slopes

The summary in Table 17 is a gross simplification of the complex land use patterns and changes occurring at each of the fourteen sites, but the overall trends are clear:

1. Northern, eastern and central Sudan have already lost the great majority of their forest cover. The removal of remaining forests is ongoing but has slowed, except in the southern border regions, where removal of the last of the major forests is progressing rapidly. Reforestation of northern and eastern states by invasive species is locally significant.

2. Darfur has lost more that 30 percent of its forests since Sudan's independence and rapid deforestation is ongoing.

3. Southern Sudan has lost some of its forests since Sudan's independence and deforestation is ongoing due to the total dependence on fuelwood and charcoal as the main sources of energy. Deforestation is worst around major towns such as Malakal, Wau and Juba. The study did not include areas distant from major towns, where it is expected that the extent of deforestation could be less severe.

The substantial difference between UNEP and FAO work is considered to reflect the difficulty in quantifying a system with extreme seasonal and annual variations, as well as classification problems due to blurred boundaries between land classes. Based on its fieldwork, UNEP considers its figures to be the best currently available, though they are probably an under-estimation given that most of the quantitative work is based on images one to seven years old, and that all factors point to a gradual increase in deforestation rates over time.

In Figures 9.3a and 9.3b, time lapse satellite images of two sites in Southern Darfur show a similar deforestation trend: the forest is being fragmented and removed in large areas, and replaced largely by traditional slash-and-burn agriculture, which has also taken over rangelands. The annual deforestation rates are calculated at 1.33 percent for Timbisquo and 1.20 percent for Um Chelluta.

Figure 9.3a   Southern Darfur deforestation – Timbisquo

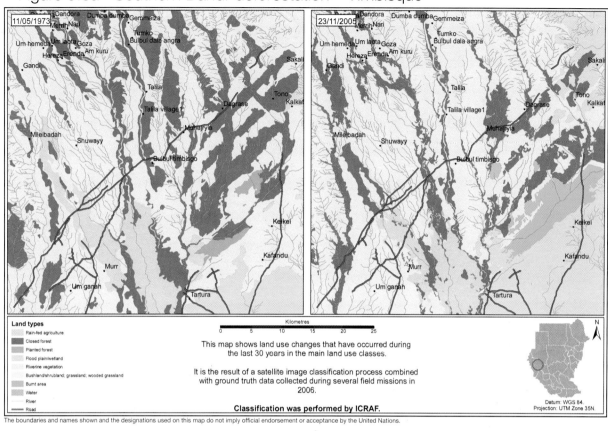

Figure 9.3b   Southern Darfur deforestation – Um Chelluta

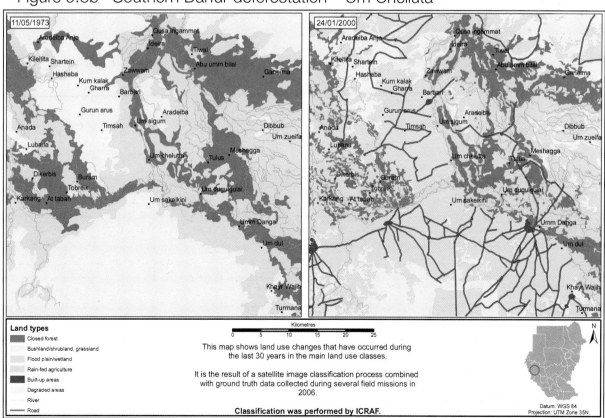

Causes of deforestation

There are several underlying causes of deforestation; these are cumulative in nature and vary considerably from region to region:

- fuelwood and charcoal extraction;

- mechanized agriculture;

- traditional rain-fed and shifting agriculture;

- drought and climate change;

- overbrowsing and fires;

- direct conflict impacts;

- commercial lumber and export industry (not a major factor); and

- traditional construction (not a major factor and not discussed).

Unsustainable rates of fuelwood extraction

As noted in previous chapters, the unsustainable extraction of fuelwood is a major problem in northern and central Sudan, as well as in refugee and displaced persons camps all over the country and particularly in Northern Darfur. The acacia groves of the Sahel have been extensively harvested for fuelwood, with a resulting rapid advance of deforestation.

The supply of charcoal to northern cities is a major business that is currently depleting the forests of central, southern and western Sudan, particularly Southern Kordofan, the northern part of Upper Nile state and eastern parts of Darfur.

According to the FNC, the charcoal and mechanized agriculture interests work closely together, with

Figure 9.4   Wau deforestation

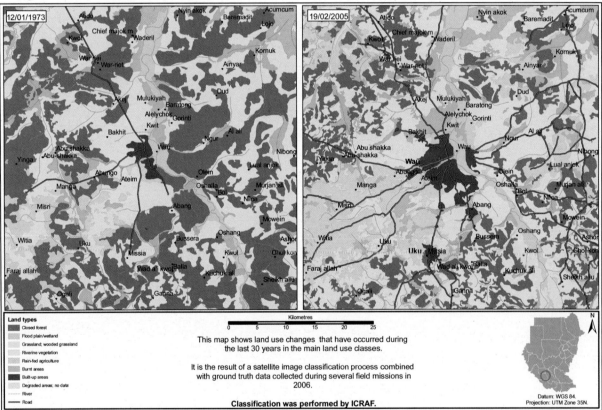

These time lapse satellite images of Wau district in Western Bahr el Ghazal show a complex pattern of intensifying land use leading to deforestation at a rate of one percent per annum and extensive forest fragmentation. Forests are replaced largely by expanding traditional slash-and-burn agriculture and new rangelands. Bare degraded land has appeared in previously forested areas, indicating either overgrazing or exhaustion from traditional cultivation

A brick kiln near Kadugli, Southern Kordofan. The remaining forests of Southern Kordofan are being consumed by the fuelwood and charcoal industries

some cases of unsuitable land being 'cleared' for agriculture in order to collect fuelwood. Together, these two industries are considered to be the primary cause of deforestation in central Sudan.

Expansion of mechanized agriculture

The expansion of mechanized agriculture in central Sudan (see Chapter 8) has occurred at the direct expense of forests. Large amounts of woodland have been cleared in the development of mechanized rain-fed farming in the eastern and central states, as well as smaller amounts in Upper Nile and Southern Kordofan states. Legal requirements to avoid the development of agricultural schemes in forest areas and to retain ten percent of forest as shelter belts have been systematically ignored. These forests were valuable chiefly as protection against desertification, but also as a source of fuel for pastoral people in those regions.

Before it was cleared for mechanized agriculture, this land in Blue Nile state consisted of low rainfall savannah and rangeland

Intensification of traditional rain-fed and shifting agriculture

When practised sustainably, traditional shifting agriculture does not result in a net loss of forest cover. However, the current unsustainable practices induced by population growth are resulting in major loss and fragmentation of forests. The ICRAF study shows that this is the main cause of deforestation in Southern Sudan and Darfur.

Wau district, Western Bahr el Ghazal. When shifting agriculture becomes unsustainable, forest cover disappears permanently

Drought, climate change and desertification

The repeated droughts of the 1970s and 1980s killed a large number of trees in the Sahel belt. Many of these areas have not been recolonized by trees since, as drier conditions and increased land use pressure have reduced the potential for seed distribution, germination and new growth. In regions such as Northern Darfur, the longer-term drop in precipitation has shifted the northern limit for several tree species a significant distance (50 to 200 km) to the south.

It is generally accepted that deforestation can promote desertification due to soil depletion, erosion and sand encroachment. At the same time, the development of hostile conditions causes gradual deforestation as trees die and are increasingly not replaced.

Pastoralist impacts: wildfires and tree browsing

The annual burning practised by pastoralist societies to renew grass and suppress shrubs and tree seedlings has a major impact on tree cover. Another issue is the use of foliage for camel fodder, which is a particular problem in areas like Southern Kordofan and Northern Darfur, where camel herders have migrated into land occupied by cattle herders and farmers. Some slow-growing species such as mangrove forests in Red Sea state have been devastated by camel browsing.

Direct conflict impacts

The scorched earth tactics used by militias in Darfur have resulted directly in localized deforestation. At present, UNEP does not have any detail on the scale of this phenomenon, and can only note its existence.

A settlement in the semi-desert north of El Fasher, Northern Darfur. The combination of drought, desertification, over-population and over-exploitation has drastically reduced forest cover in Northern Darfur

Wildfire in Blue Nile state. Fires lit by pastoralists to promote grass growth destroy existing trees and suppress sapling regrowth

Commercial lumber and export industry

In contrast to the situation in many countries, the commercial finished timber industry has not been a major factor in deforestation to date. Despite the existence of large forest resources in the south, Sudan actually imports finished timber, as poor transportation links and a lack of infrastructure have so far made commercial timber extraction difficult.

During the north-south conflict, both sides were involved in the illicit extraction of hardwoods, but the scale of extraction was limited by security, access and transportation constraints. In Southern Sudan, the main areas partially deforested due to this commercial activity are in the vicinity of Wau, Yei, Nimule and south of Torit. This trade has effectively stalled since the signing of the peace agreement.

This open woodland adjacent to a burnt village near El Geneina, Western Darfur, has been deliberately destroyed

The challenges of tackling deforestation in Sudan

At the national level, current observed rates of deforestation will reduce forest cover by over ten percent per decade. In some areas under extreme pressure, total loss has already taken place or is expected within the next ten years. There is clearly major cause for concern and an urgent need for corrective action.

Inspection of a two-year old plantation in Um Haraza, Sennar state. Reforestation has been successful in central Sudan when the FNC and state officials have been given adequate resources and mandates

The wide range of causative factors for deforestation in Sudan and the extent of regional variation indicate that solutions will have to be area-specific even while addressing national-scale demands. UNEP considers that the task of turning back deforestation in Sudan is unfortunately too large and too difficult to have a realistic chance of success in all regions.

Given the finite resources available to both GONU and GOSS, the first priorities in tackling deforestation should not be to launch large-scale investments in tree-planting or similar ventures. Despite obvious good intentions, there are many examples of destroyed communal forests and shelter belts in the northern states, where deforestation rates have only increased over time. Tree-planting on anything but a gigantic and economically non-feasible scale is unlikely to reverse this trend.

The recommended alternative approach is to analyse the situation in each region, start to resolve the underlying political, social, legal and economic issues, and only then prioritize areas and issues where some degree of success is most likely.

Many areas on the northern edge of the Sahel belt in Sudan are too degraded and too dry for large-scale reforestation to be feasible. Natural regeneration over time may be the only option

This timber bound for sale in Khartoum comes from 500 km south, near Renk in Upper Nile state

9.6 Potential conflict 'flashpoint' over the charcoal industry in Southern Sudan

The unmanaged mining of forest resources by the charcoal industry in the north-south boundary zone is one of several issues that could – in a worst-case but realistic scenario – constitute a potential trigger for renewed conflict at the local level (see Chapter 4).

At present, the charcoal industry in northern Sudan obtains its wood mainly from Southern Kordofan and riverine forests in Blue Nile and Upper Nile states. Current extraction rates are completely unsustainable, and as a result, the industry moves its operations gradually southwards each year.

UNEP predicts that within five to ten years, the northern states of Sudan will only be able to obtain sufficient supplies of charcoal from Southern Sudan and Darfur, as all other major reserves will have been exhausted. The extraction of charcoal from Southern Sudan is currently occurring outside any legal framework on resource- and benefit-sharing, and often without local agreement.

In essence, the benefits of the commercially-driven deforestation of the southern state of Upper Nile are flowing north, while the negative impacts are felt in Upper Nile state. This situation provides another catalyst for local conflict in the sensitive border zone.

9.7 Development opportunities for the timber industry in Southern Sudan

Southern Sudan's considerable forest reserves are commercially valuable and could – if managed well – support a significant wealth-creating export industry on a sustainable basis. Existing teak plantations alone could potentially generate up to USD 50 million per year in export revenue. Mahogany reserves could be the source of substantial hard currency as well. The sale of charcoal to the north is also a likely high-growth market.

Yet these resources are currently being wasted and the opportunity lost. Reserves are shrinking due to a combination of slash-and-burn clearance for agriculture, poor harvesting techniques and illegal logging. Meanwhile, a lack of governance discourages legitimate investors. The commercial timber industry needs to be radically reformed, as the trade is widely

The opportunity exists for Southern Sudan to extract much better value from each felled tree than is obtained at present. Teak plantations alone could potentially generate up to USD 50 million per year in export revenue, but the commercial timber industry is in need of reform to ensure that its practices are environmentally sustainable

perceived as badly managed in many parts of the country. Official Southern Sudan Agricultural Revitalization Programme (SSARP) statistics show that some 8,000 m³ have been exported since 2000, whereas other sources suggest that the figure is more likely to be around 90,000 m³ [9.7].

The new GOSS Ministry of Agriculture and Forestry declared a temporary ban on timber harvesting in January 2006 and intends to introduce revised timber sales procedures to reduce corruption and illegal logging, and enable the potential of Southern Sudan's forest reserves to be realized. The current harvesting ban is unlikely to remain in place for long, however, as timber is needed for the expanding local construction industries. Foreign logging concessionaires that exported teak in the past are also interested in acquiring new concessions.

Economic drivers will ensure that an export timber industry of some sort will evolve rapidly in Southern Sudan. What is at stake is the environmental sustainability of this industry, and how much benefit flows through to local populations. Political will and rapid action from GOSS, as well as support from the international community, are urgently needed. USAID, the

European Commission and others have already started to fund small-scale capacity-building programmes, but more investment is required.

9.8 Forestry sector governance

Robust legislation in the north

Legislation on the use of forests was first developed in the colonial period, with the Woods and Forests Ordinance of 1901, the Forests Ordinance of 1908, and the Forest Conservation Rules of 1917, which designated most forests as government property and established extensive forest reserves.

After independence, the authority of state and local administrations to manage forests was confirmed, and the comprehensive Forest Act of 1989 laid out a range of ownership categories and control measures. Controls over tree-cutting outside reserves were tightened by the requirement of permits. In addition, investors in agricultural schemes were obliged to conserve no less than ten percent of the total area of rain-fed projects and no less than five percent of the total area of irrigated projects to serve as shelter belts and windbreaks. Investors were also obliged to convert cleared trees into forest products. To manage forestry resources according to the Forest Act of 1989, the Forests National Corporation (FNC) was established as a semi-autonomous self-financing body in the same year. Forestry legislation was again strengthened and significantly modernized by the Forests and Renewable Resources Act of 2002.

Following the signing of the CPA and the adoption of the Interim Constitution in 2005, the responsibility for the management of forestry resources in the south was explicitly assigned to the new Government of Southern Sudan.

Northern and central Sudan enforcement issues

Northern governance issues relating to forests are simple at core: the legislation and structures are appropriate but enforcement and government investment is generally weak.

Throughout its time working with FNC officials in northern and central Sudan, the UNEP team witnessed extensive good work by the organization, but also a complete inability to enforce forestry laws due to a lack of resources and judicial support at the local level. Well-connected elements of the charcoal industry and the mechanized agriculture schemes appeared to be able to bypass the FNC and evade sanctions for obvious major violations. Minor violations are endemic and almost impossible to police.

In consultations, the FNC leadership stated that political support at the federal level was good, but called explicitly for the enforcement of existing legislation and for sound management practices to be translated to the state level. This gap between top level support and conditions on the ground indicates that the challenge will be to transform political will into practical action.

The FNC is in many respects a model organization for natural resource management in Sudan as it is self-managed, technically very competent and has a strong field presence. Its effectiveness, however, is crippled by a lack of support at the ground level. UNEP therefore considers that resolving the forestry governance issues for most of northern Sudan will be relatively straightforward, as only political will (at all levels) and appropriate investments are required. Other success factors are already largely in place.

Illegal charcoal production is a major cause of deforestation in Southern Kordofan

Darfur governance vacuum

Though the FNC is present and GONU legislation remains valid, the current situation in Darfur has led to an effective governance vacuum, with all of the associated negative implications.

Southern Sudan's current vulnerability

The situation in Southern Sudan is completely different from the rest of the country. Since 2005, the management of forests in the south falls to the GOSS Ministry of Agriculture and Forestry. The Ministry is very new and weak, and there are virtually no laws, detailed policies, or operational plans governing the forest resources of Southern Sudan.

No management activities are currently being conducted due to a lack of qualified forest managers. The Department of Forestry, in collaboration with the Kagelu Forestry Training Centre, is attempting to bridge this gap by offering refresher courses to forestry staff in the fields of silviculture, inventory and forest management, but it is expected that it will be some time before best forest practices are applied in the south. The forestry resources of Southern Sudan are thus presently extremely vulnerable to illicit exploitation.

9.9 Conclusions and recommendations

Conclusion

Sudan is in the midst of a genuine deforestation crisis. Most of the resources in northern, eastern and central Sudan have already been lost and the remainder is being depleted at a rapid pace. The large-scale timber resources of Southern Sudan are also disappearing quickly, and are generally being wasted as trees are burnt to clear land for crop-planting and to promote the growth of grass.

The sustainable use of the remaining timber resources in Southern Sudan represents a major development opportunity for the region, and requires both encouragement and the urgent development of governance to avoid potential over-exploitation.

Background to the recommendations

In simple terms, the solution to the deforestation of Sudan is to slow deforestation rates and increase replacement. In practice, however, this is anticipated to be very difficult to achieve, particularly in regions that are still in conflict or under extreme stress due to water shortages. As stated earlier, the recommended approach is to analyse the situation in each region, start to resolve the underlying political, social, legal and economic issues, and prioritize areas and issues where some degree of success is possible.

In Southern Sudan, it is likely that the timber industry will become a self-sustaining major tax and foreign exchange earner for GOSS. Industry and governance development work should therefore be regarded as an investment to jump-start an important industry. The focus should be on infrastructure, environmental and social sustainability, and governance.

Recommendations for the Government of National Unity

R9.1 Undertake an awareness-raising programme at the political level. The delivery of the latest facts and consequences of deforestation in Sudan to its leadership is a high priority. This will entail some further technical work to cover other parts of the north.

CA: AR; PB: MAF; UNP: UNEP and FAO; CE: 0.2M; DU: 1 year

R9.2 Invest in and politically support the Forests National Corporation. At present, this otherwise very capable institution cannot fulfill its mandate due to a lack of political support and funding.

CA: GI; PB: MAF; UNP: FAO; CE: 5M; DU: 3 years

R9.3 Introduce the concept and practice of modern dryland agroforestry techniques. This would entail a combination of awareness-raising, technical assistance and capacity-building, and practical action through demonstration projects in several states.

CA: TA; PB: MAF; UNP: FAO; CE: 2M; DU 5 years

R9.4 Develop a new national management plan and guidelines for mesquite and update the Presidential Decree to fit. This would entail a range of activities including assessment, cost-benefit analysis, governance and capacity-building.

CA: GROL; PB: MAF; UNP: FAO; CE: 0.4M; DU: 1 year

R9.5 Develop and implement a plan to resolve the Darfur camp fuelwood energy crisis. There are numerous options available and many piecemeal studies have been conducted, so any major programme should be preceded by a rapid options analysis and feasibility assessment. Major investment is needed to address this large-scale problem.

CA: PA; PB: UNHCR; UNP: UNEP; CE: 3M; DU: 3 years

Recommendations for the Government of Southern Sudan

R9.6 Undertake an awareness-raising programme at the political level. The delivery of the latest facts and consequences of deforestation in Southern Sudan to its leadership is a high priority.

CA: AR; PB: MAF; UNP: UNEP and FAO; CE: 0.1M; DU: 1 year

R9.7 Undertake capacity-building for the forestry sector. A large-scale multi-year programme is required.

CA: CB; PB: MAF; UNP: FAO; CE: 4M; DU: 3 years

R9.8 Develop legislation for the forestry sector. This work needs to progress from first principles, as soon as possible.

CA: GROL; PB: MAF; UNP: FAO; CE: 0.5M; DU: 2 years

R9.9 Complete a forestry inventory for the ten southern states and set up systems to monitor deforestation rates. This work could be combined with capacity-building.

CA: AS; PB: MAF; UNP: FAO; CE: 0.5M; DU: 1 year

R9.10 Regularize, reform and control the charcoal trade in Southern Sudan, with a focus on Upper Nile and Central Equatoria states. The multiple objectives include conflict risk reduction, resource management, control of corruption and the generation of tax revenue.

CA: GROL; PB: MAF; UNP: FAO; CE: 0.4M; DU: 2 years

R9.11 Introduce the concept and practice of modern agroforestry techniques. This would entail a combination of awareness-raising, technical assistance, capacity-building and practical action through demonstration projects in several states.

CA: TA; PB: MAF; UNP: FAO; CE: 2M; DU: 5 years

R9.12 Introduce the concept of forest product certification for timber export from Southern Sudan. This would entail a sustained development process to set up and embed the system into GOSS.

CA: GROL; PB: MAF; UNP: FAO; CE: 0.3M; DU: 2 years

Recommendations for the international community

R9.13 Introduce the concept and practical aspects of carbon sequestration to Sudan and attempt to integrate this into the forestry sector in the north and south. First and foremost, this would entail research to attempt to match commercial opportunities with potential carbon sinks. Suitable opportunities would then require development, support and oversight for a number of years before becoming commercially self-sustaining.

CA: GROL; PB: GONU and GOSS MAF; UNP: UNEP; CE: 0.3M; DU: 2 years

Freshwater Resources

With almost two-thirds of the Nile basin found within its borders, Sudan enjoys a substantial freshwater resource base. At the same time, 80 percent of the country's total annual water resources are provided by rivers with catchments in other countries. This leaves Sudan vulnerable to externally induced changes in water flows.

Freshwater resources

10.1 Introduction and assessment activities

Introduction

In a country that is half desert or semi-desert, the issue of freshwater availability is critical. At present, much of Sudan's population suffers from a shortage of both clean water for drinking, and reliable water for agriculture. These shortages are a result of natural conditions as well as underdevelopment. Development in this sector is surging ahead, however, and there is now an urgent need to ensure that this growth is environmentally sustainable.

Sudan has a substantial freshwater resource base (from now on referred to simply as water resources). Indeed, almost two-thirds of the Nile basin is found within its borders and its groundwater reserves are considerable. Yet there is a very broad disparity in water availability at the regional level, as well as wide fluctuations between and within years. These imbalances are a source of hardship in the drier regions, as well as a driving force for resource-based conflict in the country.

The unfinished Jonglei canal project in Southern Sudan played an important role in triggering the resumption of the north-south civil war. More recently, large-scale projects such as the Merowe dam have been strongly contested by local communities, and in the arid regions of Darfur, the current conflict also stems partly from issues of access to and use of water. The equitable use of water resources and the sharing of benefits are therefore considered key for the development of the country and the avoidance of further conflict.

In addition, there are several long-standing as well as emerging issues facing Sudan's water sector, including the challenges of providing potable water and sanitation services to a growing population, waterborne diseases, water pollution, aquatic weed infestations, the degradation of watersheds and freshwater ecosystems, and the construction of dams, which is expected to be the dominant factor that will fundamentally alter the environmental integrity of the country's rivers and wetlands over the next twenty-five years.

Wetlands throughout Sudan face a wide range of threats, including dam construction, upstream catchment degradation and oil exploration

Assessment activities

The study of freshwater resource issues in Sudan was an integral part of the general assessment, as water is a cross-cutting subject for virtually all sectors. UNEP teams visited dams, rivers, *khors* (seasonal watercourses), canals, *hafirs* (traditional small water reservoirs), wells and irrigation schemes in twenty-two states. Important sites visited include:

- the main Nile north of Khartoum through to Dongola;

- the White Nile from Juba to Bor and at Malakal, Kosti and Khartoum;

- the Blue Nile throughout Gezira, Sennar and Khartoum states;

- the Gash river at Kassala;

- the Atbara river at Atbara;

- the unfinished Jonglei canal in Jonglei state;

- major dams in central Sudan: Jebel Aulia on the White Nile, the Sennar and Roseires dams on the Blue Nile, and the Khashm el Girba on the Atbara; and

- *hafirs* in Darfur, Khartoum state, Northern Kordofan and Kassala state.

UNEP was not granted access to the Merowe dam but was able to assess the area downstream of the site.

10.2 Overview of the freshwater resources of Sudan

A large but highly variable resource

Sudan's total natural renewable water resources are estimated to be 149 km³/year, of which 80 percent flows over the borders from upstream countries, and only 20 percent is produced internally from rainfall [10.1]. This reliance on externally generated surface waters is a key feature of Sudan's water resources and is of critical importance for development projects and ecosystems alike, as flows are both highly variable on an annual basis and subject to long-term regional trends due to environmental and climate change.

As detailed in Chapter 3, the share of water generated from rainfall is erratic and prone to drought spells. In dry years, internal water resources fall dramatically, in severe cases down to 15 percent of the annual average.

The main basins

At the watershed level (the basic unit for integrated water resources management), Sudan comprises seven main basins:

- the Nile basin (1,926,280 km² or 77 percent of the country's surface area);

- the Northern Interior basins, in north-west Sudan (352,597 km² or 14.1 percent);

- Lake Chad basin, in western Sudan (90,109 km² or 3.6 percent);

- the Northeast Coast basins, along the Red Sea coast (83,840 km² or 3.3 percent);

- Lake Turkana basin, in south-eastern Sudan (14,955 km² or 0.6 percent);

- the Baraka basin, in north-eastern Sudan (24,141 km² or 1 percent); and

- the Gash basin, a closed basin in north-eastern Sudan (8,825 km² or 0.4 percent).

Table 18.   Summary data for Sudan water balance [10.1]

Statistic and measurement period or report date	Data /estimate
Water balance (1977 - 2001)	
Internal sources – rain and groundwater recharge	30 km³ per year
River inflows from other countries	119 km³ per year
Total	149 km³ per year
Water currently available for sustainable use (1999)	
Sudan share of Nile water under 1959 Sudan-Egypt treaty	20.5 km³ per year
Non-Nile streams	5.5 km³ per year
Renewable groundwater	4 km³ per year
Total	30 km³ per year
Nile treaty targets for swamp reclamation (1959)	
Proposed total additional from swamp reclamation projects	18 km³ per year
Sudan share from proposed projects	9 km³ per year

Figure 10.1 Sudan hydrological basins

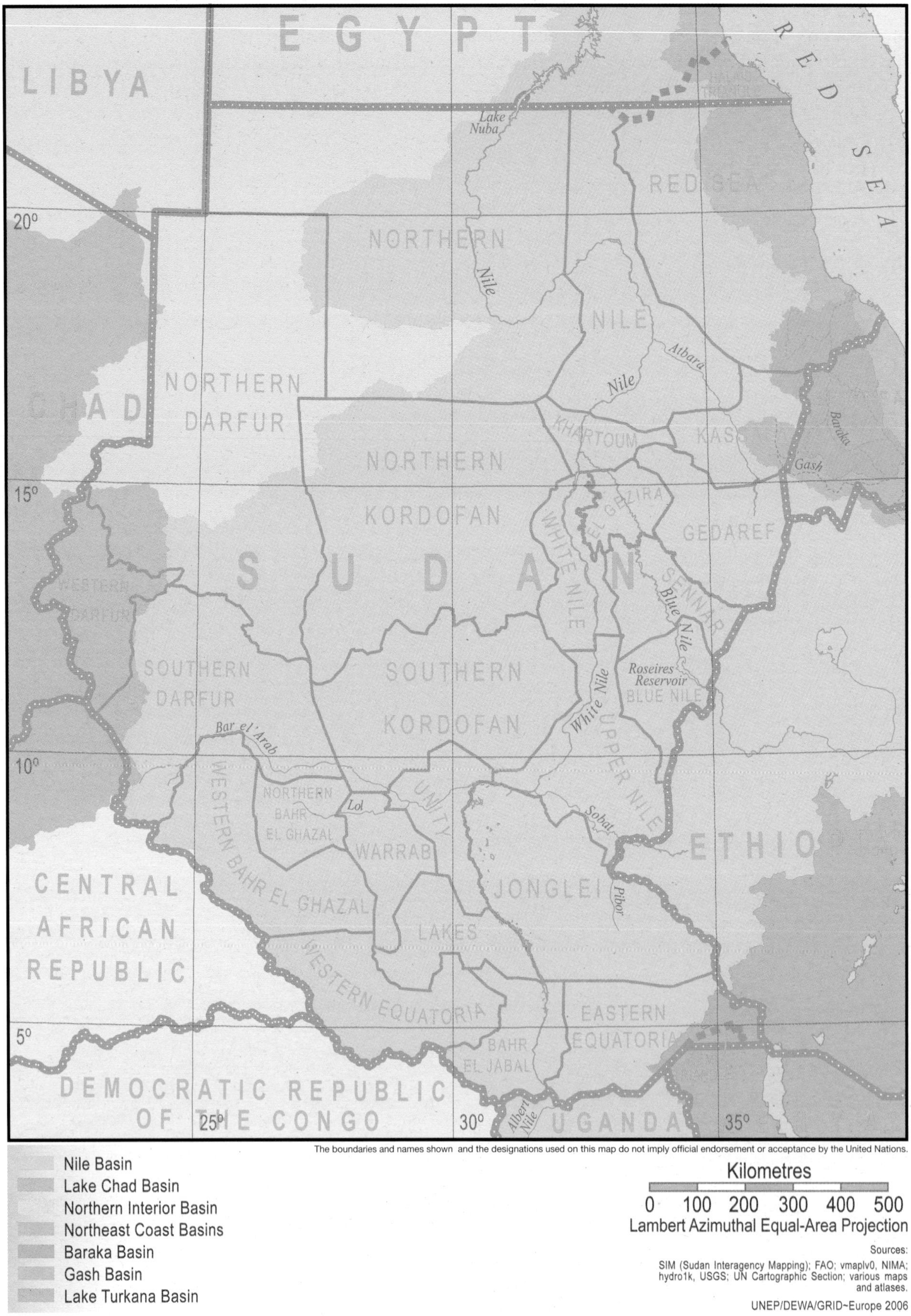
EGYPT
LIBYA
RED SEA
CHAD
SUDAN
ETHIO
CENTRAL AFRICAN REPUBLIC
DEMOCRATIC REPUBLIC OF THE CONGO
UGANDA
RED SEA
NORTHERN
NILE
NORTHERN DARFUR
NORTHERN KORDOFAN
KHARTOUM
KASSALA
GEZIRA
GEDAREF
WHITE NILE
SENNAR
WESTERN DARFUR
SOUTHERN DARFUR
SOUTHERN KORDOFAN
BLUE NILE
UPPER NILE
UNITY
NORTHERN BAHR EL GHAZAL
WARRAB
WESTERN BAHR EL GHAZAL
JONGLEI
LAKES
WESTERN EQUATORIA
BAHR EL JABAL
EASTERN EQUATORIA
Lake Nuba
Nile
Atbara
Baraka
Gash
Blue Nile
White Nile
Roseires Reservoir
Bar el'Arab
Lol
Sobat
Pibor
Albert Nile
20°
15°
10°
5°
25°
30°
35°
The boundaries and names shown and the designations used on this map do not imply official endorsement or acceptance by the United Nations.
Nile Basin
Lake Chad Basin
Northern Interior Basin
Northeast Coast Basins
Baraka Basin
Gash Basin
Lake Turkana Basin
Kilometres
0 100 200 300 400 500
Lambert Azimuthal Equal-Area Projection
Sources:
SIM (Sudan Interagency Mapping); FAO; vmaplv0, NIMA; hydro1k, USGS; UN Cartographic Section; various maps and atlases.
UNEP/DEWA/GRID~Europe 2006

The dominance of the Nile basin is evident in the fact that nearly 80 percent of Sudan lies within it, and that conversely, 64 percent of the Nile basin lies within Sudan. With the exception of the Bahr el Ghazal sub-basin, all of Sudan's drainage basins – including the main Nile sub-basins – are shared with neighbouring countries. Nile waters, as well as those of the seasonal Gash and Baraka rivers, mainly originate in the Ethiopian highlands and the Great Equatorial Lakes plateau [10.1].

The Lake Chad and Bahr el Ghazal basins are the only ones to receive important contributions from rainfall inside Sudan. These hydrological characteristics underline the importance of international cooperation for the development and sustainable management of Sudan's water resources.

Wetlands, fisheries and groundwater

Sudan boasts a significant number of diverse and relatively pristine wetlands that support a wide range of plants and animals and provide extensive ecosystem services to local populations. The principle wetlands are the Sudd – which is a source of livelihood for hundreds of thousands of pastoralists and fishermen – Bahr el Ghazal, Dinder and other Blue Nile *maya*s, the Machar marshes, Lake Abiad and the coastal mangroves. In addition, there are a large number of smaller and seasonal wetlands that host livestock in the dry season and are important for migrating birds.

The rivers and wetlands of Sudan support significant inland fisheries, which are exploited for sustenance as well as on a commercial basis. Fisheries development is generally limited and is unbalanced, as most of the resources are in the wetlands of Southern Sudan, while most of the fishing is practised in the more limited waters of central and northern Sudan.

Sudan also possesses significant groundwater resources. Indeed, one of the world's largest aquifers – the deep Nubian Sandstone Aquifer System – underlies the north-western part of the country, while the Umm Rawaba system extends over large areas of central and south Sudan, and has a moderate to high recharge potential. In Western Darfur and south-western Sudan, groundwater resources are generally limited but locally significant, due to the basement complex geology. In the coastal zone, finally, the limited groundwater is brackish to saline.

Sudan's wetlands support fisheries, which in turn support communities. Fish caught from a seasonal lake by the While Nile dries on the roof prior to being packed for local markets

Papyrus mat weaving is one of the main sources of livelihood for displaced persons and impoverished communities along the banks of the White Nile

Water consumption

Sudan consumes an estimated 37 km³ of water per year, of which 96.7 percent are used by the agricultural sector. Withdrawals by the domestic and industrial sectors amount to 2.6 and 0.7 percent respectively [10.1]. Water consumption is mainly reliant on surface waters, but groundwater extraction is rapidly growing. At present, groundwater is chiefly used for domestic purposes and small-scale irrigation in the Nile flood plain and its upper terraces, as well as in the *wadis*.

10.3 Environmental impacts and issues of the water sector

The single most critical issue related to water resources in Sudan today is the new and planned large dams and related development schemes. A number of other issues were also noted in the course of the assessment.

Large dams and water management schemes:

* impacts and issues of existing large dams;
* the Merowe dam;
* the Jonglei canal; and
* planned large dams and schemes.

Other issues:

* traditional dams;
* wetland conservation;
* invasive plant species;
* water pollution;
* groundwater exploitation;
* transboundary issues and regional issues; and
* freshwater fisheries.

10.4 Large dams and water management schemes

Existing large dams: performance problems and major downstream impacts

The situation with existing dams in Sudan can be used as a benchmark to help evaluate the balance of benefits and disadvantages of the country's proposed future dams (next section). UNEP visited all of Sudan's existing large dams: Jebel Aulia on the White Nile, the Sennar and Roseires dams on the Blue Nile, and the Khashm el Girba dam on the Atbara river.

For Sudan, the development benefits of large dams are very clear: they provide the majority of the electricity in the country and support large-scale irrigation projects. As such, they can be considered a cornerstone of development for the country.

However, like most major water and infrastructure projects, large dams also have a range of negative effects, including environmental impacts. All of the dams visited by UNEP were found to have both performance problems and visible, though variable, negative impacts on the environment. Much of the issues noted are irreversible and possibly unavoidable. Nonetheless, they provide important lessons that can help minimize impacts of future dam projects through improved design and planning.

UNEP's inspection of existing dams highlighted two principal environmental issues:

- performance problems caused in part by upstream land degradation; and

- downstream impacts due to water diversion and changes in flow regime.

Loss of active dam storage by sediment deposition

UNEP considers the performance problems of existing large dams to be cases of environment impacting infrastructure, rather than the reverse. With the exception of the Jebel Aulia dam, all of the reservoirs of Sudan's existing dams are severely affected by sediment deposition. It is estimated that 60 percent of Roseires's storage capacity, 54 percent of Khashm el Girba's, and 34 percent of Sennar's have been lost to siltation [10.3]. The construction of the Roseires dam upstream of Sennar in 1966 significantly decreased the sedimentation problem in the latter.

Islands and seasonal grasses are visible in the Sennar dam reservoir, which is now 60 percent full of sediment

At the Roseires dam reservoir, a dredger is continuously used to remove sediment from the electric turbine water inlets. Soil washed from the Ethiopian highlands is the main source of the sediment

Table 19. Existing large dams in Sudan [10.2, 10.3]

Name	Location	Year of commissioning	Purpose	Capacity (10⁹ m³)		Capacity loss
				Design	Present	
Sennar	Blue Nile	1925	Irrigation, flood control	0.93	0.37	60 %
Jebel Aulia	White Nile	1937	Hydropower	3.00	3.00	0
Khashm el Girba	Atbara river	1964	Irrigation, flood control	1.30	0.60	54 %
Roseires	Blue Nile	1966	Flood control, hydropower	3.35	2.20	34 %
Total Sudan storage capacity				**8.58**	**6.17**	**28 %**
Percentage of Sudan's storage capacity of its share				**46 %**	**33 %**	**13 %**

At Roseires, which currently accounts for 75 percent of Sudan's electricity production, sediments have reached the power intakes, affecting turbine operation and undermining electricity production. Though a proposal exists to raise reservoir storage capacity by increasing dam walls by ten metres, it is unlikely to be a sustainable solution in the long term.

Sediment accumulation is even more severe in the Khashm el Girba reservoir. Flushing is carried out during the flood peak, but this leads to massive fish kills downstream and the reservoir lake is virtually fishless as a result. Reservoirs in seasonal *wadis* are similarly affected: a significant portion of the El Rahad reservoir capacity in *khor* Abu Habil in Northern Kordofan, for instance, has been lost due to high sediment loads. The same is true for the many small check-dams in the Nuba mountains.

The root cause of the dams' performance problems is linked to upstream land degradation. The high rate of sedimentation in the Blue Nile and Atbara rivers is partly natural, and partly the end result of land degradation and soil erosion in the drainage basins of both Sudan and Ethiopia. Addressing the cause of the sedimentation would therefore require a regional-level undertaking involving substantial revegetation of the watershed and other major works. At present, dam operators are forced to attempt to address only the symptoms of this problem.

Degradation of downstream ecosystems

Sudan's existing large dams have resulted in a major degradation of downstream habitats. The three impacts of most concern are reduced annual flow, removal of annual flood peaks and increased riverbank erosion. These impacts are associated with major dam projects worldwide and are not unique to Sudan.

In simplistic terms, the removal of water and sediment (which silts up the dam reservoirs instead) has resulted in the partial destruction of downstream ecosystems. Both *maya* wetlands (swamps dominated by *Acacia nilotica*) on the Blue Nile, and Dom palm (*Hyphaene thebaica*, an endangered species in Sudan) forests along the Atbara river, have been adversely impacted by the construction of dams, which suppress the flood pulses that nourish these economically valuable ecosystems. The large-scale disappearance of the Dom palm forests in the lower Atbara is at

Prior to the construction of the Khashm el Girba dam, riparian communities relied on water pools of the Atbara river during the dry season. Annual flushing of the dam has sealed many of these ponds with sediment, leaving communities and livestock thirsty

Old plans to construct a dam at the Nile's Third Cataract, near Kerma, have recently been resuscitated as part of Sudan's major dam development programme. Environmental impact assessments and public participation need to be strengthened to ensure that environmental sustainability and social equity are fully integrated into dam building

least partly attributable to the construction of the Khashm el Girba dam. On the Blue Nile, infrequent flooding of the *maya* systems has led to a change in species composition; in some cases their survival has been threatened by hydrologic disconnectivity from the main river [10.4, 10.5].

Downstream of its juncture with the Blue Nile and the Atbara river, the main Nile is threatened by serious riverbank erosion, a phenomenon known locally as *haddam*. Dams on the Blue Nile and Atbara rivers have significantly altered daily and seasonal flows, both in terms of water and sediment flows and in terms of velocity and current direction. Riverbank erosion is discussed in more detail in Chapter 3.

A lack of environmental impact assessment and mitigation

No environmental impact assessments were carried out for the existing large dams in Sudan and their current operation is clearly not influenced by the need to limit ongoing impacts to downstream ecosystems and communities.

There is no doubt that the dams have had a major positive impact on the development of the country and that significant benefits have flowed to the recipients of the diverted waters (the large irrigation schemes). What is unclear is the overall environmental and economic balance of such projects, as the losses to downstream communities and ecosystems have not been fully accounted for. Given the cost of the dams and the observed rate of sedimentation, the economics of future dam projects in this region should be carefully examined.

The Merowe dam

The Merowe dam – which is currently the largest new dam project in Africa – was in the late stages of construction at the time of the UNEP survey. Environmental impacts (outside of construction) had therefore yet to occur, but there was no opportunity to further influence the design, for environmental or other reasons.

The Merowe dam project followed the same pattern as older dams in Sudan. The dam is set to bring massive benefits to the country through electricity generation, but the displacement of upstream communities in the dam reservoir zone has led to unrest and local conflict. What has not occurred is a full and transparent environmental, economic and social impact assessment, to weigh the positive and negative features of the project, and attempt to maximize the positives while mitigating the negatives.

UNEP has completed a very preliminary appraisal of the potential environmental impact of the dam, using the limited documentation available, field visits to the areas downstream of the dam, and the background information provided by visits to existing large dams, agricultural schemes and desert regions in Sudan in 2006 [10.6, 10.7, 10.8, 10.9] (see Case Study 10.1). This analysis shows that the impacts on the downstream communities and ecosystems may be severe and that further assessment is needed as the first step towards mitigating these impacts. Secondly, the envisioned plans for the new irrigation schemes should be reviewed based on the experiences of existing dams and schemes in Sudan.

CS 10.1   UNEP appraisal of the environmental impact of the Merowe dam

The Merowe dam, which is set to double the electricity production of Sudan [10.6], will undoubtedly contribute massively to the development of the country and provide a host of benefits. It is the first large dam project in the country to include any form of environmental impact assessment (EIA). It also features an organized resettlement plan for affected downstream populations.

However, like all new large dams worldwide, the Merowe project is surrounded by controversy related to its projected and actual social, environmental and economic impacts. UNEP, focusing on the environmental aspects only, has conducted an appraisal of the Merowe EIA process, associated documents and the actual environmental issues. The findings indicate several areas of concern.

The Merowe dam is the most upstream major development on the main Nile and is currently the largest dam development in Africa after the Aswan dam in Egypt. Reservoir impoundment will lead to the loss of 200 km of riverine farmland and habitat [10.7], permanently and radically changing the downstream ecosystem of a region that supports hundreds of thousands of people. A major new irrigation scheme is also planned.

The Merowe dam EIA license was only issued in 2005, over two years after work on the project physically started in early 2003. The EIA document was developed by a foreign consultancy working primarily on the dam design process, and had little connection to the potentially impacted communities. The report is apparently now publicly available from the Ministry but has not been disseminated, and no public hearings have been held concerning its findings.

Properly undertaken, an EIA process can provide a credible framework for the affected people to communicate their concerns and gain the trust of the project's proponents. In this case, however, the delays and closed approach undermined the entire process in terms of impact analysis and mitigation, and public buy-in.

UNEP's technical analysis and reconnaissance fieldwork downstream of the dam site indicated several significant impacts that were not addressed in the EIA:

*   **Silt loss for flood recession agriculture and dam sedimentation:** The dam will collect the fertile silt that kept the downstream riverine agricultural systems (*gerf* land) viable. This issue alone places the downstream communities at major risk. As other existing large dams, the Merowe dam is likely to be affected by high rates of sedimentation. During consultation, Ministry officials indicated that a sediment flushing routine is planned during operations, but the details and impacts of this are unclear.

*   **Riverbank erosion:** The dam's power plant is scheduled to operate at full capacity during four hours per day releasing 3,000 m³/s; during the remaining time, only two of the ten turbines will run, generating 600 m³/s [10.6]. The concentration of discharge over a short time period and the resulting strong four to five metre daily fluctuations in water levels will almost inevitably have major detrimental effects on the riverbanks and adjacent agricultural schemes.

*   **Reduced river valley groundwater recharge:** The Nile is typically full for five to six months of the year, but the dam's construction will lower the base flow considerably, which is likely to disrupt groundwater refilling over a great distance downstream of the dam. This could have significant consequences for the expanding cultivation of the upper terraces, which relies increasingly on small tube wells (*mataras*) for year-round irrigation.

*   **Questionable net gain on food production:** In combination, the above effects may seal the fate of much of the downstream farmland. While the dam project does include a planned new irrigation scheme, assessments of existing schemes in Sudan indicate that they commonly perform well below design expectations (see Chapter 8). In the case of Merowe, the proposed new irrigation areas are low fertility desert soils in a hyper-arid and extremely hot environment. The overall net gain in terms of food production should be re-examined closely based on prior dam performance and projected downstream economic losses.

*   **Blocking of fish migrations and the impact on locally endangered species** like the Nile crocodile. These issues were not addressed in the EIA.

None of the downstream scheme managers and farmers interviewed by UNEP had been presented with the findings of the dam's EIA report. Neither were they aware of any studies to assess the dam's impact on bank erosion, or consulted about its potential implications, despite the fact that they reportedly made repeated requests to the dam authorities for clarification on this issue. Ministry officials have indicated that a consultation process for downstream communities is planned.

The dam is now built and filling up. It is therefore too late to make any changes to its core design. What is possible and indeed needed, however, is an urgent follow-up impact analysis aimed at assessing what can be done to minimize the negatives and accentuate the positive impacts of this mega-project. Key areas to address include the planned flow regime and the irrigation scheme plans.

Figure 10.2   Merowe dam

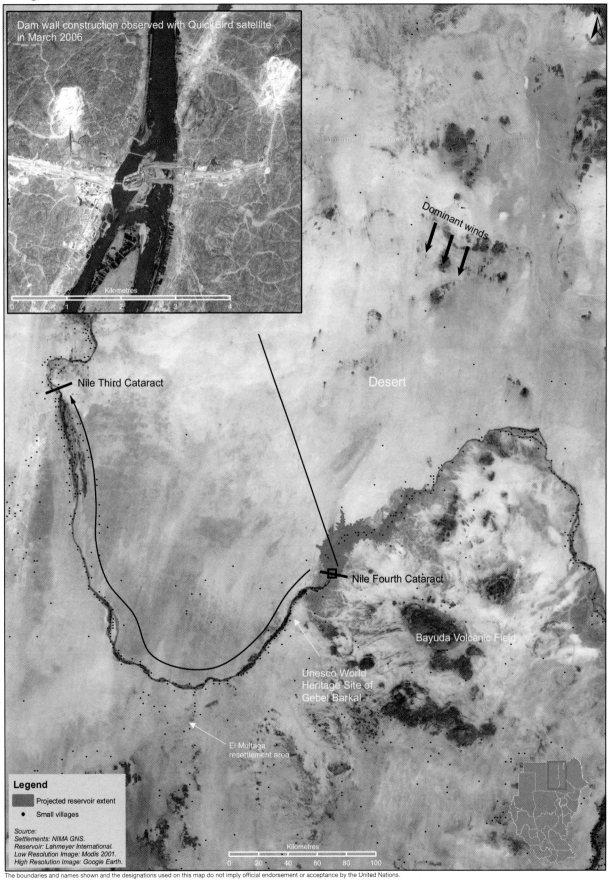

Dam wall construction observed with QuickBird satellite in March 2006

Kilometres
0   1   2   3   4

Dominant winds

Nile Third Cataract

Desert

Nile Fourth Cataract

Bayuda Volcanic Field

Unesco World
Heritage Site of
Gebel Barkal

El Multaga
resettlement area

Legend

Projected reservoir extent

•   Small villages

Source:
Settlements: NIMA GNS.
Reservoir: Lahmeyer International.
Low Resolution Image: Modis 2001.
High Resolution Image: Google Earth.

Kilometres
0   20   40   60   80   100

The boundaries and names shown and the designations used on this map do not imply official endorsement or acceptance by the United Nations.

The main channel excavator is composed of several self-propelled sections. Once the largest of its type in the world, it now lies in a derelict state in the canal bed

CS 10.2   The Jonglei canal

Launched in 1980, the construction of the Jonglei canal was interrupted by the outbreak of conflict in Southern Sudan in 1983. Though the economic motivations for the project still exist for some parties, a combination of political issues, economics and environmental concerns make the resumption of construction unlikely.

The idea of using a canal to bypass the Sudd wetlands was first conceived in the early 1900s by Egyptian and British authorities. The White Nile loses up to 50 percent of its annual flow through evaporation and evapotranspiration as it winds through the Sudd. A canal could potentially capture this water for downstream users, as well as partially drain the wetlands for agriculture [10.10, 10.11].

The project in its modern form was developed during the 1970s. The project team included multinational contractors and financiers, and had the strong support of the Khartoum government, as well as of Egypt and France. In contrast, there was little knowledge and even less acceptance of the project by local stakeholders, who were principally transhumant pastoralists and a minority population of subsistence farmers and fishermen. It is likely that the project would have resulted in a net negative impact for local communities, due to the loss of *toic* grazing land and fishing sites.

Of the canal's planned 360 km, approximately 260 km were excavated before southern Sudanese rebel military forces sabotaged the main excavator in 1983, rendering the construction too dangerous to continue. The canal excavator now lies in a derelict and corroded condition, and is probably irreparable. The canal itself does not connect to any major water bodies or watercourses, and acts only as a giant ditch and embankment superimposed on a very flat seasonally flooded plain. It is approximately eighty metres across and up to eight metres deep, including a four-metre embankment.

The canal channel has gradually filled due to erosion and lack of maintenance, reducing the angle of its slopes to a maximum of 35 degrees. It has been extensively reclaimed by vegetation, with sparse to dense woodland and scrub found along both sides. In addition, the central channel is seasonally flooded to a depth of one to two metres and supports a significant fish population and an evolving ecosystem.

The canal bank is now being used as the route for the new Juba-Malakal road, which is expected to have significant direct and indirect impacts on the environment of the canal.

The canal course cuts across the migration pathways of the white-eared kob (*Kobus kob leucotis*) and the tiang (*Damaliscus lunatus tiang*) [10.12, 10.13], and was noted to be a partial barrier to migration in the 1980s, causing concentration at preferred crossing points and increasing losses due to falls, predators, poaching and drowning. In its current condition, however, the canal is not considered to represent a significant physical barrier to larger wildlife, except in the wet season when swimming is required to cross some sections. In order to fully remove the migration barrier and avoid any inadvertent hydraulic connection to the Nile, the canal would need to be partially filled in to form land bridges at a number of points.

In its original design, the canal project would have had major negative environmental impacts on the Sudd wetlands [10.14]. The viability of the project is questionable on these grounds alone, irrespective of the numerous social, political and economic issues attached to any potential resumption of the construction. However, the principal lesson learnt from the Jonglei canal is that major ventures lacking local support are at risk, and that achieving such support requires both broad consultation and benefit-sharing.

The Jonglei canal

The Jonglei canal project – an unfinished project to build a canal to bypass the Sudd wetlands and capture the water for downstream users – was closely linked to the resumption of north-south conflict in 1983 and had strong international ties. As it was never completed, its anticipated major environmental impacts never came to pass. However, lessons learnt from this project (see Case Study 10.2) should be applied to both existing efforts in peacebuilding between north and south, and to future development plans for the Nile, as promoted by a range of local, regional and international interests.

Massive dam development in the planning stages

As of late 2006, the Government of National Unity is on the verge of launching a new and ambitious dams building programme (in addition to the Merowe dam). The importance conferred on dams is reflected in the September 2005 decision by Presidential Decree No. 217 to place the Dams Implementation Unit (formerly known as the Merowe Dam Project Implementation Unit) under the President's Office. More than two dozen dam feasibility studies are planned or currently underway. In Southern Sudan, an important hydropower programme is envisioned on the White Nile.

As the unfinished Jonglei canal is not connected to any major watercourse or water body, it is currently a 260 km-long ditch. The channel has been eroded and revegetated, and is seasonally flooded, supporting a new ecosystem

The unfinished Juba-Malakal trunk road project includes a 250 km stretch to be built on the west bank of the Jonglei canal. Approximately 100 km had been built by mid-2006, opening this remote area up for development

Figure 10.3   Nile sub-basins, dams and hydroelectric schemes

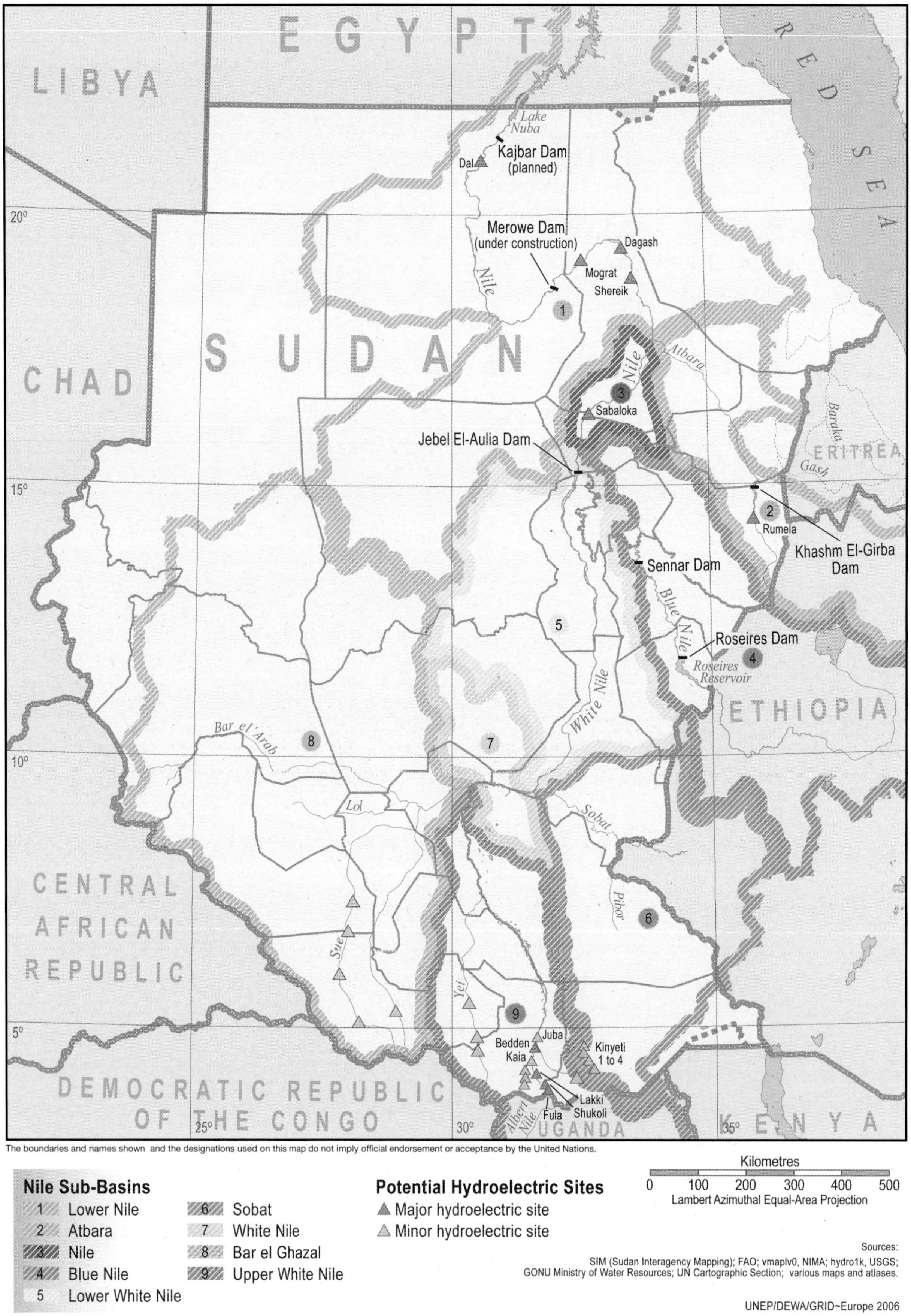

The boundaries and names shown and the designations used on this map do not imply official endorsement or acceptance by the United Nations.

Nile Sub-Basins

1	Lower Nile	6	Sobat
2	Atbara	7	White Nile
3	Nile	8	Bar el Ghazal
4	Blue Nile	9	Upper White Nile
5	Lower White Nile		

Potential Hydroelectric Sites

▲ Major hydroelectric site
△ Minor hydroelectric site

Kilometres

0   100   200   300   400   500
Lambert Azimuthal Equal-Area Projection

Sources:
SIM (Sudan Interagency Mapping); FAO; vmaplv0, NIMA; hydro1k, USGS;
GONU Ministry of Water Resources; UN Cartographic Section; various maps and atlases.

UNEP/DEWA/GRID~Europe 2006

The history of major water scheme development in Sudan is mixed. This is partly linked to the method of project development: dams and water schemes have historically been promoted by decree at the federal level, with limited or no local consultation, and no environmental impact assessments. This approach failed for the Jonglei canal in 1983 and has elicited problems for the Merowe dam project as well.

Controversy generated by major water schemes is certainly not unique to Sudan. Dams have and continue to be strongly contested in many countries. In recent years, they have been the subject of an intensive debate at the international level, most notably by the World Commission on Dams [10.15].

However, as Sudan surges ahead with its construction plans, it is in an advantageous position to re-examine its own national experience, as well as draw on the knowledge base and latest lessons learned from regional and global dam reviews, so as to avoid repeating past mistakes.

Two of the underlying strategic tenets recommended by the World Commission on Dams are 'gaining public acceptance' and 'recognizing entitlements and sharing benefits' [10.15]. For Sudan, this would require the revision of top-down approaches by which the decision to construct a dam is made by decree. Information-sharing and an open and transparent public and multi-stakeholder consultation process need to be institutionalized in Sudan's dam sector. This also implies that dams should not be regarded as an end in their own right, but rather be evaluated and discussed within the context of defined water and energy needs and the full range of available options to meet those demands.

Sedimentation of traditional small dams and water-harvesting structures

The small traditional dams inspected by UNEP did not have any of the environmental impacts of larger dams, but did have a number of performance problems. In addition, they provided clear examples of how local conflict over scarce natural resources can arise.

Traditional dugouts fed by rainwater and run-off (called *hafirs*) have played a critical role for centuries – in Darfur and Kordofan in particular – in supplying water for domestic use in villages and to pastoralists in

remote areas vulnerable to erratic rainfall variations. However, increasing siltation from topsoil erosion and drifting sands as well as poor maintenance have led either to a serious decline in the water storage capacity or to the outright loss of many *hafirs*.

Due to increasing competition over limited water supplies, many *hafirs* have become 'flashpoints' between pastoralists and farmers. The situation has been compounded by the development of horticultural schemes around *hafirs*, as witnessed in Southern Kordofan [10.16].

Lack of investment and maintenance during the conflict years led to complete or partial loss of many hafirs, such as this one at El Tooj, near Talodi in Southern Kordofan. Constructed in 1972 as part of a national campaign to eradicate thirst, the water treatment facility was targeted during the conflict and local communities have been drinking untreated water ever since

A small dam complex in Darfur, with a banked catchment area, storage dams and associated small-scale irrigated agriculture

10.5 Sustainable use and conservation of wetlands

An important national resource under pressure

UNEP has found that most of Sudan's major wetlands are currently facing significant conservation threats.

During the long north-south conflict, wetlands in the south were adversely affected by uncontrolled hunting and poaching. With peace, the country's wetlands in all areas are under mounting pressure from development plans. The most significant issues are major infrastructure projects such as oilfields, dams and water engineering projects, roads, housing schemes, conversion for agriculture and settlement, as well as resource over-exploitation by a growing population. Other emerging threats include invasive alien species, namely water hyacinth and mesquite. This all points to the necessity of developing strategic action plans and building national capacity aimed at the wise use of wetlands.

Issues related to the Sudd are covered in Case Study 10.3, while the remaining mangrove wetlands – which are in steep decline and in urgent need of protection – are discussed in Chapter 12. The Machar marshes are very remote and were not visited by UNEP, but the Governor of Upper Nile state reported that the construction of roads for oil exploration constituted a major risk for the marshes. As for the Bahr el Arab wetlands, the principle threat is considered to be habitat degradation by land clearance for agriculture, overgrazing and fires.

Degradation of the Blue Nile wetlands

The *maya* ecosystems of the Blue Nile are badly degraded and in continuing decline. UNEP visited seven *maya*s (swamps dominated by *Acacia nilotica*) along the Blue Nile and found them all to be degraded by accelerated siltation. Several, such as Um Sunut and Kab in Gezira state and El Azaza in Sennar state, were effectively disconnected from the main river. The main causes of this decline are upstream dam construction and catchment changes. Other issues include extensive felling of riverine forests, damage from overgrazing and wildlife poaching.

Mayas like this one in Dinder National Park play a critical role in supporting wildlife populations during the dry season

Table 20.   Status of the six most significant wetlands in Sudan [10.2, 10.17, 10.18, 10.19. 10.20]

Wetland	State(s)	Approximate size	Ecosystem integrity
Sudd	Jonglei, Unity, Upper Nile	57,000 km²	Generally in very good condition
Machar marshes	Upper Nile	6,500 km²	Status unknown
Blue Nile mayas, including Dinder	Blue Nile, Sennar	Discontinuous (< 1,000 km²)	Moderately to heavily degraded
Bahr el Arab	Northern Bahr el Ghazal, Warrab, Unity	Discontinuous	Status unknown
Lake Abiad	Southern Kordofan	5,000 km²	Moderately degraded
Red Sea mangroves	Red Sea state	Linear and discontinuous (< 100 km²)	Badly degraded and shrinking

The plant biota of the Sudd range from submerged and floating vegetation in the open water to swamps dominated by papyrus. Over 350 plant species have been identified in the wetland

CS 10.3   The Sudd wetlands

Sudan has some of the most extensive wetlands in all of Africa and until recently, only a small percentage of this important habitat had any legal protection. In June 2006, however, the Sudd wetlands were listed as a site under the Ramsar Convention.

The Sudd is the second largest wetland in Africa, and the ecosystem services it provides are of immense economic and biological importance for the entire region. In the rainy season, the White Nile and its tributaries overflow to swell the Sudd swamps situated between the towns of Bor in the south and Malakal in the north. The swamp habitats themselves cover more than 30,000 km², while peripheral ecosystems such as seasonally inundated woodlands and grasslands cover a total area some 600 km long and a similar distance wide. The flooded area varies seasonally and from year to year, due to variations in rainfall and river flows. Its greatest extent is usually in September, shrinking in the dry season.

The plant biota of the Sudd range from submerged and floating vegetation in the open waters to swamps dominated by *Cyperus papyrus*. In addition, there are extensive phragmites and typha swamps behind the papyrus stands. Seasonal floodplain grasslands up to 25 km wide are dominated by wild rice *Oryza longistaminata* and *Echinochloa pyramidalis*. Over 350 plant species have been identified, including the endemic *Suddia sagitifolia*, a swamp grass [10.17].

The swamps, floodplains and rain-fed grasslands of the Sudd also support a rich animal diversity, counting over 100 species of fish, a wide range of amphibians and reptiles (including a large crocodile population) and 470 bird species [10.17]. The swamps host the largest population of shoebill (*Balaeniceps rex*) in the world: aerial surveys in 1979-1982 counted a maximum of 6,407 individuals. Hundreds of thousands of birds also use the Sudd as a stopover during migration; migratory species include the black-crowned crane (*Balearica pavonina*), the endangered white pelican (*Pelecanus onocrotalus*) and the white stork (*Ciconia ciconia*).

In addition, more than 100 mammal species have been recorded. Large mammals have always been hunted by local communities as an important food source. Given the present widespread availability of modern weaponry, however, the current status of large mammals, including elephants, needs to be reassessed urgently. Historically, the most abundant large mammals have been the white-eared kob (*Kobus kob leucotis*), the tiang (*Damaliscus lunatus tiang*) and the Mongalla gazelle (*Gazella rufifrons albonotata*), which use the floodplain grasslands in the dry season [10.21]. The endemic Nile lechwe (*Kobus megaceros*) and the sitatunga (*Tragelaphus spekii*) are resident, and it is anticipated that there are still significant populations of hippopotami (*Hippopotamus amphibius*).

The ecosystem services performed by this immense wetland, which extend far downstream, include flood and water quality control. Other services within the ecosystem itself are year-round grazing for livestock and wildlife, fisheries, and the provision of building materials, among many others. The Sudd is inhabited principally by Nuer, Dinka and Shilluk peoples, who ultimately depend on these ecosystem services for their survival. The central and southern parts of the Sudd have small widely scattered fishing communities. Up to a million livestock (cattle, sheep and goats) are kept in the area, herded by the pastoralists to their permanent settlements in the highlands at the beginning of the rains in May-June and down to intermediate elevations during the dry season. Crops include sorghum, maize, cowpeas, groundnuts, sesame, pumpkins, okra and tobacco.

There are three protected areas in the Sudd: Shambe National Park, and the Fanyikang and Zeraf game reserves. In June 2006, an area totaling 57,000 km² was declared Africa's second largest Ramsar site [10.17].

The Sudd and its wildlife are currently at risk from multiple threats, including oil exploration and extraction, wildlife poaching, pastoralist-induced burning and overgrazing, and clearance for crops. The resumption of the Jonglei canal project would also put the wetland at significant risk. Listing the Sudd as a protected site under the Ramsar Convention is an important but mainly symbolic initiative that now needs to be consolidated with practical measures to help conserve this critical natural asset.

10.6 Invasive plant species

Infestations on land and water

The watercourses of Sudan are afflicted with two invasive species: water hyacinth, which threatens the Nile basin watercourses, and mesquite, which has infested many of the seasonal *khors* and canals of northern Sudan. Mesquite is covered in detail in Chapter 8.

Water hyacinth

The most problematic aquatic weed in Sudan is water hyacinth (*Eichhornia crassipes*), a native plant of South America that was officially declared an invasive pest in 1958 [10.22]. Water hyacinth forms dense plant mats which degrade water quality by lowering light penetration and dissolved oxygen levels, with direct consequences for primary aquatic life. The weed also leads to increased water loss through evapotranspiration, interferes with navigation and fishing activities, and provides a breeding ground for disease vectors such as mosquitoes and the vector snails of schistosomiasis.

Water hyacinth (Eichhornia crassipes) grows rapidly; until recently, it invaded the entire stretch of the White Nile from Juba to Jebel Aulia

Workshops of the Ministry of Agriculture's Water Hyacinth Control Division at Jebel Aulia lie idle as funding from donor agencies has dried up. The northern limit of hyacinth infestation is now reportedly between Kosti and Duweim, although its presence was cited in the Jebel Aulia dam reservoir in June 2006, for the first time in seven years

The Jebel Aulia dam has served as a barrier to the spread of the invasive water hyacinth

A 1,750 km stretch of the White Nile, from its upper reaches near Juba to Duweim (some 70 km south of Khartoum), is infested. The hyacinth spread used to extend to the Jebel Aulia dam, but a causeway at Duweim is apparently acting as a precarious barrier to downstream propagation. In Sudan, control measures initially relied on large-scale applications of chemicals. An estimated 500 tonnes of the herbicide 2, 4-D were applied to the White Nile annually [10.22]. This practice has now ceased, but it may have had significant long-term impacts on aquatic life and human health; these have not yet been assessed. Mechanical and biological control methods have also been used in Sudan, though a comprehensive evaluation of the success of these efforts has not been carried out to date.

Hyacinth control measures were hampered during the conflict years; as a result, efforts focused on sensitive locations such as near the Jebel Aulia dam. Today, there are no control operations underway at all. The role of the Plant Protection Department of the Ministry of Agriculture, which is responsible for hyacinth control, is currently limited to monitoring infestations, and it has no capacity to respond to the spread.

In the south, the impact of water hyacinth on the Sudd is completely unknown, although it is anticipated to be considerable, given that these wetlands comprise a large number of oxbow lakes and slow-moving channels which are ideal conditions for weed growth. The scale of infestation can be gauged every wet season, when up to 100 metre-long rafts of detached weed float down the White Nile downstream of the Sudd.

10.7 Water pollution

A major but largely unquantified issue

While water pollution is clearly a significant issue in Sudan, it has not been adequately quantified. Indeed, the sector is characterized by a lack of historical data and investment. Systematic surface water quality monitoring programmes in Sudan are limited to three sites: the main Nile at Dongola, the Blue Nile at Soba (near Khartoum), and the White Nile at Malakal. Other sites and groundwater are tested on an ad hoc basis. Monitoring data is publicly available but limited in scope.

This lack of information makes it difficult to adequately assess water quality and the likely changes that may take place in the future. With this in mind, UNEP noted three principal water quality issues:

- diffuse pollution from agrochemicals and sewage;
- point source industrial pollution; and
- high levels of suspended sediments.

Biological water pollution

Biological water pollution from sewage and waterborne infectious agents is the most serious threat to human health in Sudan. The limited monitoring that has occurred so far has confirmed bacteriological contamination of the Nile and shallow groundwater aquifers in Khartoum state and elsewhere in northern Sudan. There is very limited laboratory data for Southern Sudan but the waterborne disease statistics clearly show that it is a major problem. This is discussed in more detail in Chapter 6.

Given that fertilizer usage in Sudan is minimal by world standards, laboratory analysis of Nile waters only detected very low levels of nitrates. However, high nitrate levels were recorded at individual wells near concentrations of livestock [10.2].

Pesticide pollution

Non-point source pollution is a cause for serious concern in the major irrigated schemes, particularly in Gezira and its Managil extension, Rahad and the country's five major sugar estates, where large-scale agrochemical applications continue despite overall declining usage trends. Various studies (mainly university graduate theses) have found serious pesticide contamination

The lack of a storm water drainage system in Khartoum causes major flooding, as observed here in August 2006. As the flood waters recede, pools of stagnant water increase the risk of spreading waterborne diseases, particularly in crowded areas like IDP camps

A local resident collects drinking water from the Nile. Biological water pollution from sewage and waterborne infectious agents is the most serious threat to human health in Sudan

The fast-growing cities of Southern Sudan are in desperate need of sewage systems

Pumping stations supply drinking water from irrigation canals that are susceptible to contamination from aerial pesticide application, such as this one in Deim el Masheihk on the Managil extension of the Gezira scheme

in the Gezira canals, as well as in boreholes in the Qurashi (Hasahesa) area and the Kassala horticulture zone. Accidental aerial spraying and pesticide drift reportedly lead to frequent fish kills in irrigation canals; these fish are sometimes collected for consumption [10.2].

Derelict and inadequate pesticide storage facilities and disposal measures, as observed in warehouse schemes at Hasahesa, Barakat and El Fao, as well as in stores of the Plant Protection offices in Gedaref, also pose a serious water pollution hazard. Complaints about the strong smell and contaminated spill during the rainy season have been received from Gedaref University, located downstream of the pesticide warehouse.

There is also a growing trend to apply pesticides in rain-fed mechanized agriculture schemes, which may lead to widespread contamination of both surface and groundwater, including the water points used by nomads. For example, herbicide application (mainly the persistent organochlorine 2, 4-D) in mechanized schemes is standard practice in Gedaref state [10.23] and is expanding in Dali and Mazmum in Sennar state, as well as in Habila in Southern Kordofan. Given the persistent nature of many pesticides and their biological magnification in the food-chain, long-term monitoring of surface and groundwater should be implemented, particularly in the states of Gezira, Sennar, White Nile and Gedaref, which host the main irrigated schemes.

Industrial effluent

Water pollution from industry is mostly limited to specific 'hot spots' such as North Khartoum, Port Sudan and Wad Medani. Given the current boom in industrial investment, however, it is an issue of growing concern. The majority of industrial facilities do not have dedicated water treatment facilities. Effluent is typically released either into the domestic sewage system (where one exists), or directly into watercourses or onto land.

For example, wastewater from the industrial area of North Khartoum (Bahri) flows untreated into the sewage treatment plant of Haj Yousif. Release of untreated industrial wastewater into watercourses or onto land is common practice, as was observed by the UNEP team in the Bagair industrial area, and at Assalaya and Sennar sugar factories, which dispose of their wastewater directly into the White and Blue Nile respectively. A major fish kill was reported in the Blue Nile in March 2006, following an accidental spill of molasses from the north-west Sennar sugar factory [10.2].

There are some positive developments, however, as a few large enterprises, such as the Kenana Sugar Company and some oil companies, have installed or are in the process of installing wastewater treatment plants [10.24]. This is a particularly critical issue for the oil industry, which is expected to generate large and increasing amounts of wastewater as the oilfields mature.

Suspended solids from eroded catchments

The heaviest water pollution load in Sudan is probably caused by suspended sediment. Recorded levels of suspended solids in rivers and reservoirs in the wet season range from 3,000 ppm to over 6,000 ppm, which corresponds to highly turbid/muddy conditions. While many of Sudan's rivers and streams are naturally turbid, the problem has been amplified by the high rates of soil erosion due to deforestation and vegetation clearance, overgrazing, dams, haphazard disposal of construction materials, and mining.

High levels of suspended sediment have adverse impacts on drinking water quality as well as on aquatic life, and in Sudan, have led to considerable economic losses due to the siltation of dams and irrigation canals. The impact is particularly visible in the Atbara river and the Blue Nile, whose catchments are seriously degraded by poor land management practices. In 2000, government sources estimated the total sediment load of the Blue Nile to be 140 million tonnes per annum [10.2].

Locals collect polluted effluent from the north-west Sennar sugar factory, for use in brick-making. The untreated effluent flows directly into the Blue Nile. This led to significant fish kills in the summer of 2006

Poor management of an experimental well drawing on fossil water from the NSAS has led to the creation of a wetland in the desert

10.8 Groundwater exploitation

A largely untapped but also unmanaged resource

On a national scale, Sudan makes limited use of its groundwater, but it is a critical resource at the local level, particularly in the northern and central regions, and in Darfur. Data on the use and quality of groundwater, however, is rarely collected and extraction is generally completely unmanaged. There is anecdotal evidence of unsustainable extraction rates, but in the absence of monitoring data, the situation only becomes apparent when the wells run dry.

UNEP has focused on three examples of this general problem:

* the exploitation of the Nubian aquifer (discussed in the following section on transboundary issues);
* the use of upper terrace and other shallow aquifer systems; and
* the use of groundwater in the humanitarian aid community in Darfur.

The richness of groundwater resources in Sudan was recently evidenced in a piezometric survey at Gaab el Sawani, which showed the static water level to range from 1 to 6 m above ground level

Figure 10.4   Groundwater resources of Sudan

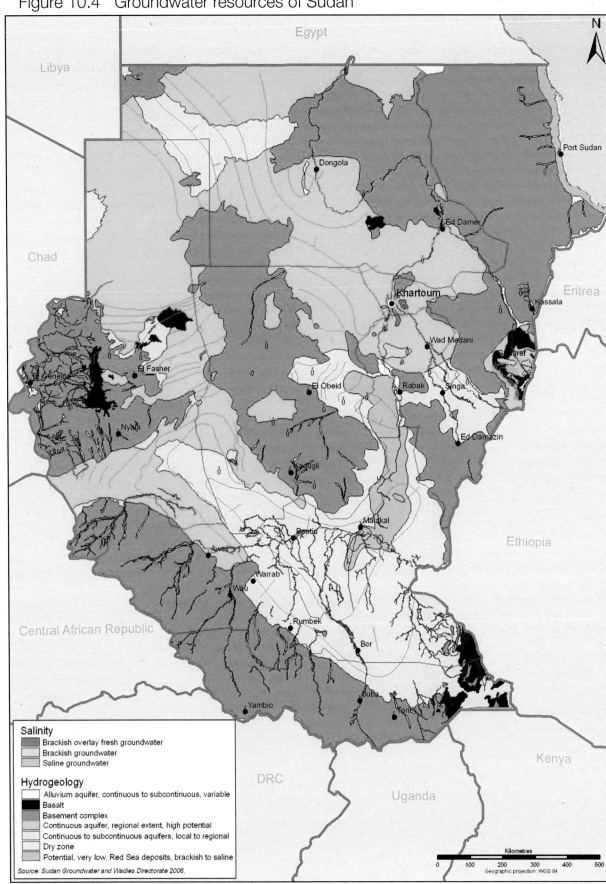

Salinity

- Brackish overlay fresh groundwater
- Brackish groundwater
- Saline groundwater

Hydrogeology

- Alluvium aquifer, continuous to subcontinuous, variable
- Basalt
- Basement complex
- Continuous aquifer, regional extent, high potential
- Continuous to subcontinuous aquifers, local to regional
- Dry zone
- Potential, very low, Red Sea deposits, brackish to saline

Source: Sudan Groundwater and Wadies Directorate 2006.

The boundaries and names shown and the designations used on this map do not imply official endorsement or acceptance by the United Nations.

Use of upper terrace and other shallow aquifer systems

There is little published data available on Sudan's shallow groundwater resources such as the Umm Rawaba formation, which is reportedly an excellent source of near-surface groundwater. Overall, however, there is growing investment and reliance on groundwater resources in Sudan, particularly on the use of *mataras* (irrigation wells) in the Nile floodplain and adjoining upper terraces, as well as in the *wadis*. There are reports of falling aquifer levels in Wadi Nyala and Kassala, and of seawater incursion in the shallow groundwater of the Red Sea coastal zone [10.2].

The sustainability of *mataras* in the upper terraces and *wadis* is questionable, and there are many anecdotal reports of declining groundwater levels that require scientific verification [10.2, 10.25]. For example, in Lewere in the Nuba mountains, groundwater levels have allegedly dropped from 3 to 70 metres, while in Atmoor, levels were said to have fallen by up to 10 metres.

Groundwater use in the humanitarian relief effort in Darfur

The humanitarian aid effort in Darfur has led to the drilling and establishment of hundreds of wells and water points since 2003. Many of these deep boreholes are located in or near displaced persons camps, and have high flow pumps installed to service populations of several thousand. These camps are commonly established in arid regions where groundwater is the only reliable source of water for up to ten months of the year. Given that the camps may stay in place for many more years, there is a clear need to ensure that groundwater extraction rates are sustainable. As of mid-2006, however, no organized groundwater level monitoring was taking place in camps in Darfur (see Chapter 5).

A recent groundwater vulnerability study of four large camps in Darfur indicated that camp wells extracting water solely from the basement complex aquifer were probably unsustainable in the medium term (two years) and that alternatives were needed [10.26].

The rapid expansion of shallow irrigation wells, locally known as mataras, in the Nile upper terraces needs to be sustainably managed to avert aquifer depletion

It is critically important that the water supply wells drilled in IDP and refugee camps do not run dry. Groundwater level monitoring should commence to allow the calculation of sustainable yields

10.9 Transboundary and regional issues

A need for cooperation over shared resources

Careful management and a high level of awareness are required for a number of transboundary and regional issues in the water sector in Sudan to avoid project failure or worse, catalysing regional disputes or even conflict.

Water projects and the CPA

In the Comprehensive Peace Agreement and subsequent Interim Constitution, the federal government (Government of National Unity) was granted specific sole authority over the management of Nile waters and Nile basin water resources. The Government of Southern Sudan and state governments were given separate powers related to water supply projects. GONU thus clearly has the mandate for any new major water project.

Given that the White Nile borders or flows through five of the ten states of Southern Sudan, northern state water projects may affect the southern states and vice versa. Therefore, it is considered critical that the GONU and GOSS conduct open and regular dialogue on Nile waters and development issues in order to not undermine the CPA. As of mid-2006, this was reported to be occurring, though not on a formal or regular basis.

Upstream watershed conditions, climate change and future projects in Sudan

The quantity, timing and quality of most of the Nile, Gash and Atbara river waters flowing through Sudan depend not on Sudan but on upstream countries, principally Ethiopia (Blue Nile, Atbara, Gash), Uganda (White Nile), and Tanzania and Kenya which border Lake Victoria (White Nile). These four countries all face a range of environmental problems including large-scale deforestation and land degradation. In addition, Uganda has recently increased water extraction from Lake Victoria for hydroelectric power, contributing to a significant drop in the lake's level. As a result, the currently observed changes in Nile flow rates (levels appear to be declining

overall but variability is increasing) and turbidity are expected to increase over time.

Climate change will also affect the performance of the existing and planned major water resource management projects in Sudan. Both rainfall and river flows are expected to be affected within the next thirty years, and some impacts may already be occurring (see Chapter 3).

Large-scale water development demands a high level of flow predictability to ensure confidence for the large capital investment required. Accordingly, Sudan needs to better understand upstream catchment environmental issues and the likely impacts of climate change, and adjust its plans to suit.

Management of the shared Nubian Sandstone Aquifer System

The vast Nubian Sandstone Aquifer System (NSAS) represents the largest volume of freshwater in the world. It is estimated at 150,000 km^3 or nearly 200 years of average Nile flow. This deep artesian aquifer underlies approximately 376,000 km^2 of north-west Sudan (17 percent of the NSAS total area of 2.2 million km^2). It is shared with Chad, Egypt and Libya, and is primarily comprised of non-renewable or 'fossil' water some 20,000 years old [10.27]. A smaller basin of the NSAS, which is known as the Nubian Nile aquifer, receives recharge from the Nile river. The direction of groundwater flow in the NSAS is generally from south-east to north-east. Hence, Sudan and Chad are in an upstream position providing minor recharge to Egypt and Libya downstream.

The aquifer remains largely untapped in both Sudan and Chad. In contrast Libya and Egypt, through the Great Man-Made River and the South Valley Development projects respectively, are now actively pumping water for ambitious agricultural schemes [10.27, 10.28]. Large-scale irrigated agriculture with fossil water in a hyper-arid environment is a controversial issue due not only to potential wastefulness but also to the risk of soil salinization. Despite increasing pressure to mine the NSAS to meet the demands of a growing population, the need for wise and sustainable use of this precious resource, based on sound scientific knowledge and a regionally agreed strategy, cannot be overstated.

A catch from the White Nile. At present, the freshwater fisheries of Southern Sudan are only lightly exploited

To this end, a GEF project involving the four basin countries was launched in 2005. Its primary objective is to develop an NSAS water resource database and to promote technical exchange of information and expertise, as well as provide capacity-building for local staff. The project also aims to create a framework for a legal convention and institutional mechanism for shared management of the Nubian Aquifer System [10.29].

10.10   Freshwater fisheries: an unbalanced but promising resource

The freshwater fisheries of Sudan are an important source of sustenance for millions of riverine dwellers, and support a small informal commercial sector.

In the northern states near the major cities, resources are reportedly fished to saturation, with stable or dropping catches [10.20]. In the absence of hard water quality monitoring data, the reason for such catch reductions cannot be accurately determined, but localized overfishing and sedimentation are likely causes.

While there is no catch data for the freshwater fisheries of Southern Sudan, field observations and discussions with fisheries experts working on the White Nile indicated that the resource is clearly under-exploited, principally due to a lack of capacity in the local fishing sector.

As with any natural resource extraction, the sustainability of fisheries will only be achieved through good management, starting with data collection to assess the scale and health of the resource.

10.11   Water sector environmental governance

The ministerial-level structure for water governance is straightforward, as both the Government of National Unity and the Government of Southern Sudan have ministries for water resources management. In practice, however, governance is more complex, as water is a cross-cutting sector with other major ministries.

Laying nets in the White Nile at Bor, Jonglei state. The challenge for fishermen in this region is not catching enough fish, but preserving the catch so that it can be transported and sold outside of the area

CS 10.4   Development of fisheries in Southern Sudan

The Muntai Fisheries Training Centre based in Padak in Jonglei state is a positive example of sustainable development tied to better use of natural resources. The centre, which focuses on the transfer of skills to local artisanal fisherman, is part of an agricultural development project funded by USAID. A particular focus is placed on obtaining better value for fish catches and reducing wastage through the use of preservation techniques such as smoking and drying.

The wide variety of species and the large size of many fish indicate that the fishery potential of the White Nile is probably underexploited. The centre proposes to conduct catch surveys and commence development of fishery policies and by-laws in parallel with the capacity-building process.

Officials reported that the fishing community was actually only a small percentage of the local Dinka community, but that this minority was in some respects significantly better off than the majority of pastoralists, as they had both food security and a reliable source of income. The Dinka people are still food aid recipients, depend heavily on cattle-rearing and are expecting an influx of returnees to significantly increase local population density. In this context, sustainable initiatives to broaden the food base and promote rural business are most welcome.

This is particularly the case for major GONU projects such as the Merowe dam, for which a special dams unit was developed that overlays the responsibilities of the ministries for water resources, agriculture, energy, industry and environment. In Southern Sudan, the GOSS ministry is currently in the institution-building phase, and issues such as inter-ministerial mandates on cross-cutting issues have yet to be fully addressed.

The most significant governance issue for the water sector is considered to be its culture of development through mega-projects rather than sustainable development principles. At the working level, the water sector suffers from a lack of enforceable working regulations, standards or enforcement capacity, with particular gaps noted for water pollution and groundwater.

An irrigation canal headman. Pilot projects to establish water user associations in the Gezira scheme have shown reduced operational costs and more efficient on-farm water management

The introduction of improved smoking methods has raised the income of fishermen in the Bor region by expanding the market and increasing the price of fish

10.12 Conclusions and recommendations

Conclusion

At present, the national approach to water resources management in Sudan is based largely on resource exploitation and biased towards mega-projects. The water resources sector currently also faces a range of serious environmental challenges, which will require innovative management approaches as well as significant investments to rehabilitate degraded systems and strengthen technical capacity. In light of Sudan's ambitious dam-building programme, perhaps the most challenging task will be to develop a new decision-making framework for water projects that is based on equity, public participation and accountability.

Background to the recommendations

Substantial development of the water resources of Sudan is anticipated in the next decade. Such development should not be discouraged, but should be designed, constructed and operated in a more sustainable manner.

The two key themes of the recommendations are to strengthen national capacity for water resources management, and to introduce the philosophy and practical aspects of Integrated Water Resource Management (IWRM) to Sudan.

As the investment for most new and major water schemes will come from or be controlled by the Government of National Unity, the GONU Ministry of Irrigation and Water Resources is considered the appropriate counterpart for most of the capacity-building and advocacy proposed here, though some effort should be placed with equivalents in the Government of Southern Sudan and at the state level. Assistance to the Darfur states is a particular priority as substantial investments in this sector are anticipated as soon as the security situation allows.

Recommendations for the Government of National Unity

R10.1 Strengthen technical capacity in sustainable water resource management. This will entail significant investment in training and equipment for data collection, analysis and corrective action planning. All existing dam operations would be covered, as well as project planning for dams, groundwater and irrigation schemes. Priority targets for assistance would be the Dams Implementation Unit and the Ministry of Irrigation and Water Resources.

CA: CB; PB: MIWR and DIU; UNP: UNEP; CE: 2M; DU: 2 years

R10.2 Develop integrated water resources management (IWRM) plans for degraded basins. Priority should be given to the Blue Nile and Atbara river basins, Darfur, Khor Abu Habil in Northern Kordofan, and the Nuba mountains in Southern Kordofan. One of the key targets of these plans should be to propose integrated measures aimed at reducing river siltation levels and downstream riverbank erosion.

CA: GROL; PB: MIWR and DIU; UNP: UNEP; CE: 1M; DU: 2 years

R10.3 Develop and embed guidelines on dams in environmental law. The guidelines should include public consultations, and options and ecosystem integrity assessments. A legislative mandate prohibiting the initiation of any dam construction activities prior to the issuance of an EIA permit, and stipulating public participation throughout the dam project cycle as well as disclosure and timely distribution of all environmental information about the dam should be developed.

CA: GROL; PB: MIWR and DIU; UNP: UNEP; CE: 0.1M; DU: 2 years

R10.4 Conduct an additional environmental assessment of the Merowe dam project and develop specific mitigation measures for the operation of the facility. Key issues include the analysis and mitigation of downstream impacts and absorbing environmental lessons learnt from existing dams and irrigation schemes.

CA: AS: PB: MIWR and DIU; UNP: UNEP; CE: 0.5M; DU: 2 years

R10.5 Establish a national water quality monitoring programme for both surface and groundwater to include key physical, chemical and biological parameters. Include a tailor-made water quality monitoring programme for pesticide residues in the large-scale irrigation schemes. Inventory and assess water pollution 'hot spots'.

CA: AS; PB: MIWR and DIU; UNP: UNEP; CE: 5M; DU: 2 years

R10.6 Develop a capacity-building programme and implement pilot projects on water conservation and management aimed at local user groups including water use associations. Priority should be given to the main irrigation schemes.

CA: CB; PB: MIWR and DIU; UNP: UNEP; CE: 2M; DU: 2 years

R10.7 Strengthen the capacity of regulatory authorities in groundwater data collection and management. This entails the development of a robust licensing system.

CA: CB; PB: MIWR and DIU; IP: UNEP; CE: 1M; DU: 2 years

Recommendations for the Government of Southern Sudan

R10.8 Build capacity for sustainable water resource management, using IWRM as a founding philosophy. Capacity-building should include groundwork to assist the establishment of the ministry itself, and should initially focus on impact assessment and mitigation for planned water supply and power generation projects in the ten southern states.

CA: CB; PB: MWRI; UNP: UNEP; CE: 1M; DU: 2 years

R10.9 Develop and implement an integrated management plan for the Sudd wetlands. The cost estimate covers plan development and the first two years of implementation.

CA: GROL; PB: MWRI; UNP: UNEP/Ramsar Convention; CE: 1M; DU: 2 years

The Assistant Director of the Roseires dam explains the challenges of operating a facility that is of national significance for both power generation and irrigation

Wildlife and Protected Area Management

Birds of prey settle for the night on the flood plains of the White Nile in Jonglei state. While the past few decades have witnessed a major decline in wildlife in Sudan, remaining populations can still be considered internationally significant.
© Nick Wise

Wildlife and protected area management

11.1 Introduction and assessment activities

Introduction

As late as 1970, Sudan boasted some of the most unspoilt and isolated wilderness in east Africa, and its wildlife populations were world-renowned. While the past few decades have witnessed a major assault on both wildlife and their habitats, what remains is both internationally significant and an important resource opportunity for Sudan.

Ecosystems, issues, and the institutional structures to manage wildlife and protected areas differ markedly between north and south in Sudan. In the north, the greatest damage has been inflicted by habitat degradation, while in the south, it is uncontrolled hunting that has decimated wildlife populations. Many of the issues in the following

sections are hence addressed separately for the two areas of the country. It should be noted that the most important remaining wildlife and protected areas in northern Sudan are on the coastline or in the Red Sea; these are covered in Chapter 12.

This chapter focuses on wildlife and protected areas as a specific sector. It is acknowledged that the larger topic of biodiversity has not been adequately addressed in this assessment. While the importance of conserving biodiversity is unquestionable, a significant difficulty for action on this front – in Sudan as elsewhere – is the lack of government ownership: no single ministry is responsible for this topic. As a result, the observed implementation of recommendations under the label of biodiversity is poor.

Although it has not been included as a specific sector in this assessment, the biodiversity of Sudan was studied and reported on in 2003 by a programme funded by the Global Environment Facility (GEF) under the auspices of the Convention on Biological Diversity (CBD) [11.1].

White-eared kob and zebra migrating through Boma National Park in 1983

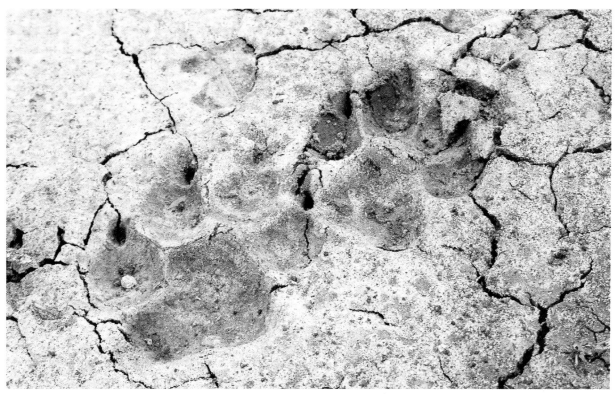

Lion tracks in Padak county, Jonglei state. In the absence of formal survey data for much of the country, the evidence for wildlife populations is often anecdotal and qualitative

Assessment activities

The investigation of issues related to wildlife and protected areas in Sudan was conducted as part of the overall assessment. Two commissioned desk studies – one by the Boma Wildlife Training Centre, the other by the Sudanese Environment Conservation Society (SECS) – summarized the extent of existing knowledge for the south and north respectively [11.2, 11.3]. UNEP was able to visit one major site in the north (Dinder National Park), as well as a number of smaller reserves. The protected areas of Southern Sudan and Darfur were inaccessible due to security and logistical constraints. However, information was obtained from interviews and other sources in the course of general fieldwork in Southern Sudan.

Due to historical and ongoing conflicts, the available data on wildlife is highly skewed, with most recent information limited to northern and central states. This lack of up to date field data is a core problem for Southern Sudan's protected areas, but major studies by the Wildlife Conservation Society are underway in 2007 to correct this.

11.2 Overview of the wildlife and habitats of Sudan

The arid and semi-arid habitats of northern Sudan have always had limited wildlife populations. In the north, protected areas are mainly linked to the Nile and its tributaries, and to the Red Sea coast, where there are larger concentrations of wildlife. In contrast, the savannah woodlands and flooded grasslands of Southern Sudan have historically been home to vast populations of mammals and birds, especially migratory waterfowl. This abundance of wildlife has led to the creation of numerous national parks and game reserves by both British colonial and independent Sudanese authorities.

There is a large volume of literature on the wildlife of Sudan as recorded by casual observers who travelled through or lived in Sudan during the 19th and first half of the 20th centuries. A 1940s account, for instance, describes large populations of elephant, giraffe, giant eland, and both white and black rhino across a wide belt of Southern Sudan. Because of the civil war, however, few scientific studies of Sudan's wildlife have been conducted, and coverage of the south has always been very limited.

© PHIL SNYDER

The migration of white-eared kob across the flood plains of Southern Sudan is one of the least known but most spectacular wildlife wonders of the world. Hundreds of thousands of animals move in a seasonal search for dry ground, new pasture and water (inset). Kob are perfectly adapted to the flood plain environment of Southern Sudan and have been hunted by local people for centuries

CS 11.1 The management of migratory wildlife outside of protected areas: the white-eared kob

One of the distinctive features of the wildlife population of Southern Sudan is that much of it is found outside of protected areas. This presents a range of challenges for conservation and management, as illustrated by the case of the white-eared kob antelope.

White-eared kob (*Kobus kob leucotis*) are largely restricted to Southern Sudan, east of the Nile, and to south-west Ethiopia [11.19, 11.20]. These antelope are dependent on a plentiful supply of lush vegetation and their splayed hooves enable them to utilize seasonally inundated grasslands. The spectacular migration of immense herds of white-eared kob in search of grazing and water has been compared to that of the ungulates in the Serengeti.

Substantial populations of white-eared kob occur in Boma National Park, the Jonglei area and in Badingllo National Park [11.20]. The paths of their migration vary from year to year, depending on distribution of rainfall and floods (see Figure 11.1). A survey and documentary film made in the early 1980s followed the herds of the Boma ecosystem as they moved between their dry and wet season strongholds that year, and found that the herds moved up to 1,600 km per year, facing a range of threats as they migrated through the different seasons, ecosystems and tribal regions [11.5].

The principle threats to the kob are seasonal drought, excessive hunting pressure and now the development of a new aid-funded rural road network cutting across their migration routes. The sustainable solution to excessive hunting is considered to be its containment and formalization rather than its outright prohibition, a measure which is both unachievable and unenforceable. White-eared kob represent an ideal opportunity for sustainable harvesting: they have a vast habitat, are fast breeders and are far better adapted to the harsh environment of the clay plains and wetlands than cattle. The spectacular nature of the kob migration may support some wildlife tourism in future but it is unrealistic to expect tourism revenue to provide an acceptable substitute for all of the livelihoods currently supported by hunting.

Minimizing the impact of the new road network will require some innovative thinking to integrate animal behaviour considerations into road design and development controls. Dedicated wildlife-crossing corridors, culverting and underpasses are all options that could reduce road accident-related animal deaths, while banning hunting within set distances of the new roads may help to control vehicle-assisted poaching.

Figure 11.1   Kob migration

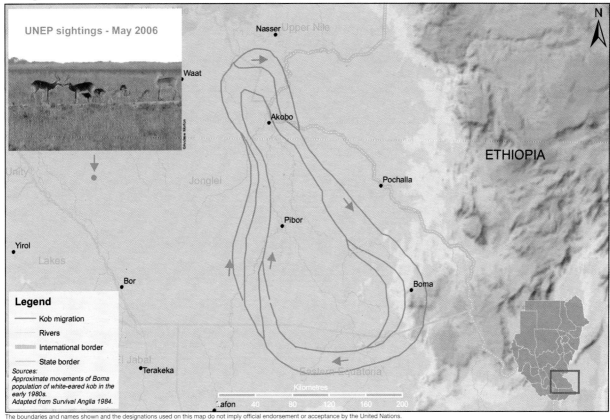

The boundaries and names shown and the designations used on this map do not imply official endorsement or acceptance by the United Nations.

As a result of this lack of technical fieldwork, virtually all up to date evidence of wildlife distribution in Southern Sudan outside of a few protected areas is anecdotal and cannot be easily substantiated. Nonetheless, this type of information is considered to warrant reporting in order to assess priorities for more substantive assessments. Key information from 2005 and 2006 includes the sightings of elephants in the northern part of the Sudd wetlands, and the sighting of very large herds of tiang and white-eared kob in Jonglei state. It is of note that both of these sightings took place outside of legally protected areas (see Case Study 11.1).

The only other recent data available on Southern Sudan is from ground surveys of Nimule, Boma and Southern National Park, carried out by the New Sudan Wildlife Conservation Organization (NSWCO) in 2001. The results of these surveys and other information provided to UNEP by the Boma Wildlife Training Centre indicate that many protected areas, in Southern Sudan at least, have remnant populations of most species.

Tiang, Bokor reedbuck and white-eared kob near the main road in Mabior, Jonglei state. Wildlife in Southern Sudan are found as much outside as inside protected areas

Wildlife habitats and occurrence by region

The regional environments of Sudan defined in Chapter 2 can be used as a basis for the description of current wildlife habitats and populations:

- arid regions (coastal and arid region mountain ranges, coastal plain, stony plains and dune fields);
- the Nile riverine strip;
- the Sahel belt, including the central dryland agricultural belt;
- the Marra plateau;
- the Nuba mountains;
- savannah;
- wetlands and floodplains;
- subtropical lowlands;
- the Imatong and Jebel Gumbiri mountain ranges; and
- subtidal coastline and islands – covered in Chapter 12.

The delimitations of the various areas in which wildlife are present are derived from a combination of ecological, socio-economic, historical and political factors. It should be noted, however, that the boundaries between certain regions are ill-defined, and that many animals migrate freely across them.

Arid regions. The mountains bordering the Red Sea, as well as those on the Ethiopian border and in Northern Darfur, are host to isolated low density populations of Nubian ibex, wild sheep and several species of gazelle [11.3]. Larger predators are limited to jackal and leopard. Due to the lack of water, wildlife in the desert plains are extremely limited, consisting principally of Dorcas gazelle and smaller animals. Life centres on *wadis* and oases, which are commonly occupied by nomadic pastoralists and their livestock.

The Nile riverine strip. The Nile riverine strip is heavily populated and as such only supports birdlife and smaller animals (including bats).

The Sahel belt, including the central dryland agricultural belt. In the Sahel belt, the combination of agricultural development and roving pastoralists effectively excludes large

Empty landscapes: the UNEP team travelled through the Nuba mountains without seeing or hearing any reports of remaining wildlife

wildlife, although the region does host migratory birds, particularly in the seasonal wetlands and irrigated areas. With the important exception of Dinder National Park, the expansion of mechanized agriculture has eliminated much of the wild habitat in the Sahel belt.

The Marra plateau. The forests of Jebel Marra historically hosted significant populations of wildlife, including lion and greater kudu [11.3]. Limited surveys in 1998 (the latest available) reported high levels of poaching at that time. Due to the conflict in Darfur, there is only negligible information on the current status of wildlife in this region.

The Nuba mountains. The wooded highlands of the Nuba mountains historically held large populations of wildlife, but all recent reports indicate that the civil war led to a massive decline in numbers and diversity, even though forest cover is still substantial. The UNEP team travelled extensively through the Nuba mountains without any sightings or reports of wildlife.

Savannah. The bulk of the remaining wildlife of Sudan is found in the savannah of central and south Sudan, though the data on wildlife density in these regions is negligible.

Historical reports include large-scale populations of white and black rhino, zebra, numerous antelope species, lion, and leopard. In addition, aerial surveys carried out in the woodland savannah of Southern National Park in November 1980 revealed sizeable population estimates of elephant (15,404), buffalo (75,826), hartebeest (14,906) and giraffe (2,097) [11.4]. The number of white rhino in Southern National Park was estimated to be 168, which then represented a small but significant remnant population of an extremely endangered subspecies of rhino. In 1980, aerial surveys carried out in Boma (mixed savannah and floodplain habitats) indicated that the park was used by large populations of a wide variety of species as a dry season refuge, with the exception of the tiang, whose numbers increased considerably during the wet season [11.5].

Wetlands and floodplains. The vast wetlands and floodplains of south Sudan, which include the Sudd and the Machar marshes, are an internationally significant wildlife haven, particularly for migratory waterfowl. These unique habitats also support many species not seen or found in large numbers outside of Sudan, such as the Nile lechwe antelope, the shoebill stork and the white-eared kob.

Subtropical lowlands. The subtropical lowlands form the northern and western limits of the central African rainforest belt and thus host many subtropical closed forest species, such as the chimpanzee.

The Imatong and Jebel Gumbiri mountain ranges. The wetter microclimates of these isolated mountains in the far south of Southern Sudan support thick montane forest. There is only negligible information available on wildlife occurrences in these important ecosystems.

The flooded grasslands of Southern Sudan support very large bird populations, including black-crowned cranes (Balearica pavonina) (top left), pink-backed pelicans (Pelecanus rufescens) (top right), cattle egrets (Bubulcus ibis) (bottom left), and saddle-billed storks (Ephippiorhynchus senegalensis) (bottom right), seen near Padak in Jonglei state

Globally important and endangered species in Sudan

Sudan harbours a number of globally important and endangered species of mammals, birds, reptiles and plants, as well as endemic species.

In addition, there are a number of species listed as vulnerable by IUCN, including sixteen species of mammals, birds and reptiles: hippopotamus (*Hippopotamus amphibius*); cheetah (*Acinonyx jubatus*); African lion (*Panthera leo*); Barbary sheep (*Ammotragus lervia*); Dorcas gazelle (*Gazella dorcas*); red-fronted gazelle (*Gazella rufifrons*); Soemmerring's gazelle (*Gazella soemmerringei*); African elephant (*Loxodonta africana*); Trevor's free-tailed bat (*Mops trevori*); horn-skinned bat (*Eptesicus floweri*); greater spotted eagle (*Aquila clanga*); imperial eagle (*Aquila heliaca*); houbara bustard (*Chlamydotis undulata*); lesser kestrel (*Falco naumanni*); lappet-faced vulture (*Torgos tracheliotos*); and African spurred tortoise (*Geochelone sulcata*) [11.12].

The Mongalla gazelle is not endangered but has a relatively small habitat. Rangeland burning such as has recently occurred here is favourable to this species, as it thrives on short new grass

Table 21. Globally endangered Species occurring in Sudan [11.6, 11.7, 11.8, 11.9, 11.10, 11.11, 11.12]

Common name	Scientific name	Red List category
Mammals		
Addax*	*Addax maculatus*	CR A2cd
African ass	*Equus africanus*	CR A1b
Dama gazelle	*Gazella dama*	CR A2cd
Nubian ibex	*Capra nubiana*	EN C2a
Grevy's zebra*	*Equus grevyi*	EN A1a+2c
Rhim gazelle	*Gazella leptoceros*	EN C1+2a
African wild dog	*Lycaon pictus*	EN C2a(i)
Chimpanzee	*Pan troglodytes*	EN A3cd
Birds		
Northern bald ibis	*Geronticus eremita*	CR C2a(ii)
Sociable lapwing	*Vanellus gregarius*	CR A3bc
Basra reed warbler	*Acrocephalus griseldis*	EN A2bc+3bc
Saker falcon	*Falco cherrug*	EN A2bcd+3b
Spotted ground-thrush	*Zoothera guttata*	EN C2a(i)
Reptiles		
Hawksbill turtle	*Eretmochelys imbricata*	CR A1bd
Green turtle	*Chelonia mydas*	EN A2bd
Plants		
Medemia argun	*Medemia argun*	CR B1+2c
Nubian dragon tree	*Dracaena ombet*	EN A1cd

CR = critically endangered; EN = endangered; * questionable occurrence in Sudan

11.3 Overview of protected areas

Variable protection

A significant number of areas throughout Sudan have been gazetted or listed as having some form of legal protection by the British colonial or the independent Sudanese authorities. In practice, however, the level of protection afforded to these areas has ranged from slight to negligible, and many exist only on paper today. Moreover, many of the previously protected or important areas are located in regions affected by conflict and have hence suffered from a long-term absence of the rule of law.

Protected areas of northern Sudan

According to the information available to UNEP, northern Sudan has six actual or proposed marine protected sites [11.13], with a total area of approximately 1,900 km², and twenty-six actual or proposed terrestrial and freshwater protected sites, with a total area of approximately 157,000 km² [11.1, 11.2, 11.14, 11.15, 11.16, 11.17].

Table 22.   Protected areas of northern Sudan (including marine areas)

Map reference	Protected area (* proposed)	Type (* proposed)	Km²	Habitat(s)	Key species
Marine protected areas					
30 53	Dongonab Bay	National park/ Ramsar site*/ Important bird area	3,000	Marine/tidal	Dugong, marine turtles, white-eyed gull
32	Sanganeb	National park/ Ramsar site*	260	Marine	Coral, marine fish
42	Suakin Archipelago*	National park/ Important bird area/ Ramsar site*	1,500	Marine	Marine turtles, crested tern
	Khor Kilab	National park*	2	Marine	Coral
	Abu Hashish	National park*	2	Marine	Coral
	Shuab Rumi	National park*	4	Marine	Coral
Terrestrial protected areas					
39 59	Radom	National park/ MAB reserve/ Important bird area	12,500	Savannah woodland	Buffalo, giant eland, leopard, hartebeest
35 52 58	Dinder	National park/ MAB reserve/ Ramsar site/ Important bird area	10,000	Savannah woodlands and flooded grasslands (mayas)	Reedbuck, oribi, buffalo, roan antelope, red-fronted gazelle
36	Jebel Hassania*	National park	10,000	Semi-desert	
43	Wadi Howar*	National park	100,000	Desert	
19	Jebel Gurgei Massif*	Game reserve	100		
	Rahad*	Game reserve	3,500		
26	Red Sea Hills*	Game reserve	150		
27	Sabaloka	Game reserve	1,160	Semi-desert	
28	Tokor	Game reserve	6,300	Semi-desert	
49	Erkawit Sinkat	Wildlife sanctuary	120	Semi-desert	
50	Erkawit	Wildlife sanctuary	820	Semi-desert	
3	Jebel Bawzer (Sunut) Forest	Bird sanctuary/ Ramsar site*	13	Semi-desert	
8	Lake Nubia	Bird sanctuary	100	Freshwater lake	Pharaoh eagle owl, crowned sandgrouse
2	Jebel Aulia Dam*	Bird sanctuary	1,000	Freshwater lake	
7	Lake Kundi*	Bird sanctuary	20	Freshwater lake	
6	Lake Keilak*	Bird sanctuary	30	Freshwater lake	
1	El Roseires Dam*	Bird sanctuary	700	Freshwater lake	
4	Khashm el Girba Dam*	Bird sanctuary	100	Freshwater lake	
9	Sennar Dam*	Bird sanctuary	80	Freshwater lake	
45	Jebel Elba*	Nature conservation area	4,800		
46	Jebel Marra Massif*	Nature conservation area/ Important bird area	1,500	Savannah grassland and woodland	Greater kudu, red-fronted gazelle
5	Lake Abiad	Bird sanctuary	5,000	Freshwater lake	Ruff, black-crowned crane

Figure 11.2 Protected areas of Sudan

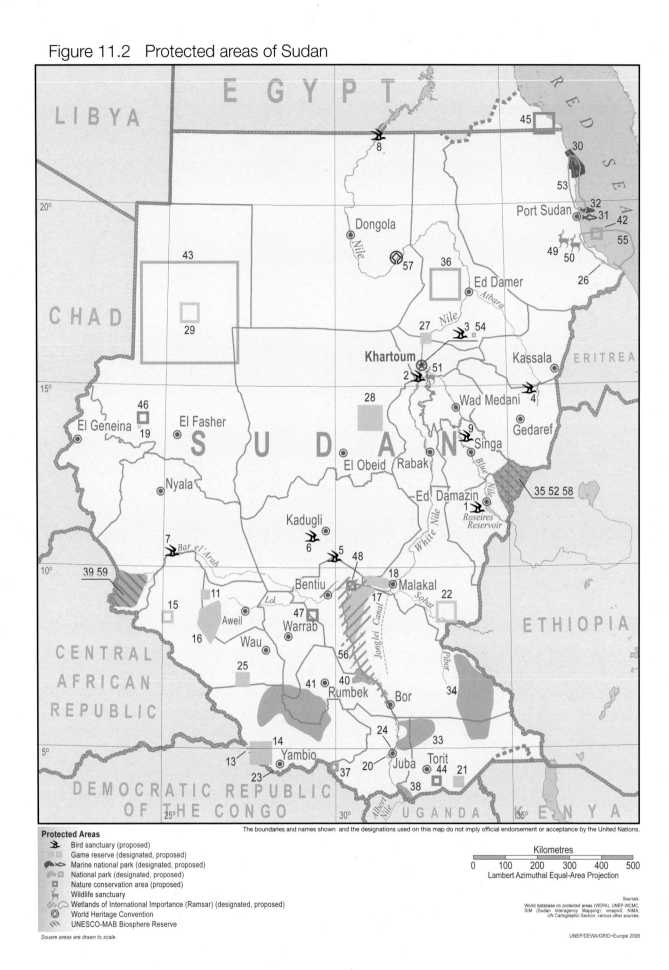

Index to Protected Areas map

National designations

Site number, Site name ([P]Proposed, [U]Unknown location), Area (ha)

Bird sanctuary:

1. El Roseireis Dam[P]	70'000	5. Lake Abiad[P]	500'000	8. Lake Nubia[P]	10'000
2. Jebel Aulia Dam[P]	100'000	6. Lake Keilak[P]	3'000	9. Sennar Dam[P]	8'000
3. Jebel Bawzer Forest (Sunut Forest)[P]	1'234	7. Lake Kundi[P]	2'000		
4. Khashm El-Girba Dam[P]	10'000				

Game reserve:

10. Abroch[P][U]	150'000	17. Ez Zeraf	970'000	24. Mongalla	7'500
11. Ashana	90'000	18. Fanikang	48'000	25. Numatina	210'000
12. Barizunga[P][U]	200'000	19. Jebel Gurgei Massif	10'000	26. Red Sea Hills	15'000
13. Bengangai	17'000	20. Juba	20'000	27. Sabaloka	116'000
14. Bire Kpatuos	500'000	21. Kidepo	120'000	28. Tokor	630'000
15. Boro[P]	150'000	22. Mashra[P]	450'000	29. Wadi Howar[P]	400'000
16. Chelkou	550'000	23. Mbarizunga	1'000		

Marine national park:

30. Dongonab Bay	300'000
31. Port Sudan[P]	100'000
32. Sanganeb	26'000

National park:

33. Badinglo	1'650'000	37. Lantoto[P]	76'000	41. Southern	2'300'000
34. Boma	2'280'000	38. Nimule	41'000	42. Suakin Archipelago[P]	150'000
35. Dinder	1'000'000	39. Radom	1'250'000	43. Wadi Howar[P]	10'000'000
36. Jebel Hassania[P]	1'000'000	40. Shambe	62'000		

Nature conservation area:

44. Imatong Mountains[P]	100'000	46. Jebel Marra massif[P]	150'000	48. Lake No[P]	100'000
45. Jebel Elba[P]	480'000	47. Lake Ambadi[P]	150'000		

Wildlife sanctuary:

49. Erkawit Sinkat	12'000
50. Erkawit	82'000
51. Khartoum	1'500

International conventions and programmes

Site number, Site name, Area (ha)

Wetlands of International Importance (Ramsar):

52. Dinder National Park	1'000'000
53. Dongonab Bay-Marsa Waiai[P]	280'000
54. Jebel Bawzer Forest (Sunut Forest)[P]	1'234
55. Suakin-Gulf of Agig[P]	1'125'000
56. Sudd	5'700'000

World Heritage Convention:

57. Gebel Barkal and the Sites of the Napatan Region

UNESCO-MAB Biosphere Reserve:

58. Dinder National Park	1'000'000
59. Radom National Park	1'250'000

A baboon in Dinder National Park, Sennar state. The level of actual protection is highly variable but generally weak throughout Sudan. Poaching is a problem in all major parks

Nominally protected areas thus cover approximately ten percent of northern Sudan, with three sites – Wadi Howar, Dinder and Radon – accounting for a large portion of this figure. While this is significant and worthy of support, the actual level of protection provided and ecosystem integrity are more important than sheer size.

Wildlife authorities interviewed by UNEP in northern Sudan reported consistent problems with protected area management, ranging from poaching to livestock encroachment and land degradation. Many sites were so degraded

from their original condition as to potentially warrant de-listing. The UNEP investigation of Dinder National Park, for example, found that this major site was not only badly damaged and under severe stress, but was also being starved of the requisite funds for proper management (see Case Study 11.2).

Overall, terrestrial and freshwater sites in northern Sudan were found to be very degraded and on a continuing decline. Marine protected areas were generally in better condition due to a low level of development pressure.

Protected areas of Southern Sudan

Given that the legally protected areas of Southern Sudan were in a conflict zone for over two decades, they have not been managed or effectively protected. During the war, the presence of the military gave some areas under SPLA control a measure of protection, but these were also used to supply bushmeat.

With the recent addition of the Sudd wetlands – which were listed as a site under the Ramsar Convention in 2006 – Southern Sudan comprises twenty-three sites, for a total area of 143,000 km² or approximately 15 percent of the territory.

Again, this large figure is positive, but the condition of these areas and the level of actual protection are of more import.

The level of actual protection provided to these twenty-three sites is considered by UNEP to be negligible but rising as the GOSS wildlife forces start to build capacity and mobilize. The condition of the areas is more difficult to gauge, but all available evidence points to a massive drop in the numbers of large wildlife due to poaching.

The most reliable evidence comes from Boma National Park, which was surveyed three times,

Table 23. Protected areas of Southern Sudan

Map reference	Protected area (* proposed)	Type (* proposed)	Km²	Habitat(s)	Key species
33	Badingilo (incl. Mongalla game reserve)*	National park/ Important bird area	8,400	Flooded grasslands and woodlands	Elephant, buffalo, giraffe
34	Boma	National park/ Important bird area	22,800	Savannah woodlands, grasslands, swamps	White-eared kob, tiang, reedbuck
37	Lantoto*	National park	760	Tropical forest	Chimpanzee, elephant
38	Nimule	National park/ Important bird area	410	Savannah and riverine woodlands	Elephant, cheetah
40	Shambe	National park (within Ramsar site)	620	Flooded savannah and riverine forest	Nile lechwe, buffalo
41	Southern	National park/ Important bird area	23,000	Savannah woodland	Giant eland, elephant, rhino
11	Ashana	Game reserve/ Important bird area	900	Savannah woodland	Elephant, giant eland
13	Bengangai	Game reserve/ Important bird area	170	Tropical forest	Elephant, bongo, buffalo
14	Bire Kpatuos	Game reserve	5,000	Tropical forest	Bongo, yellow-backed duiker
15	Boro*	Game reserve	1,500	Savannah woodland	Elephant
16	Chelkou	Game Reserve	5,500	Savannah woodland	Elephant, giant eland, buffalo
17	Ez Zeraf	Game reserve (within Ramsar site)	9,700	Flooded grassland and woodland	Nile lechwe, sitatunga, hippo
18	Fanikang	Game reserve (within Ramsar site)	480	Flooded grassland and woodland	Nile lechwe
20	Juba	Game reserve/ Important bird area	200	Savannah grassland and woodland	Heuglin's francolin, Arabian bustard
21	Kidepo	Game reserve/ Important bird area	1,200	Savannah grassland and woodland	Elephant, heuglin's francolin
22	Mashra*	Game reserve	4,500	Flooded grassland	Elephant
23	Mbarizunga	Game reserve	10	Tropical forest	Bongo, bushbuck, yellow-backed duiker
25	Numatina	Game reserve	2,100	Savannah woodland	Elephant, giant eland, roan antelope
7	Lake Kundi	Bird sanctuary	20	Freshwater lake	Yellow-billed stork, black-crowned crane
44	Imatong mountains	Important bird area/ Nature conservation area	1,000	Montane forest and woodland	Blue duiker, bushbuck
47	Lake Ambadi	Nature conservation area	1,500	Freshwater lake	
48	Lake No	Nature conservation area	1,000	Freshwater lake	
56	Sudd	Ramsar site/ Important bird area	57,000	Rivers, lakes, flooded grasslands and savannah	470 bird species, 100 mammal species and 100 fish species

Table 24.   Comparison of population estimates of larger ungulates in the years 1980 and 2001 in Boma National Park [11.2]

Species	2001 Count (wet season)	1980 Count (wet season)	1980 Count (dry season)
White-eared kob	176,120	680,716	849,365
Lesser eland	21,000	2,612	7,839
Roan antelope	1,960	2,059	3,085
Mongalla gazelle	280	5,933	2,167
Tiang	Not seen	116,373	25,442
Lelwel hartebeeste	5,600	8,556	47,148
Zebra	Not seen	24,078	29,460
Buffalo	Not seen	2,965	11,179
Giraffe	Not seen	4,605	9,028
Waterbuck	Not seen	620	2,462
Grant's gazelle	Not seen	1,222	1,811
Elephant	Not seen	1,763	2,179
Lesser kudu	Not seen	654	170
Oryx	Not seen	1,534	396
Cattle	7,980	7,056	93,815

twice in 1980 (in the dry and wet seasons) and once in 2001 [11.2]. As shown in Table 24, the wildlife populations recorded in 2001 had dropped dramatically, but there were still significant numbers of most species, with the exception of elephant, giraffe, zebra and buffalo. In scientific terms, the two surveys are not directly comparable. Nonetheless, the fact that viable populations of several species of wildlife still existed in Boma in 2001 is important for the future of wildlife and protected areas in Sudan.

A key figure to note is the cattle count, which documents the extent of encroachment into the park by pastoralists.

11.4 Wildlife and protected area management issues

There are four issues facing the wildlife and protected area management sector, which are cumulative in effect:

- habitat destruction and fragmentation;
- park encroachment and degradation;
- commercial poaching and bushmeat; and
- wildlife tourism (or lack thereof).

Habitat destruction and fragmentation

Habitat destruction and fragmentation from farming and deforestation is the root cause of most

biodiversity loss in northern and central Sudan. Vast areas of savannah and dryland pasture have been replaced with agricultural land, leaving only limited shelter belts or other forms of wildlife refuge. The intensity of mechanized agricultural development has forced pastoralists to use smaller grazing areas and less suitable land, leading to the degradation of the rangelands and increased competition between livestock and wildlife.

The net result is that larger wildlife have essentially disappeared from most of northern and central Sudan, and can only be found in the core of the protected areas and in very low numbers in remote desert regions.

In Southern Sudan, the lack of development has resulted in much less habitat destruction, but the intensification of shifting agriculture is causing large-scale land use changes across the region, particularly in the savannah. The floodplains are less affected, but the continued burning will negatively impact some species, while benefiting others, such as the antelope.

An additional important issue in Southern Sudan is the impact of ongoing and planned development like the creation or rehabilitation of rural trunk roads. This is a particular concern for Jonglei state, where the new road cuts directly across the migration route of the white-eared kob (see Case Study 11.1).

Park encroachment and degradation

Livestock is present in most of the legally protected terrestrial areas of Sudan, irrespective of their legal status. In some cases, pastoralists used the area long before the legal status came into effect; in others, the site has been invaded during the last thirty years. Pastoralists and their herds are now well entrenched in many major parks, creating competition for water and fodder, leading to land

Habitat destruction and fragmentation is the root cause of biodiversity loss in northern and central Sudan. The expansion of mechanized agriculture has deforested large areas and removed the shelter belts that host wildlife populations

degradation through burning and overgrazing, and facilitating poaching. Encroachment has partly destroyed the integrity of Dinder National Park [11.3], and now represents a major challenge for the developing wildlife sector in Southern Sudan.

A particular risk for Southern Sudan is armed conflict in the parks, as the wildlife forces (over 7,300 men as of late 2006) mobilize and start to confront pastoralists and poachers. Modern non-confrontational approaches entailing community engagement will be required if the wildlife sector in Southern Sudan is to avoid damaging gun battles between locals and rangers. The semi-resident population of pastoralists and bushmeat hunters from the Murle tribe in Boma National Park – who have become accustomed to living in the park and are heavily armed – illustrates this problem.

Commercial poaching and bushmeat

The ready availability of firearms has been the most significant factor in the reduction of wildlife in Southern Sudan, and has also compounded the problems of habitat destruction in northern and central Sudan. Uncontrolled and unsustainable levels of hunting have devastated wildlife populations and caused the local eradication of many of the larger species including elephant, rhino, buffalo, giraffe, eland and zebra.

Tiang are extensively hunted in the flood plains of Southern Sudan

The infrastructure and staff capacity of Dinder National Park were greatly improved thanks to a grant from the Global Environment Facility, but sufficient and sustainable government funding is urgently needed now that GEF support has come to an end (left)

The core of the park is comprised of wetlands that are critically important as reliable sources of water in the dry season (top right)

Although many have been poached, the park still supports a significant population of larger mammals. Warthogs are very common in the park's wetlands (bottom right)

CS 11.2 Dinder National Park: an ecosystem under siege

Dinder National Park is the most important terrestrial protected area in the northern states of Sudan. Located on the Ethiopian border, straddling Blue Nile and Kassala states, it is approximately 10,000 km² in size. The most important features of the park are a series of permanent and seasonal wetlands known locally as *mayas*, which are linked to streams running off the Ethiopian highlands to the east.

The habitat and wildlife of Dinder National Park can currently be described as badly degraded and under serious threat from a number of ongoing problems, including encroachment, habitat degradation and poaching.

Until the 1960s, the area surrounding Dinder was relatively uninhabited. Since then, however, migration and land use changes have resulted in development around the park, to the extent that some forty villages now exist along its borders. Large-scale mechanized agriculture to the north and west has not only pushed traditional agricultural communities to the edge of the park, but by taking over most of the land previously used for grazing, has also led pastoralists to invade the park in large numbers. Livestock compete with wildlife for fodder and water, and transmit diseases such as rinderpest and anthrax, while burning degrades the grassed woodland habitat. Poaching is also a major problem, as is the felling of trees for firewood by trespassers and fires set in the course of honey extraction.

Between 2002 and 2006, the park benefited from a USD 750,000 Global Environment Facility (GEF) grant that resulted in increased capacity for the wildlife force and a well thought out management plan with a strong emphasis on community involvement in the conservation of the park. This funding ceased in early 2006 and the future preservation of the park hangs in the balance. Without further injection of funding by the government or the international community, it is very likely that the gains achieved by the GEF grant will be lost and that degradation will continue.

Figure 11.3   Dinder National Park

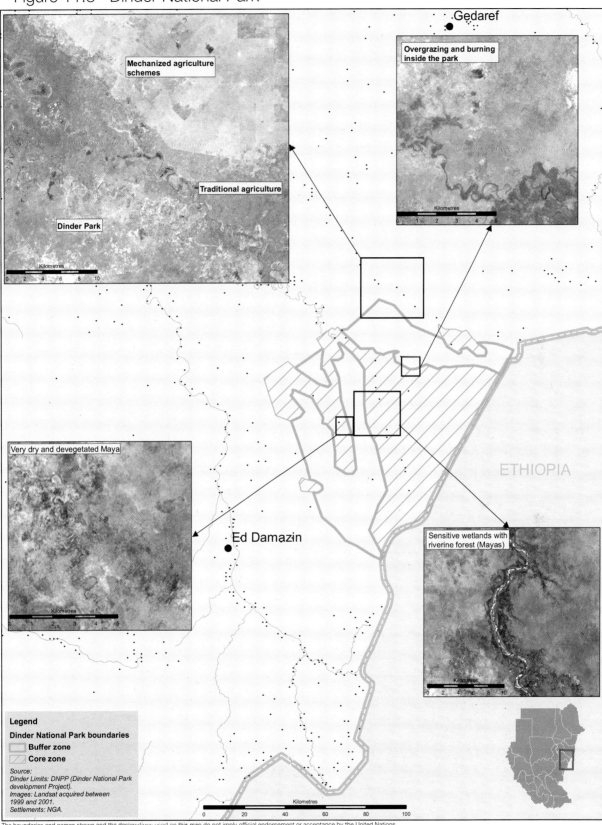

Mechanized agriculture schemes

Traditional agriculture

Dinder Park

Overgrazing and burning inside the park

Very dry and devegetated Maya

Sensitive wetlands with riverine forest (Mayas)

Gedaref

ETHIOPIA

Ed Damazin

Legend

Dinder National Park boundaries

Buffer zone

Core zone

Source:
Dinder Limits: DNPP (Dinder National Park development Project).
Images: Landsat acquired between 1999 and 2001.
Settlements: NGA.

The boundaries and names shown and the designations used on this map do not imply official endorsement or acceptance by the United Nations.

The harvesting of animals in Sudan takes two general forms: commercial poaching for non-meat products, and the bushmeat culture and industry. The two forms are often combined, but each has different cultural, ecological and legal aspects and needs to be tackled in a different manner.

Commercially oriented poaching for non-meat products, such as ivory, skins and live animals for pets, was historically a major industry but is now reduced due to a steep drop in the targeted wildlife populations. This form of harvesting is completely illegal in Sudan, with the sole exception of the continued existence of a small-scale commercial trophy hunting business in the Red Sea hills.

Important poaching targets are now almost exclusively found in Southern Sudan and include elephants, snakes, leopards, parrots, chimpanzees and tortoise, with the live animal trade being most important for the latter three species and classes. Ivory poaching was and still is a significant problem that needs to be addressed as a matter of priority in order to safeguard the remaining few elephants in the country (see Case Study 11.3). Protecting the limited number of chimpanzees still present is also considered a vital task for the wildlife forces of Southern Sudan (see Case Study 11.4).

Bushmeat (meat harvested by hunting wild animals) has always been part of the Sudanese

The collection of baby animals to serve as pets is common in Southern Sudan. The long-term survival rate of such individuals is very low. A Patas monkey in Jonglei state (top), a servile cat in Aweil, Northern Bahr el Ghazal (bottom left) and a hyena in Rumbek, Lakes state (bottom right)

In Sudan, the demand for ivory comes principally from tourists and foreign workers who are perhaps unaware of the global ban on ivory trading

CS 11.3   The illegal ivory trade in Sudan and the regional extinction of the African elephant

Sudan has been a centre for elephant hunting and ivory trade for centuries. Since 1990, however, it has been illegal under the Convention on International Trade in Endangered Species (CITES) to export ivory. Killing elephants or selling ivory from animals killed after 1990 is also illegal in Sudan. Given that most of the old (pre-1990) unmarked stock was in all likelihood used up long ago, any current ivory trade is no doubt illegal.

Nonetheless, the ivory trade and poaching of elephants in Sudan continue to this day, with export through illegal international trade networks. The international NGO Care for the Wild conducted a detailed investigation of the issue in 2005, and follow-up reconnaissance and interviews by UNEP in mid-2006 largely confirmed the findings.

During the war years, the main agents of the ivory trade were the military forces of the north that benefited from their unmonitored access to the south and the borders with the Central African Republic (CAR) and the Democratic Republic of Congo (DRC). The drastic reduction in elephant populations within Sudan and the gradual withdrawal of the northern forces from Southern Sudan have probably reduced direct military involvement, but private raiders remain in business. There have been consistent reports of heavily armed horsemen from Northern and Southern Kordofan, as well as Southern Darfur, coming into Southern Sudan, CAR and DRC on ivory-poaching trips. The latest report was received by UNEP from a government official in Western Bahr el Ghazal in July 2006.

The main centre of the ivory trade is Omdurman, a city across the river from Khartoum. The 2005 NGO report quotes 50 souvenir shops, 150 craftsmen and up to 2,000 items in individual shops. The main customers were reported to be Asian expatriates. UNEP visits to shops in Omdurman in July 2006 also revealed substantial amounts of ivory on sale and confirmed the presence of foreign ivory buyers.

The illegal ivory trade is a critical force driving the regional extinction of the African elephant. In order for the elephant to have a chance of survival in Sudan and elsewhere in central Africa, this trade needs to be shut down by tackling both the supply and the demand. There is no doubt that this will be a very arduous task.

Completely cutting off the supply through anti-poaching measures in the south will be extremely difficult due to the overall lack of governance in the region, the wide availability of firearms and the multiple national borders. At the same time, addressing the demand will be a particularly sensitive and politically challenging task. Possible but controversial measures to stop the demand include shutting down the carving industry through national legislation, or exerting diplomatic pressure on Asian governments to enforce the CITES convention on their own citizens traveling to Sudan, through a combination of persuasion and enforcement.

diet, with the exception perhaps of the most ancient agricultural societies based along the Nile. It partly sustained the SPLA during the conflict and was a critical fallback food source for millions of Sudanese in times of crop and livestock failure. During periods of famine, southern Sudanese reported eating any and all types of wild fauna, from buffalo to field mice.

The current issue with the bushmeat 'industry' is a combination of a lack of control and a lack of data. Indeed, there is very limited control on the continued harvesting of important food species such as the white-eared kob, but there is also no data available to assess whether current rates of harvesting are sustainable.

It is unrealistic to expect a blanket ban on bushmeat to be enforceable in Southern Sudan at this time. What is needed instead is the establishment of a system and culture of sustainable harvesting, where local hunters and communities take the bulk of the responsibility for the care of such resources.

Wildlife tourism

The main problem with wildlife tourism in Sudan is that it does not exist on a commercial scale. In 2005, the total number of foreign visitors to Dinder National Park and the marine parks was less than one thousand. Protected areas are hence not commercially self-sustaining and need constant subsidization, creating an evident issue of prioritization for one of the world's poorest countries.

There is currently no wildlife tourism industry whatsoever in Southern Sudan either, and the prospects for rapid growth are slight due to insecurity and a lack of infrastructure. Accordingly, the habitual issue of controlling the impacts of tourism does not yet apply to Sudan.

Crocodile and python skin accessories are popular in markets in Khartoum, but there is no data on the impact of this trade on reptile populations in Sudan

This young chimpanzee – named Thomas by wildlife rangers – was confiscated from a trader in Yei, Central Equatoria, in April 2006. He is shown here with his current keeper, the Undersecretary to the Government of Southern Sudan, Ministry of Environment, Wildlife Conservation and Tourism. His fate is uncertain as chimpanzees are completely unsuitable as pets and there are no rehabilitation or holding facilities in Sudan. The Ministry is searching for solutions, both for Thomas and for chimpanzee conservation in general

CS 11.4   Chimpanzee hunting and live capture in Southern Sudan

The chimpanzee (*Pan troglodytes*) is found in relatively undisturbed tropical forest regions in central and western Africa; the forests of the far southern edge of Sudan represent the eastern limit of its habitat.

Like all of the great apes, the chimpanzee is in danger of extinction. Throughout its range, the species is subject to a variety of threats, including habitat loss and fragmentation, the bushmeat industry, and live capture. While all of these issues are important in Sudan, the predominant problem is the bushmeat trade and the resulting live capture of animals. Typically, a mother and other family members are shot for meat, and the juveniles are captured alive for later sale as pets.

Sudan has been invited to sign the Kinshasa Declaration supporting the Great Apes Survival Project (GRASP) but, as of June 2007, has yet to do so.

11.5 Wildlife and protected area sector governance

Governance structure

The governance structure and legal situation of the wildlife and protected area management sector are complex and partially dysfunctional. The 2005 Interim National Constitution explicitly places management of the wildlife of Southern Sudan under the authority of the GOSS. At the same time, a number of international treaties such as the Convention on International Trade in Endangered Species (CITES) and the Ramsar Convention are managed at the federal level. This creates some confusion for the management of sites and issues in Southern Sudan.

Government of National Unity

In the Government of National Unity, wildlife and protected area management are the responsibility of the Ministry of Interior, as wildlife forces are part of the country's unified police forces. The controlling ordinance is the 1986 Wildlife Conservation and National Parks Ordinance. While there are numerous deficiencies in the structures and legislation which hamper practical governance, a principal problem is under-investment in the forces, resulting in a very low level of capacity in the field.

Government of Southern Sudan

Wildlife and protected area management in Southern Sudan are the responsibility of the Wildlife Conservation Directorate of the GOSS Ministry of Environment, Wildlife Conservation and Tourism. Like many of the new GOSS institutions, this structure is still extremely weak in capacity due to shortages in skilled manpower, equipment and accommodation. It does, however, have moderate amounts of funding and is receiving limited capacity-building.

While there is currently no GOSS legislation on wildlife and protected area management, the SPLM had a working Commission on Wildlife, and issued a number of directives for areas under its control.

A particular and unusual challenge for the new ministry is the requirement from GOSS to absorb large numbers of troops demobilized from the Unified Forces and directed to civilian sectors such as the police, wildlife forces, prisons and fire brigades. As of late 2006, the projected size of the wildlife force was over 7,300, which would probably make it the world's largest. If not well managed, training, managing and financing such a large force is expected to be major problem for the ministry that could distort the operations of the unit and distract it from its core role as the focal point for environmental governance (including wildlife) in Southern Sudan.

On a positive note, the Wildlife Conservation Society, an international NGO, announced in November 2006 that it was forming a multi-year partnership with the GOSS to build capacity in the wildlife forces and progress sustainable management of wildlife resources via a series of practical projects. One of the early activities planned is a major aerial survey of the protected areas to count wildlife populations and assess habitat conditions. The first stage of the fieldwork was completed in early 2007.

Innovative and sustainable solutions are needed to stem the decline of wildlife of Southern Sudan. These juvenile ostriches taken from the wild as chicks and raised in an aid compound in Padak will grow too big, powerful and dangerous to be kept as pets. The long-term fate of these particular individuals is sealed, but the species can be preserved in the region

11.6 Conclusions and recommendations

Conclusion

The issues relating to wildlife and protected area management are notably different in the north and south of Sudan. Economic pressures underlie the destruction of northern and central Sudan's wildlife, as well as the degradation of its protected areas. In a period of conflict and extreme poverty, investment in this sector was not a priority for the predecessors of the Government of National Unity. However, the new wealth provided by oil revenue will hopefully allow a gradual turnaround of this situation.

In Southern Sudan, the limited short- to medium-term prospects for wildlife tourism imply the need for alternative revenue streams to finance wildlife management. Potential alternatives include sustainable game ranching and the formalization of the bushmeat industry.

With the exception of three park areas (Dinder, Sanganeb and Dongonab Bay), the data on the wildlife and protected areas of Sudan is insufficient to allow the development of management plans. Before detailed planning can take place, more in-depth assessments will need to be carried out.

Background to the recommendations

The following recommendations are structured to fit the post-CPA institutional arrangements. They are aimed at pragmatic solutions for economic sustainability and prioritization of expenditure. For Southern Sudan, the need for comprehensive capacity-building within the wildlife management sector is clear. As of early 2007, GOSS is in receipt of assistance from both USAID and the Wildlife Conservation Centre; moreover, it has capacity for self-improvement via the Boma Wildlife Training Centre. However, it should be noted that the wildlife sector is unique in that is has a high potential for attracting partnerships with international NGOs and thus has better funding prospects than many other environmental sectors.

Recommendations for the Government of National Unity

R11.1 Reform and rationalize institutions, laws and regulations. The institutions, laws and regulations related to wildlife and protected area management at all levels of government need to be rationalized and improved. Due to the overlapping nature of many of the existing institutions, laws and regulations, this would, in the first instance, need to be done as a joint exercise by GONU, GOSS and state governments.

CA: GROL; PB: MI and MEPD; UNP: UNEP and INGOs; CE: 0.5M; DU: 3 years

R11.2 Invest in the management of Dinder National Park. This would entail implementation of the current management plan, which is both adequate and up to date.

CA: GI; PB: MI and MEPD; UNP: UNEP and INGOs; CE: 3M; DU: 5 years

R11.3 Shut down the illegal ivory carving and trading industry. This is a clear governance issue with north-south peace implications that can be addressed without causing significant economic hardship on the national scale.

CA: GROL; PB: MI; UNP: UNEP and CITES; CE: nil; DU: 1 year

Recommendations for the Government of Southern Sudan

R11.4 Develop interim strategies and plans for the management of protected areas and wildlife including the surveying of all protected areas. Detailed long-term plans, policies and legislation cannot be rationally developed or implemented due to the current lack of information and governance capacity. Interim measures are needed.

CA: PA; PB: MEWCT; UNP: UNEP and INGOs; CE: 4M; DU: 2 years

R11.5 Develop focused plans for the management of Nimule National Park, the Sudd Ramsar site (including its elephant population) and the conservation of chimpanzees and migratory antelopes including the white-eared kob. These four items have common features (international support, practicality and conservation urgency) that make them targets for early practical action.

CA: GROL; PB: MEWCT; UNP: UNEP and INGOs; CE: 2M; DU: 2 years

Marine Environments and Resources

Port Sudan, which hosts the largest sea freight terminal in the country, typifies the situation for marine resources in Sudan: economic development is occurring at the expense of the environment, and the surrounding lagoons are suffering from land-based pollution and modification due to the indiscriminate building of infrastructure.

Marine environments and resources

12.1 Introduction and assessment activities

Introduction

The coral reefs of the Sudanese territorial waters in the Red Sea are the best preserved ecosystems in the country. To date, these precious assets have been largely protected by the lack of development, but the economic and shipping boom focused on Port Sudan and the oil export facilities is rapidly changing the environmental situation for the worse.

At present, the state of the coastal environment is mixed: while steady degradation is ongoing in the developed strip from Port Sudan to Suakin, good conditions prevail elsewhere along the coast. On and above the tideline, the symptoms of overgrazing and land degradation are as omnipresent in Red Sea state as elsewhere in dryland Sudan.

The preservation and sustainable development of the marine resources of Sudan will require an integrated approach. For this reason, all of the issues specifically related to marine and coastal environments are collated and discussed here, though several cut across sectors covered in other chapters of this report.

Assessment activities

For this assessment, UNEP drew upon a significant available databank on the marine resources of Sudan [12.1, 12.2, 12.3, 12.4, 12.5]. In addition, a UNEP field mission covered the coastal strip from 100 km north of Port Sudan to the Tokar delta. Fieldwork included an extensive investigation of the Port Sudan area.

UNEP has been involved in the assessment and management of the natural resources of the Red Sea since the 1980s in its role as a supporter and participant in the Regional Organization for the Conservation of the Environment of the Red Sea and the Gulf of Aden (PERSGA). PERSGA-sponsored projects have included surveys of the coral reefs and other important marine habitats of Sudan.

While it did not extend to the habitat's condition, UNEP's assessment of the marine environment of Sudan was considered adequate to cover and provide an update on the main environmental issues.

A typical shoreline north of Port Sudan, with sparse vegetation on a sandy-silty beach, a sheltered zone and the fringing reef (indicated by the breaking waves in the distance)

Figure 12.1   Sudan coastline

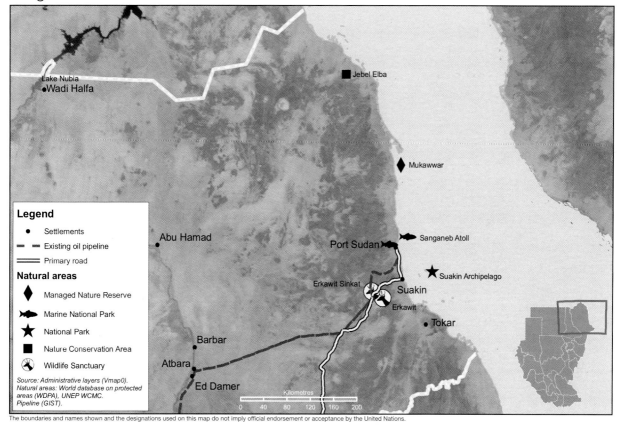

The boundaries and names shown and the designations used on this map do not imply official endorsement or acceptance by the United Nations.

12.2 Overview of marine and coastal environments and resources

The Red Sea

The Sudanese Red Sea is famous for its attractive and mostly pristine habitats, particularly its coral reefs. Three distinct depth zones are recognized: shallow reef-studded shelves less than 50 m deep, deep shelves 500 to 1,000 m deep, and a central trench more than 1,000 m deep, reaching a maximum of 3,000 m off the city of Port Sudan. The Red Sea is home to a variety of pelagic fish including tuna, but the overall fish density is relatively low due to limited nutrient input. The sea hosts important populations of seabirds and turtles, as well as mammals such as dugong, dolphins and whales.

Coastline and islands

The coastline of Sudan on the Red Sea is some 750 km long, not including all the embayments

and inlets [12.2]. Numerous islands are scattered along the coast, the majority of which have no water or vegetation. The dominant coastal forms are silty beaches, rocky headlands and salt marshes, commonly bordered with mangroves. Fringing coral reefs are very common and water clarity is generally high due to the lack of sedimentation.

Average precipitation in the coastal areas is extremely low, ranging from 36 mm per year at Halaib to 164 mm per year at Suakin, so that the desert extends right to the tide mark. The only exception is the Tokar delta, which receives substantial run-off from seasonal streams originating in the Ethiopian and Eritrean highlands.

The islands and most of the coastline are relatively undisturbed and host important feeding and nesting sites for a variety of seabirds. The three most ecologically important habitats are coral reefs, mangroves and seagrass beds.

Coral reefs

Three types of coral reefs are found in Sudanese waters:

- fringing reefs, which lie 1-3 km offshore;
- patch reefs, which lie up to 15 km offshore, separated from the fringing reef by deep and wide channels; and
- pillar reefs or atolls, found 20 km or more offshore, such as the Sanganeb atoll.

The coral reefs of Sudan are considered to be in moderate to good health, despite an extensive cover of algae over some fringing reefs. Some die-back/coral bleaching has occurred, particularly in the upper ten metres [12.3, 12.7].

Mangroves

Mangrove stands are a key coastal habitat, which provide forage, wood products and breeding grounds for fish. Extensive stands were originally found in areas where the seasonal streams (*khors*) reach the coast, as these produce the brackish and sediment-rich conditions necessary for mangroves to thrive. Mangroves stands are currently under severe pressure along the entire coastline from a combination of overgrazing and over-cutting, and in some regions, wholesale destruction due to coastal industrial development.

Seagrass beds

Seagrass beds are found in shallow coastal waters, around mangroves and between the low tide line and fringing reefs. They are highly productive habitats that provide grazing for dugong, and support fish and trochus shellfish.

12.3 Environmental impacts and issues

A high quality environment under pressure

The Sudanese marine and coastal environment is in relatively good condition overall, with isolated badly degraded areas. The region, however, is subject to a mounting list of environmental impacts linked to urban and industrial development, and to overgrazing. The principal environmental issues are:

- coastal habitat destruction by development;
- oil industry spill risks;
- passing ship pollution;
- pollution from land-based sources;
- risk of importing invasive species in ballast water;
- fisheries management;
- mangrove cutting and overgrazing; and
- marine protected areas and tourism.

Soft coral at Sanganeb. The coral reefs of Sudan are in very good to moderate condition away from the major urban areas. They are partly protected by their isolation and the lack of run-off from the desert

A major extension to the Port Sudan harbour, known as the Green Port, is going ahead in an area surrounded by seagrass beds and coral reefs. It is now necessary to focus on planning port operations to minimize ongoing impacts

Coastal habitat destruction by development

Development along the Red Sea coast is largely limited to a 70 km strip extending from Port Sudan to Suakin. This zone includes the two cities, the major ports, the oil terminals, saltworks, a shrimp farm and the new Red Sea Economic Free Trade Zone.

The damage to coastal habitats due to construction within this strip is extensive and in some cases both completely unnecessary and probably uneconomic in the long term. In some areas such as the main commercial port of Port Sudan, habitat destruction is unavoidable: though regrettable, local environmental damage is outweighed by the scale of the economic benefit. In other cases, however, the benefits of development are questionable.

Twenty kilometres south of Port Sudan, productive mangroves have been destroyed by saltworks construction; saltwater access canals and banks have cut through mangrove stands, disrupted groundwater flows and sediment deposition patterns. Approximately eight kilometres south of Port Sudan at Kilo Tammania, mangroves have been destroyed by the poor design of an outfall access road and recreation area [12.2].

As discussed in Chapters 7 and 13, industrial development in Sudan occurs in the absence of an effective environmental impact and management culture. This is clearly apparent in the Port Sudan region.

Figure 12.2   Port Sudan and coral reef

Coral reef destroyed for port development

The boundaries and names shown and the designations used on this map do not imply official endorsement or acceptance by the United Nations.

Oil industry spill risks

The risk of oil spills from the relatively new Bashir crude oil export terminal is discussed in detail in Chapter 7. The risks are considered to be moderate and the reported response measures close to international standards. The new Alkheir petroleum and gas export terminal is also considered to represent a moderate risk.

However, the loaded crude oil and product tanker traffic leaving the two terminals and traveling east to the Indian Ocean remains a considerable risk, due to the navigational hazard presented by the numerous fringing and patch reefs. In 2004, a freight vessel, the MV Irrens, grounded on the reef at the Wingate anchorage area some 10 km east of the Alkheir terminal [12.2].

Passing and docked ship bilge water and oil pollution

The Red Sea is a major shipping transit route, connecting the Indian Ocean with the Suez Canal. The ports of Sudan host a range of vessels, from

Ships passing and entering ports in Sudan currently have no place to deposit oily waste, such as that generated by clearing bilges and fuel tanks. In the absence of facilities and controls, the risk is that ships jettison this oil at sea

small coastal tenders to bulk grain carriers. In the absence of controls and facilities for receiving oily waste from bilges, ships discharge this effluent into the sea. This results in chronic oil pollution around the ports, but also along the coast, as discharges from passing ships drift landwards.

Pollution from land-based sources

The industrial facilities and utilities of Port Sudan are a major source of land-based pollution for the Red Sea. They include two power stations, a desalination plant and the harbour dockyard. Other facilities in the area, such as a tire factory, a tannery, and an oil seed factory, are now closed down.

Electrical power stations A and C were found to be dumping substantial quantities of waste oil onto open ground in adjacent vacant land (station C is described in more detail in Case Study 7.1). In addition, the desalination plant was found to be at the origin of a significant pollution by hypersaline effluent (see Case Study 12.1). The harbour dockyard, which has no oily water treatment facility, was another expected source of pollution, but was not inspected. Other parts of the harbour, including the main warehouse, were investigated and found to be relatively clean, except for one open warehouse filled with unwanted pesticides and other chemicals.

The Port Sudan landfill is located at the head of a seasonal watercourse. Every wet season, the run-off draws pollution from the site to the coastal lagoons

Figure 12.3 Port Sudan power station and salt flats

The boundaries and names shown and the designations used on this map do not imply official endorsement or acceptance by the United Nations.

UNEP also visited a small oil refinery located five kilometres south of Port Sudan (see Figure 12.3). Site personnel reported that an oil-water separator was used for water treatment, and that the treated effluent and cooling water were discharged to sea, although this could not be verified by UNEP due to access restrictions. The refinery grounds and surrounds were markedly cleaner than the adjacent electrical power station C.

Additionally, the harbour lagoons are polluted by litter, waste oil and sewage from wet season run-off from the *khor* Kilab, which borders the old industrial area of Port Sudan. This area contains numerous small factories and vehicle repair workshops that dump used oil and other waste into the stream bed throughout the year.

Finally, the main Port Sudan landfill, which is located in the head of the *khor*, is a source of surface and groundwater contamination. The run-off from the dump also eventually ends up in the harbour. The landfill is covered in detail in Case Study 6.4.

Risk of importing invasive species in ballast water

No port in Sudan has facilities for receiving ballast water, which is instead discharged by the ships either in the harbour or in the approaches. This practice carries the risk of importing invasive species (larvae, parasites and infectious agents) from where the ship last docked and took in the ballast.

Fisheries management

Marine fisheries and mariculture industries in Sudan are currently underdeveloped. They are also poorly controlled and subject to repeated proposals for expansion from foreign investors.

The artisanal fleet on Sudanese waters is comprised exclusively of locally made wooden boats and small fiberglass tenders. Fishing methods include hand lines, and bottom set and pelagic gill nets, with 80 percent of the catch coming from hand lines. Prior to 2005, an Egyptian shrimp trawling fleet operated offshore of the Tokar delta, but it was

Cargo ships carry seawater as ballast, which is drawn in or discharged when cargos are loaded and unloaded. When this occurs thousands of kilometres away from the intake point, there is a risk of introducing alien species into the local marine environment

This lagoon in the centre of the city of Port Sudan is already burdened with urban pollution and shoreline development. Unless a solution for the saline effluent is found, the lagoon is expected to become a biologically dead zone

Reverse osmosis units separate seawater into two streams: freshwater for consumption and a high salinity effluent which needs to be disposed of in an appropriate manner to avoid environmental damage

CS 12.1   The impact of pollution from the Port Sudan desalination plant

This desalination-based freshwater production plant in Port Sudan provides an unfortunate case study in the importance of locating industrial facilities correctly in order to optimize benefits to local citizens and minimize environmental impacts.

The plant, which was built in 2004, plays a vital role in the provision of freshwater to the city. Based on a reverse osmosis process that is powered by diesel, it has a combined freshwater output of 7,500 m³ per day and an effluent discharge of 2,500 m³.

The facility is located on the shoreline of a shallow and moderately polluted saltwater lagoon that was an important if declining fishing ground until 2004, but is now surrounded by urban development. The original plant design envisaged extracting water from the lagoon, but health concerns forced a late revision in the form of a 4 km pipeline to convey seawater in from the coastline. The effluent from the plant, however, is currently discharged directly into the lagoon as per the original design.

The salinity of the effluent is approximately four times that of seawater, and it contains traces of chlorine and anti-scaling agent. The local authority reported that a major fish kill occurred during plant commissioning and there are current complaints from local residents regarding skin rashes, although the link between this public health problem and the increased salinity is unclear at this stage.

What is clear is that the combination of a nearly closed system and ongoing saline inputs will in time result in a hypersaline and ecologically dead (and most probably anaerobic) lagoon in an urban area. While the local authorities were very much aware of this problem at the time of UNEP's visit, there was no agreement on the solution due to the high cost of all options proposed to date.

banned by the Red Sea State Governor during the 2005-2006 season, apparently due to a licensing dispute. At present, no legal offshore fishing is conducted by foreign vessels, though the potential for illegal fishing is high as there is effectively no monitoring.

The fisheries industry is constrained by a lack of investment in facilities to handle the catch, as well as by a limited domestic market. The daily fish catch is monitored by the local fisheries authority and estimated to be approximately 1,100 tonnes per year [12.2, 12.8]. Most of the fish is consumed locally. There is a small export market to Saudi Arabia and Egypt for fresh coral fish and shark, and some 200 to 300 tonnes of trochus shellfish are exported – mainly to Europe – per year.

Though historically significant, mariculture and the collection of wild pearl oysters in the Red Sea region ended in the 1990s. It may or may not be revived. Shrimp farming has just commenced, with one farm located 35 km south of Port Sudan, but this venture is struggling to establish local and export markets.

The key environmental issue for the fisheries and mariculture industries is the lack of effective governance. This leaves the environment highly vulnerable to overfishing and uncontrolled mariculture expansion.

At present, the domestic marine fisheries industry is very limited. Most of the catch is consumed locally. A small volume of high-quality fish is exported to other Gulf countries

Camels grazing on mangroves 20 km south of Suakin. The impact of such grazing can be seen in the absence of foliage below three metres. This stand also shows signs of extensive timber-cutting

Spinner dolphins offshore of Suakin. The marine tourism industry in Sudan still operates on a small scale, catering mainly to scuba divers, but the quality and quantity of marine life holds promise for the long-term growth of the industry. Protection and control measures need to be improved to ensure that this growth occurs without harm to the environment

© RED SEA ENTERPRISES

Mangrove cutting and overgrazing

Mangrove leaves are edible for camels and are thus vulnerable to grazing damage in periods of scarcity. Most of the accessible mangrove stands visited by UNEP had the characteristic clipped look resulting from overgrazing. Mangroves can also supply wood for fuel and construction, and unsustainable cutting has clearly been a problem in the accessible stands.

Marine protected areas and tourism

There are two declared marine protected areas in Sudan: Sanganeb Marine National Park and Dongonab Bay (with Mukawar Island). Sanganeb Marine National Park is described in detail in Case Study 12.2.

Dongonab Bay National Park lies 125 km north of Port Sudan and covers 60 km of coastline and

a shallow bay with a wide diversity of marine habitats, including coral reefs and seagrass beds that support a large population of endangered dugong. The park also has a significant resident human population in a number of small fishing villages, and hosts a salt plant.

In addition, four high-value habitats have been proposed as marine protected areas:

- Suakin Archipelago, which comprises coral reefs surrounding a number of sandy islands approximately 20 km south-east of Suakin; these are important nesting sites for marine turtles and sea birds;

- Khor Kilab Bird Sanctuary, a 2 km² estuarine area on the south side of Port Sudan harbour;

- the Abu Hashish area, a 5 km² area on the eastern side of the new Green port, containing numerous coral reefs; and

- Shuab Rumi, a 4 km² area of coral reefs 50 km north of Port Sudan.

To this list, UNEP would add **all** of the remaining mangrove stands along the Sudanese Red Sea coastline, as this habitat is now under severe pressure and disappearing rapidly in some areas.

At present, the marine tourism industry is centred mainly on Sanganeb and to a lesser extent on Shuab Rumi. The Dongonab area is relatively remote and rarely visited. For the most part, tourism consists of international diving holidays, with visitors flying to Port Sudan and residing on large hotel boats, which travel to anchor at the various diving sites for a few days at a time. There is also some limited local recreation along the coastline.

The major environmental issue related to marine tourism is the lack of handling facilities at the dive sites and ports. For example, dive boats are forced to anchor on the reefs, causing damage, because they do not have mooring buoys. Tourism operators are highly aware of this problem, but do not have the legal mandate to install the necessary equipment, as that rests with the Sea Ports Corporation. An additional issue is the limited capacity for governance of the parks and tourism in all places.

© RED SEA ENTERPRISES

CS 12.2   Sanganeb National Park: a microcosm of high reef biodiversity

The Sudanese coast harbours the most diverse coral reefs in the Red Sea. The small Sanganeb Atoll, arguably the only true atoll in the Red Sea, is situated approximately 30 km north-east of Port Sudan. It lies close to the centre of Red Sea marine biodiversity, where conditions are optimal for coral growth and reef development.

Sanganeb's physical features include an outer rim that encloses three central lagoons, areas of back reefs, and shallow water reef flats dominated by massive colonies of porites, gonisatrea and montipora. Outside this outer rim, the reef drops vertically, interrupted by terraces, to the seabed some 800 m below. The drop from the reef flats to the reef slopes hosts a spectacular diversity of coral and fish species.

The coral fauna of the Sanganeb Atoll, which may well prove to be among the richest in the Red Sea, inhabits a number of different bio-physiographic reef zones. To date, a total of 124 cnidarians have been recorded. The atoll also hosts significant populations of *Trochus dentatus* (giant spider conch) and sea-cucumbers, which are commercially exploited elsewhere in Sudan.

Over 251 coral reef fish species have so far been recorded and this number may rise to more than 300. Populations of larger species such as humphead parrotfish (*Bolbometopon muricatum*), bumphead wrasse (*Cheilinus undulatus*), and groupers, which are vulnerable to overfishing throughout their ranges, appear healthy in Sanganeb. The open waters around the atoll include a large number of pelagic fish species such as tuna, barracuda, sailfish, manta rays and sharks. Sailfish are reported to spawn in the Sanganeb lagoon.

The atoll was declared a National Park in 1993 and is currently one of two marine protected areas in Sudan (the other is the Dongonab Bay and Mukawar Island National Park, gazetted in 2005). Management plans for both sites were developed by the Regional Organization for the Conservation of the Environment of the Red Sea and Gulf of Aden (PERSGA) in 2003. Sanganeb additionally lies within one of two proposed Ramsar sites along the Sudanese coast, and is on Sudan's tentative list for UNESCO World Heritage status. At present, the park covers an area of approximately 22 km², but there are proposals to create an additional buffer zone that would increase the area to approximately 260 km².

Sudan's Wildlife Conservation General Administration signed an agreement with the international NGO the African Parks Foundation to implement the existing management plans for both Sanganeb and Dongonab Bay National Parks [12.6]. In June 2006, the Foundation and IUCN undertook a baseline biodiversity survey of both parks.

The atoll has considerable potential as a major destination for diving tourism, but the infrastructure to support and manage increased tourism has yet to be put in place.

12.4 Marine and coastal environmental governance

Governance structure

The governance structure for the Sudanese Red Sea coastline, territorial seas, islands and associated marine protected areas is very complex and in consequence, fragmented.

Sudanese ports are managed by the Sea Ports Corporation, which is part of the federal Ministry of Transport. The important exception is the arrangement at the Bashir Oil Terminal port facilities, which also come under the management of the Ministry of Energy and Mining. Marine fisheries are governed by the Marine Fisheries Administration, which is part of the federal Ministry of Agriculture and Forestry. The marine protected areas are under the responsibility of the Headquarter of Wildlife Conservation in the federal Ministry of Interior, and wildlife conservation services staff are actually managed by the Ministry of Interior, as they are part of the country's united police force.

At the state level, the governor and the local government of ministers and advisors have significant and broad-reaching authority, which overlaps with the federal mandate to a large extent.

Red Sea state is unusual in that it has a working body specifically for marine environment protection – the newly formed Marine Environmental Protection Authority (MEPA). In addition, the State Council for Environment (SCE) provides an oversight and coordination role. Finally, the NGO sector is also active in Port Sudan.

Legislation and coordination

Appropriate and up to date legislation and guidance is lacking for the direction of the various authorities. Fisheries legislation, for example, is based largely upon acts drafted by the British in the 1930s. A number of important legal documents have been developed more recently, but have yet to be ratified or implemented by the federal authorities. The new state-sponsored SCE is anticipated to improve coordination between the various actors, though it is constrained by legislation to be largely advisory.

Figure 12.4 Sanganeb National Park

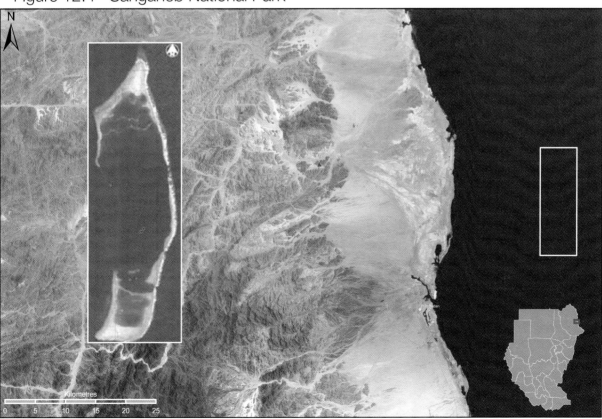

The boundaries and names shown and the designations used on this map do not imply official endorsement or acceptance by the United Nations.

Governance performance

While the Red Sea region has a number of interested and responsible parties for environmental protection, the complex governance structure and ensuing fragmentation of responsibility hamper practical performance by the authorities.

In addition, a severe lack of financial resources affects all governance operations (except for the Bashir Oil Terminal and the Sea Ports Corporation), and legislative deficiencies hinder both the authorities and civil society. For instance, many of the major facilities are managed at the federal level, which makes enforcement of legislation at the state level problematic.

12.5 Conclusions and recommendations

Conclusion

Compared to many parts of Sudan, the coastal and marine environments are still in very good condition. The marine habitats have global as well as national significance and are currently the most important foreign tourist attractions in Sudan.

The environmental issues faced by the region will require an integrated approach to have any chance of successful resolution. The multiple competing uses and threats for shared resources such as shipping channels, estuaries, coral reefs and pelagic fisheries cannot be addressed in isolation.

The general level of environmental awareness and interest among Red Sea state stakeholders is impressive and higher than that seen in many other parts of Sudan. However, this interest needs to be converted into practical action, in the first instance by transferring more authority to the local level.

Background to the recommendations

The two key themes for the recommendations are: integration, based on the concept of Integrated Coastal Zone Management (ICZM), and devolution of responsibility to the Red Sea state level.

Young men on duty on national service picking up litter from the tidal lagoons of Port Sudan. The level of interest in the environment in Red Sea state is among the highest in all of Sudan

The objective of Integrated Coastal Zone Management (ICZM) is to establish sustainable levels of economic and social activity in coastal areas while protecting the coastal environment. It brings all those involved in the development, management and use of the coast together in a framework that facilitates the integration of their interests and responsibilities.

In support of the devolution of powers, the 2005 Interim Constitution grants states the authority to manage their natural resources. This general clause needs to be strengthened for the unique coastal and marine environment, with more detail on the division of powers for a range of issues such as fisheries, coastal development, land-based marine pollution sources and tourism. This process would correct the current imbalance due to the fact that much of the interest in environmental management resides in Red Sea state while the mandate for management resides largely at the federal level.

Recommendations for the Government of National Unity

R12.1 Ratify and enforce existing prepared legal instruments for the marine environment. Documents that are ready but not yet translated into law or firm standards include the Sudanese Maritime Law and the National Oil Spill Contingency Plan.

CA: GROL; PB: GONU Assembly; UNP: UNEP; CE: 0.1M; DU: 2 years

R12.2 Develop legislation and statutory guidance covering offshore fisheries. This should cover issues such as prohibited areas and the granting of licenses to both domestic and international operators.

CA: GROL; PB: MAF; UNP: FAO; CE: 0.3M; DU: 2 years

R12.3 Adequately fund the marine fisheries inspection and data collection services operating out of the Red Sea ports to enable monitoring of catches and offshore fisheries including foreign vessels.

CA: GI; PB: MAF; UNP: FAO; CE: 3M; DU: 2 years

R12.4 Adequately fund the two marine protected areas of the Red Sea that have existing management plans and follow through with those plans to develop self-sustaining revenue streams for those areas. Sanganeb Marine National Park is the priority site.

CA: GI; PB: MI, UNP: UNEP; CE: 5M; DU: 5 years

Recommendations for the Red Sea State Government

R12.5 Enforce existing EIA legislation on planned developments on the coastline, including the Red Sea Free Trade Zone. This will require more direct involvement of the Red Sea State Government in support of the Marine Environment Protection Authority.

CA: GROL; PB: RSS MEPA; UNP: UNEP; CE: 0.1M; DU: 2 years

R12.6 Enforce existing water pollution legislation on industrial and utilities plant discharges into the Red Sea. This will require more direct involvement of the Red Sea State Government in support of the Marine Environment Protection Authority.

CA: GROL; PB: RSS MEPA; UNP: UNEP; CE: 0.1M; DU: 2 years

R12.7 Advocate and progress federal/state power-sharing on marine environmental issues. Set out and restructure the power-sharing arrangements for coastal and marine natural resources management to allow direct liaison and resolution at the state level.

CA: GROL; PB: RSS MEPA; UNP: UNEP; CE: 0.1M; DU: 3 years

R12.8 Introduce the concept of Integrated Coastal Zone Management through revised master-planning for the whole coast with a focus on the areas of Port Sudan, Suakin and Tokar.

CA: GROL; PB: RSS MEPA; UNP: UNEP; CE: 0.4M; DU: 3 years

Environmental Governance and Awareness

Under UNEP sponsorship, several consultation meetings were held between the environment ministries of the Government of National Unity and the Government of Southern Sudan to discuss national action plans for environmental management.

Environmental governance and awareness

13.1 Introduction and assessment activities

Introduction

Environmental governance and awareness are at a crossroads in Sudan. For several decades, the priorities of a war economy and a range of escalating environmental issues overran incremental progress in these areas. Now, two major events have radically reshaped the governance context and helped create the conditions for positive change.

First, the Comprehensive Peace Agreement (CPA) [13.1] and the Interim Constitution [13.2] have made much of the existing governance structures and legislation obsolete, creating a major opportunity for reform. Second, the injection of oil revenue has greatly boosted the financial resources of both the Government of National Unity (GONU) and the Government of Southern Sudan (GOSS), enabling such reform to be translated into concrete action.

This chapter provides an overview of the national structures, legislation and culture related to environmental management and awareness, with a focus on how to integrate or 'mainstream' environmental considerations into government and society in Sudan.

Assessment activities

Not only was the review of environmental governance in Sudan an integral part of UNEP's work in the country, but the assessment process itself was modelled to concurrently assist in the development of improved governance and a higher level of environmental awareness.

A detailed institutional assessment was conducted for the GONU, GOSS and selected state governments, including Khartoum, Red Sea, Gezira, Sennar, White Nile and Bahr el Jabal (Central Equatoria) [13.4]. This entailed a legal and practical review of all current and relevant treaties and legislation (including the CPA, the DPA, and the GONU and GOSS Constitutions) and follow-up interviews with government officials in both executive bodies and in over twenty ministries at the three working levels – national, regional and state.

The role of civil society was also evaluated, through extensive interaction with NGOs and the tertiary education sector, as represented by the many academics involved in the assessment process.

13.2 Overview of environmental governance structures

A complex and evolving national context

The main feature of environmental governance in Sudan is that it has not been able to keep pace with the evolving national context, as driven by a series of major changes, such as the cessation of the north-south conflict, the associated peace agreement and Interim Constitutions, the development of the oil industry, the escalation of the Darfur crisis and the partial resolution of the Eastern Front conflict. Underlying these events have been the creeping processes of population growth, climate change and land degradation. The net result today is a governance structure and culture that no longer fit the country's current circumstances.

Conflict and peace, the CPA and the 2005 National and GOSS Interim Constitutions

The cessation of hostilities between north and south opened up the country to the rule of civilian law and radically altered its political structure.

The Interim Constitution of the Republic of Sudan adopted on 6 July 2005 reflects the Comprehensive Peace Agreement (CPA) of January 2005 and defines a new set of rules for governance in general, and for environmental governance in particular. The two main elements of this new policy context are a high level of decentralization of powers to the states, and the creation of a Government of Southern Sudan (GOSS).

Table 25.   Powers and responsibilities set out in the 2005 Interim National Constitution
relating directly or indirectly to environmental governance

Schedule (A) National powers	
Section	**Title**
15	National lands and national natural resources
19	Meteorology
23	Intellectual property rights, including patents and copyright
25	Signing of international treaties on behalf of the Republic of Sudan
27	National census, national surveys and national statistics
29	International and interstate transport, including roads, airports, waterways, harbours and railways
30	National public utilities
33	Nile Water Commission, the management of Nile waters, transboundary waters and disputes arising from the management of interstate waters between northern states and any dispute between northern and southern states
Schedule (B) Powers of the Government of Southern Sudan	
2	Police, prisons and wildlife services
6	Planning for Southern Sudan government services including health, education, and welfare
9	The coordination of Southern Sudan services or the establishment of minimum Southern Sudan standards or the establishment of Southern Sudan uniform norms in respect of any matter or service referred to in Schedule C or Schedule D, read together with Schedule E, with the exception of Item 1 of Schedule C, including but not limited to, education, health, welfare, police (without prejudice to the national standards and regulations), prisons, state public services, such authority over civil and criminal laws and judicial institutions, lands, reformatories, personal law, intra-state business, commerce and trade, tourism, environment, agriculture, disaster intervention, fire and medical emergency services, commercial regulation, provision of electricity, water and waste management services, local government, control of animal diseases and veterinary services, consumer protection, and any other matters referred to in the above Schedules
10	Any power that a state or the National Government requests it to exercise on its behalf, subject to the agreement of the Government of Southern Sudan or that for reasons of efficiency the Government of Southern Sudan itself requests to exercise in Southern Sudan and that other level agrees
14	Public utilities of the Government of Southern Sudan
19	Any matter relating to an item referred to in schedule D that cannot be dealt with effectively by a single state and requires Government of Southern Sudan legislation or intervention including, but not limited to the following: (1) natural resources and forestry (2) town and rural planning (3) disputes arising from the management of interstate waters within Southern Sudan
Schedule (C) Powers of states: regarding environmental governance, most powers – executive and legislative – are at state level	
8	State land and state natural resources
13	The management, lease and utilization of lands belonging to the state
17	Local works and undertakings
21	The development, conservation and management of state natural resources and state forestry resources
23	Laws in relation to agriculture within the state
27	Pollution control
28	State statistics, and state surveys
31	Quarrying regulations
32	Town and rural planning
36	State irrigation and embankments
40	State public utilities
Schedule (D) Concurrent powers: The National Government, the Government of Southern Sudan and state governments shall have legislative and executive competencies on any of the matters listed below	
1	Economic and social development in Southern Sudan
3	Tertiary education, education policy and scientific research
4	Health policy
5	Urban development, planning and housing
6	Trade, commerce, industry and industrial development
7	Delivery of public services
12	River transport
13	Disaster preparedness, management and relief, and epidemics control
15	Electricity generation, and water and waste management
17	Environmental management, conservation and protection
19	Without prejudice to the national regulation, and in the case of southern states, the regulation of the Government of Southern Sudan, the initiation, negotiation and conclusion of international and regional agreements on culture, sports, trade, investment, credit, loans, grants and technical assistance with foreign governments and foreign non-governmental organizations
23	Pastures, veterinary services, and animal and livestock disease control
24	Consumer safety and protection
25	Residual powers, subject to schedule E
27	Water resources other than interstate waters
31	Human and animal drug quality control
32	Regulation of land tenure, usage and exercise of rights in land.
Schedule (F) Resolution of conflicts in respect of concurrent powers: If there is a contradiction between the provisions of Southern Sudan law and/or a state law and/or a national law, on the matters referred in Schedule D, the law of the level of government which shall prevail shall be that which most effectively deals with the subject matter of the law, having regard to:	
1	The need to recognize the sovereignty of the nation while accommodating the autonomy of Southern Sudan or of the states
2	Whether there is a need for national or Southern Sudan norms and standards
3	The principle of subsidiarity
4	The need to promote the welfare of the people and to protect each person's human rights and fundamental freedoms

The Ministry of Environment and Physical Development, in Khartoum

The need to preserve a measure of equality between states while awarding a high level of autonomy to Southern Sudan was addressed by granting all states a high level of autonomy, and creating a specific regional level of government – the GOSS – in the south. This model, characterized by a somewhat asymmetrical (between north and south) but overall decentralized system of governance, was adopted by the Interim Constitution.

UNEP has analysed the impact and new legal status quo of the 2005 Interim National Constitution; Table 25 on the previous page sets out its interpretation of national, regional, state and concurrent powers related to environment.

In terms of environmental governance, the impact of these changes is evident in the south, but not yet in the north and east.

In December 2005, the GOSS adopted its own regional Constitution, which echoes the key terms of the Interim National Constitution and adds detail, including substantial text on natural resource management [13.3]. On the Eastern Front, the peace process is still in its early stages, so the implications for environment and natural resource management are not clear at this stage. Finally, the Darfur Peace

Agreement (DPA) does not include significant detail on the environment and, as of June 2007, is not being implemented due to ongoing conflict.

GONU federal structure

The structure of environmental governance in the GONU is characterized by a multiplicity of small units linked to environment but not closely linked to each other. The key units are the Ministry of Environment and Physical Development (MEPD), the Higher Council for Environment and Natural Resources (HCENR), a number of state-level councils and other bodies, and departments or units in line ministries such as the Ministry of Agriculture and Forestry.

The Ministry of Environment and Physical Development was created in 2003. The MEPD's mandate, which covers surveying, construction, urban planning and now environment, is derived from the Environmental Framework Act of 2001. However, no actual environmental mandate for the MEPD is specified in the legislation, as the legislation pre-dates the establishment of an environment portfolio within the ministry. The MEPD's Department of Environmental Affairs (DEA) only has approximately ten staff members.

The Higher Council for Environment and Natural Resources was established in law by the 2001 Environmental Framework Act. Its mandate focuses on policy coordination for all sectors that have a role in the protection of the environment or use of natural resources, but no role in implementation. It was conceived as a ministerial-level forum supported by a secretariat. The Minister of Environment serves as the chairman of the HCENR. As of late 2006, however, the actual Higher Council has never been formally convened. All of its activities have been carried out by the secretariat, managed by the Secretary-General.

A key function of the HCENR to date has been that of focal point for international liaison and agreements. So far, virtually all of the international conventions, multilateral environmental agreements (MEAs) and Global Environment Facility (GEF) projects have been managed by this body. The HCENR employs 50 to 60 staff, of which approximately 20 are career civil servants. The rest are funded on short-term

contracts connected to MEA or GEF projects [13.4].

Several other ministries have important environment-related portfolios. In some ministries, this translates into dedicated departments; in others, environmental issues are in theory integrated into normal business.

The Ministry of Tourism and Wildlife (MTW) manages all wildlife issues in the northern and central states, and also plays an important role in the management of marine protected areas. In the Ministry of Agriculture and Forestry (MAF), the Forests National Corporation (FNC) comprises a great deal of practical expertise in forest management and conservation. The Ministry of Irrigation and Water Resources (MIWR) has a functioning environmental unit, though major realignment is now underway following the attachment of the Dams Implementation Unit to the President's Office. Finally, a unit within the Ministry of Industry (MoI) undertakes and partly evaluates the environmental impact assessments provided by projects [13.4].

The Ministry of Environment, Wildlife Conservation and Tourism, in Juba

GOSS regional structure

The design of the Government of Southern Sudan, which was created in the wake of the CPA, is nearly complete. Key posts have been established and awarded, but the development of the civil service is still in the early stages.

Within the GOSS ministerial structure, coordination and leadership on environment and wildlife issues are the mandate of the Ministry of Environment, Wildlife Conservation and Tourism (MEWCT). The MEWCT has over 600 allocated staff positions at the regional and state level, and over 7,300 allocated positions for the wildlife forces (see Chapter 11). The MEWCT had a budget of USD 4 million in 2006, excluding most of the costs of the wildlife personnel. Almost all of the MEWCT staff is newly appointed and relatively inexperienced in civil servant tasks. The exception is the wildlife sector, where the GOSS has inherited some of the expertise developed by the SPLM during the conflict period [13.10].

As is the case for GONU, several other GOSS line ministries have environmental responsibilities, including the Ministry of Agriculture and Forestry (MAF), the Ministry of Animal Resources and Fisheries (MARF), the Ministry of Water Resources and Irrigation (MWRI), and the Ministry of Industry and Mining (MIM).

State government structures

While the Interim National Constitution allocates fairly uniform responsibilities to all states, the environmental governance situation, in practice, varies greatly between the north, south and Darfur.

The Environmental Framework Act provides a mandate for state-level environmental administration and legislation, which was reinforced by the Interim Constitution in 2005. Several northern states (Red Sea, Gezira, Sennar, White Nile, Gedaref, Nile and Khartoum) have established environmental administrations that range from individual part-time efforts to well organized councils on environment involving several line ministries at the state level. Red Sea state is the most advanced in this respect, as it has both a coordinating council and a new Marine Environmental Protection Authority.

Interviews with these state-level units revealed that there was no universal model and that their origins were state-based, resulting from individual initiatives, personal political support, or decrees from governors or state ministerial decisions [13.4].

In contrast, state governments in the south have virtually no environmental administrations or capacity whatsoever. Similar to the GOSS in general, southern state governments are currently still growing. In principal, however, environmental issues enjoy a high level of support from the interviewed governors.

The three Darfur states are essentially in the same position as the southern states in terms of institutional capacity for environmental issues, but have even less capacity to act due to the conflict. The level of political support was not established in this assessment.

13.3 Overview of environmental and natural resource legislation

Environmental aspects of the 2005 Interim National Constitution

At the level of general principles, environmental protection is a national objective, which is not subject to interpretation by other levels of government.

In Chapter 2 of the Constitution, Article 11 states that for the State of Sudan as a whole, the conservation of the environment, and of biodiversity in particular, should be pursued, and that the State should ensure a sustainable utilization of natural resources, including by prohibiting actions that would adversely affect the existence of specific species. Article 17 reaffirms that it is the responsibility of Sudan as a whole to fulfil its international obligations. Chapter 3 adds that it is the duty of every Sudanese citizen to preserve the natural environment [13.2].

The Interim Constitution radically changes the relative authority of the various actors and stakeholders in the field of environment by transferring significant powers from the national to the state level and, in the case of GOSS, to the regional government.

The Environmental Framework Act of 2001

In 2001, the President of the Republic of Sudan signed an environmental framework law that is still in force today [13.4]. The Environmental Framework Act, referred to hereafter as the 'Act', has five chapters and twenty-nine articles:

- Chapter 1: Preliminary regulations;

- Chapter 2: the Higher Council for Environment and Natural Resources;

- Chapter 3: Policies and general trends for the protection of the environment, evaluation and environmental follow-up;

- Chapter 4: Violations, penalties and punishments; and

- Chapter 5: General rules, standards and methods of combating pollution.

Five general environmental objectives are stated in the Act, leaving it up to sector ministries to achieve these goals while performing their tasks or implementing their policies:

- the protection of the environment and its natural balance, and the conservation of its components and social and cultural elements, in order to achieve sustainable development for future generations;

- the sustainable use of resources;

- the integration of the link between environment and development;

- the empowerment of the authorities responsible for the protection of the environment; and

- the activation of the role of the concerned authorities and prevention of relaxation or disposal of duties.

Generally speaking, the law is more detailed for the protection of natural resources than for pollution control and regimes. According to Article 18, environmental impact assessments are required for projects likely to have a negative impact on the environment.

The MEPD has been asked to review and redraft the 2001 Act and all legislation to reflect the new legislative mandates of the MEPD and the HCENR under the 2005 Interim Constitution. This process will be far-reaching, not only because it will need to clarify the division of labour between MEPD and HCENR, but also because the Interim Constitution deeply affects the geographical division of powers, as indicated above.

GONU sector legislation

The GONU has a large body of sectoral legislation with linkages to environmental governance, which virtually all predates the CPA and 2005 National Constitution. Key acts and associated line ministries include:

- Ministry of Tourism and Wildlife: the Wildlife Conservation and National Parks Act (1986);

- Ministry of Agriculture and Forestry: the Forests Act (1989);

- Ministry of Agriculture and Forestry: the Pesticides Act (1994);

- Ministry of Animal Resources: the Freshwater Fisheries Act (1954) and the Marine Fisheries Act (1937);

- Ministry of Irrigation and Water Resources: the Water Resources Act (1995);

- Ministry of Health: the Environmental Health Act (1975) (water and air pollution); and

- Ministry of Industry: the Petroleum Wealth Act (1998).

Another area of governance with strong links to environmental governance is land tenure. This topic is not covered by any single line ministry, but important legislation includes the Unregistered Lands Act (1970) and the Civil Transactions Act (1984). The implications of deficiencies in land tenure are covered in Chapter 8.

GOSS legislation

As of early 2007, the process of legislation development within GOSS is still in its early stages. The legal basis for environmental governance is therefore effectively absent in Southern Sudan at this time.

In the interim period, the GOSS judiciary and ministries have taken the approach of using directives from the GOSS President, governors and ministers as temporary control measures. Though there are numerous SPLM policy documents and directives from the time of the conflict, these are not automatically translated into GOSS legislation and so are not legally valid.

In theory, the potential exists for the GOSS to use GONU legislation – including the Environmental Framework Act – as interim measures for governance of issues within the GOSS mandate, but this may be difficult to implement in practice.

State legislation

Red Sea state is the only state in Sudan to have developed a state-level framework law, known as the State Environmental Law of 2005. Other northern states have formalized their individual approaches to environmental governance via governor or state minister decrees and directives, and through reference to the GONU Environmental Framework Act of 2001.

International agreements

Sudan is a party to the following global and regional multilateral environmental agreements (MEAs):

- the Convention on Biological Diversity (CBD - 1992);

- the Cartagena Protocol on Biosafety (2000);

- the African-Eurasian Waterbird Agreement (AEWA - 1999);

- the Convention on International Trade in Endangered Species of Wild Fauna and Flora (CITES - 1973);

- the African Convention on the Conservation of Nature and Natural Resources (Africa Convention - 2003);

- the Ramsar Convention on Wetlands (1971);

- the Convention Concerning the Protection of the World Cultural and Natural Heritage (UNESCO WHC - 1972)

- the United Nations Convention to Combat Desertification (UNCCD - 1994)

- the United Nations Framework Convention on Climate Change (UNFCCC - 1994);

- the Vienna Convention for the Protection of the Ozone Layer (1985) and the Montreal Protocol on Substances that Deplete the Ozone Layer (1987);

- the Basel Convention on the Control of Transboundary Movements of Hazardous Wastes and their Disposal (1989);

- the Bamako Convention on the Ban of the Import into Africa and the Control of Transboundary Movement of Hazardous Wastes within Africa (1991);

- the Stockholm Convention on Persistent Organic Pollutants (POPs - 2001);

- the Rotterdam Convention on the Prior Informed Consent (PIC) Procedure for Certain Hazardous Chemicals and Pesticides in International Trade (1998);

- the United Nations Convention on the Law of the Seas (1982) and the Convention on the International Maritime Organization (1958); and

- the Regional Convention for the Conservation of the Environment of the Red Sea and the Gulf of Aden (PERSGA - 1982).

Funding supplied to Sudan in the period 2002 - 2006 to support the implementation of MEAs was approximately USD 5 million in total (see Chapter 14) [13.11, 13.12, 13.17, 13.18, 13.19, 13.20].

The 2001 Environment Act gives the HCENR the mandate to specify the channels assigned to implement the MEAs. In most cases, the HCENR has designated itself as the focal point. Many of the MEA support projects have a project coordinator hosted by the HCENR, and most activities are conducted at the federal level in Khartoum. Following the realignment of powers set out in the 2005 Interim Constitution, the national implementation mechanisms required by most MEAs will now fall largely under the responsibility of the states.

Aside from progress reporting, compliance with the agreements is variable, but overall at a low level.

13.4 Environmental education and civil society

Environmental education and awareness

Environmental education and awareness in Sudan are relatively limited, but gradually increasing.

Environmental science is a popular subject in the country's universities, and environmental studies programmes have multiplied over the years. Due to a lack of funding and equipment, as well as to a certain extent the lack of a culture of experimental science, environmental science is taught almost purely theoretically.

Environmental education at the primary and secondary school level is not institutionalized, but individual efforts at environmental curriculum development and outreach are taking place under the management of national NGOs [13.4].

National environmental NGOs

Building on a tradition of environmental societies dating back to the early 20th century, Sudan has several solid non-governmental organizations, within and outside Khartoum. Since the adoption of the Environmental Framework Act in 2001, NGOs have become important stakeholders in environmental affairs.

At present, the majority of NGO activities are focused on the northern states and the Red Sea. Environmental NGOs are present in Southern Sudan and Darfur as well, but are either very new or constrained by ongoing conflict.

Many of the activities funded by international partners have been implemented through NGOs such as the Sudanese Environment Conservation Society (SECS). Environmental NGOs were part of the technical team for this assessment, and completed a range of desk studies and field missions. They also played an active role in the Khartoum and Juba NPEM workshops in 2006 (see Section 13.8).

The South Sudan National Environment Association, which was founded in Boma in 2006, is the first national environmental NGO to be established in Southern Sudan

SECS has established several community-managed forests to provide firewood to the communities and act as shelter belts around villages and buffer zones against desert encroachment

CS 13.1   The Sudanese Environment Conservation Society

The Sudanese Environment Conservation Society (SECS) is a non-governmental and non-profit organization established in 1975 with a mandate to raise environmental awareness among different communities and advocate on issues related to environment. It is open for membership to all Sudanese who can serve its mandate, and has more than 120 branches all over the country.

The Society's activities are organized under three main programmes: Institutional Development and Capacity-Building, Environmental Rehabilitation and Environmental Education. It has established several working groups and networks throughout the country, including the Poverty Network, Desertification, Biodiversity, Environmental Law, Human Rights, Landmines, POPs, Climate Change, Women's groups, and others. SECS also hosts other programmes funded by the Nile Basin Initiative's micro-grants component, Nile Basin Discourse and the Darfur Joint Assessment Mission. Finally, SECS is a focal point in Sudan for IUCN, Bird International, UNDP, FAO, UNEP, and UN HABITAT.

At the grassroots level, SECS develops and implements practical and replicable environmental projects that contribute to the alleviation of poverty in rural and sub-urban areas. For example, the Society has established several community-managed forests, including a twenty-hectare forest in El Dein, Southern Darfur and a five-hectare forest in Sabnas, White Nile state. These community forests supply fuelwood, and can act as shelter belts around villages and buffer zones in areas afflicted by desert encroachment.

SECS has also supplied thirty schools in Khartoum state with natural water coolers, prompting other organizations to adopt the technology and supply universities, colleges, and prisons with the same. Moreover, to reduce the dependence on fuelwood and charcoal as the only source of energy for cooking, SECS has championed the introduction of Butane gas cookers and has distributed over 1,100 Butane gas cylinders in the villages of Gammoia (Khartoum state), Dinder (Blue Nile), El Rahad (Northern Kordofan), and Sabnas (White Nile) to date.

Over the years, the Society's activities have generated a vast amount of knowledge. Reports and other documents are available at the SECS library, which is open to students and researchers. Several academic institutions have also been established to address environmental issues and train researchers, such as the Institute of Environmental Studies at the University of Khartoum, the Faculty of Natural Resources at the University of Juba, and Environmental Studies at Ahliya University. SECS collaborates closely with these institutions by sharing information, as well as supporting and participating in their various activities.

Environmental data collection, management and dissemination

As highlighted throughout this report, not only is there relatively little solid environmental data available on Sudan (at both the national and international levels), but much of the existing data is obsolete.

The UNEP assessment found no institutionalized system of environmental data management or organized process for the dissemination of data to the public. Collection is limited to isolated work by individual ministries and academics. Most of the available data is linked to forestry, agriculture and health, and there is only limited information on water resources, industry, wildlife, climate and environmental governance. What does exist is generally not easily accessible to the public due to cost issues. Confidentiality constraints are not considered to be a major concern, except for isolated controversial projects and areas.

13.5 Overview of environmental governance and awareness issues

UNEP has compiled a comprehensive list of issues affecting environmental governance and awareness in Sudan. The list below focuses on central issues and opportunities only; sectoral issues are covered in Chapters 6 to 12, and governance issues relating to international aid are discussed in Chapter 14. Note that many subjects are cross-cutting and overlapping:

Social, development and investment issues:

- priorities in a post-conflict country;
- large-scale development mindset;
- lack of enforcement;
- limited governance capacity; and
- scarcity of environmental data.

Structural and legislative deficiencies:

- the CPA and Interim Constitution;
- GONU structure including international agreements;

- GONU legislation;
- GOSS structure;
- GOSS legislation;
- GONU and GOSS line ministries; and
- states.

Environmental governance and peacebuilding:

- the need and topics for north-south dialogue; and
- the NPEM process.

13.6 Social, development and investment issues

Priorities in a post-conflict country

The length and continuity of regional conflicts in Sudan put the country on a war footing for almost fifty years, with obvious impacts on its economy and governance culture. The destabilizing effects of conflict aside, Sudan remains a very poor country with an extremely limited tax base (though this is now starting to change due to oil revenue).

As a result of this uniquely unfortunate history, environmental conservation and sustainable development have not been financial or political priorities for the Government of Sudan. This is reflected in the annual budgets for all areas of environmental governance and natural resource management, which have never been adequately funded.

The promising exception to this situation is the allocation of USD 4 million by GOSS to the Ministry of Environment, Wildlife Conservation and Tourism in the 2006 budget. This scale of funding sets a very positive precedent, which must be encouraged.

Large-scale development mindset

In Sudan, the government has historically tended to rely upon a limited number of very large-scale investment projects or programmes to boost development. For some time, this tendency was exacerbated by investment and aid policies from the international community, which favoured large-scale infrastructure and agricultural development.

UNEP teams covered many of these large development projects in the course of their assessment, including large dams and the Jonglei canal (see Chapter 10), oil production (see Chapter 7), and the Gezira and New Halfa irrigation schemes, numerous sugar plantations and major rain-fed agricultural schemes in central Sudan (see Chapter 8).

These different programmes were found to have a number of negative features in common with respect to the environment: they were all conceived and supported at the highest political level; they often proceeded to the construction phase relatively quickly and without comprehensive analysis of economic, social and environmental sustainability; and they caused extensive and often unexpected environmental damage. The Jonglei canal is the best known example of the high risks and costs of this type of approach for project developers, local populations and the environment (see Case Study 10.2).

While environmental impact assessment documents were produced for the more recent projects, they were never publicly released or integrated into the planning and design process, and therefore had a negligible effect in terms of impact mitigation or community acceptance.

Significant improvements in environmental governance and sustainable development will not be possible without tackling the core issue of this effective immunity of major project developers from environmental considerations.

A more appropriate model for environmentally sensitive projects can be drawn from best international practice. Typically, the project development process includes a paced sequence of environmental, social and economic impact assessments and public consultations – before the project starts. This process can help both community acceptance and environmental sustainability.

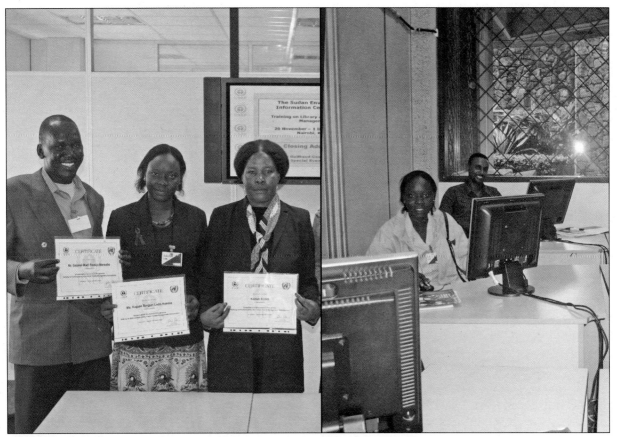

A UNEP training course on environmental information management was held for Sudanese government and NGO staff in Nairobi in late 2006. A significant investment in data collection, management and dissemination should be an early part of any programme to improve environmental governance in the country

Limited governance capacity

Environmental governance authorities in both GONU and GOSS have insufficient capacity to adequately implement existing mandates. For GONU, this is principally due to under-investment in the sector, while GOSS is completely new and therefore still weak.

The UNEP assessment found the human resource capacity to be high in many instances, with experienced and competent personnel throughout government ministries and the civil service. Just as importantly, the tertiary education system produces significant numbers of graduates in environmental subjects. The overriding constraint on the civil service's capacity is insufficient funding, which translates into deficiencies in knowledge, staff numbers, equipment, accommodation and operating expenses.

UNEP considers that given sufficient time and funding, building capacity in the Sudanese civil service to help achieve improved environmental governance is entirely possible and relatively straightforward. For such work to be sustainable, however, it would need to have significant counterpart funding from the GONU and GOSS, and avoid 100 percent international aid funding (see Chapter 14).

Lack of enforcement

Existing GONU laws have deficiencies (see next section), but are nonetheless perfectly usable for a wide range of applications, from EIA provisions to wildlife poaching to pollution control. Unfortunately, enforcement of the existing environmental legislation is extremely limited at all levels. The development of capable institutions – even if backed by improved legislation – will not result in any real improvement unless the culture of non-enforcement is addressed concurrently, starting at the highest level.

Scarcity of environmental data

The pervasive scarcity of solid quantitative data on all aspects of the environment of Sudan constrains rational planning for resource management and conservation. Besides, the absence of strong and credible signals that real problems exist – which can only be provided by up to date data – makes it difficult to even raise awareness at the government level. A significant investment in data collection, management and dissemination should therefore be an early part of any programme to improve environmental governance in Sudan.

13.7 Structure and legislative issues

Legislative complexity and overlap

The Comprehensive Peace Agreement is a landmark achievement that has brought peace to most of Sudan. The resulting governance situation, however, is highly complex. This is particularly apparent in the environmental governance and natural resource management elements of the 2005 Interim National Constitution: as shown in Table 25, there is a great deal of overlap and potential for confusion. The Schedule (F) Resolution of Conflicts in Respect of Concurrent Powers appears sensible in principle, but is expected to be very slow and complicated in practice in the event of a dispute.

GONU core structure (including international agreements)

The current GONU structure for environmental governance is problematic and considered to be a major obstacle for reform, irrespective of potential funding and legislative improvements.

At present, the various arms of government with an environmental mandate are poorly connected or not connected at all, and have duplicate mandates and insufficient resources, leading to unproductive competition and conflict. Given that the principal coordinating body, the Higher Council for Environment and Natural Resources has never actually met, high-level leadership is lacking.

The international community's environmental sector has played a role in this situation, and may have inadvertently worsened it (see Chapter 14). Indeed, the MEA and GEF funding processes have helped perpetuate an ad hoc fundraising and externally driven project-based mindset within GONU, which in turn has significantly hindered the capacity-building and reform of the responsible organizations, such as the Ministry of Environment and Physical Development.

UNEP considers substantive reform of the GONU environmental governance structure to be a pre-requisite for lasting improvement in this sector. The scope of the reform should address the following subject areas:

- the structures and interfaces of the MEPD, the HCENR secretariat and the HCENR;

- the development of coherent units within MEPD to focus on a range of coordination and policy topics including:
  - multilateral environmental agreements;
  - economic sector-specific environmental governance (for the oil industry, for example);
  - outreach and assistance to the regional and state levels; and

- the development of an Environmental Protection Authority or similar body to implement and enforce legislation.

The international convention secretariats will also need to cooperate in this process and ensure that best use is made of available resources to implement the conventions.

GONU legislation

GONU legislation in the field of environment and natural resource management has many deficiencies: it is obsolete, incomplete and unclear in parts, and as a result, difficult to enforce.

GONU officials are already aware of the deficiencies in the existing legislation and are starting to work on a revision of the Environmental Framework Act of 2001. This work needs to be strongly supported and followed through with a substantive programme of legislative development that tackles underlying details, such as the provision of statutory guidance and integration into different economic sectors, like industry and agriculture.

GOSS core structure

The GOSS core structure for environmental governance is considered to be appropriate and well designed at the ministerial level. Three major issues, however, need to be resolved in order to progress further in organizational development and capacity-building:

- organizing the large number of wildlife forces (7,300) and maintaining a balance in the ministry between the three directorates of environment, wildlife and tourism;

- determining the role of the ministry in practical issues such as the implementation of practical policies and the enforcement of environmental legislation; if appropriate, a semi-autonomous Environmental Protection Authority or similar unit may need to be developed; and

- determining the relationship between GOSS and southern states on environmental governance, in order to progress associated capacity-building and legislative development.

GOSS legislation

Given the GOSS's complete lack of environmental legislation, it is clear that a vast amount of development work is required. The principle issue of concern is timing, as the experience of other post-conflict countries has shown that this process can take several years to do well. Leaving Southern Sudan without any environmental controls during the post-conflict period is considered to be an unacceptable risk for its environment. Accordingly, some interim measures and risk-based prioritization are recommended:

- Develop an interim set of working guidelines on priority topics and issue them as a directive from the Ministry;

- Focus first on structuring framework legislation to allow work on underlying legislation to start; and

- Work concurrently on finalizing the framework legislation and the priority sector legislation.

The priority sectors are:

- environmental impact assessment and project development permitting;

- urban planning and environmental health, including waste management; and

- oil industry environmental legislation (in cooperation with GONU).

GONU and GOSS line ministries

Environmental authorities in both GONU and GOSS face the challenge of mainstreaming environmental considerations into other line ministries. This will require focused programmes to increase inter-ministerial coordination, and the development of new (or improvement of existing) sector-specific environmental legislation. It should be noted that some line ministries have strong units and/or experienced personnel working on environmental issues, while others have neither staff nor resources. Solutions will therefore need to be tailored to each ministry.

States

As a result of the 2005 Interim National and GOSS Constitutions, all of Sudan's twenty-five states now have a legal mandate for natural resource management that reaches well beyond their current capacity. They are in need of general assistance, particularly in the areas of operating expenses, human resources capacity-building and the development of state-level legislation.

In order to avoid a high level of variation between states and the unnecessary duplication of effort, GONU and GOSS federal-level bodies should provide a coordinated programme of assistance, in the form of a development 'package' that could be rapidly rolled out to all states.

13.8 Environmental governance and peacebuilding

The NPEM process

The government-led process of developing a National Plan for Environmental Management (NPEM) constitutes a good example of proactive work to improve environmental governance and practical cooperation between north and south on substantive governance issues. The process commenced in late 2005 and the first working draft was released in early 2007 [13.5]. The underlying objective or final product of the NPEM is envisaged to be an environmental action plan or series of plans that set out the priorities for Sudan in terms of corrective action and targeted investment in environmental issues.

Given that the NPEM objectives are close to those of the UNEP assessment process, they have effectively been combined. One clear difference between the two processes, however, is the form and ownership of the final documentation: UNEP is responsible for this report, while the national plans must by default be owned by the government.

If it is successfully concluded, the most likely final documentation of the NPEM will be a national-level plan presented to the GONU parliament in 2007 and a matching regional document presented to the GOSS parliament in 2007 or 2008. It is anticipated that both this process and the guidance included in the final documents will significantly assist the development of environmental governance in Sudan.

The process has also provided a platform for open and detailed dialogue between technical professionals, civil servants and politicians from northern and southern states. Two key events were held in July 2006 in Khartoum and November 2006 in Juba, respectively. Over forty papers covering environmental issues from all parts of the country were presented and discussed at these workshops, which were attended by over 300 people.

The principal added value of the NPEM model is that it is less formal and therefore less politically charged than the CPA-instigated commissions, but that it nonetheless provides an organized forum for debate on sensitive topics with the support of neutral international parties, such as UNEP and the Nile Basin Initiative.

Expanding the NPEM model to other issues and regions

As discussed in Chapter 4 and elsewhere in the report, several environmental issues represent potential 'flashpoints' that could lead to renewed conflict:

* the environmental impacts of the development of the oil industry (Chapter 7);

* the southward migration of northern pastoralists due to land scarcity and degradation (Chapters 3 and 8);

- tree-felling for the charcoal industry in the north-south boundary zone (Chapter 9);

- new and planned dams and major water projects, including any revival of the Jonglei canal project (Chapter 10);

- ivory and bushmeat poaching (Chapter 11).

The NPEM style of technical dialogue could be extended to these topics to further assist the process of peacebuilding in Sudan.

13.9 Conclusions and recommendations

Conclusion

The CPA, the Interim National Constitution and the Interim GOSS Constitution have significantly changed the framework for environmental governance in Sudan. Given that the GOSS and states now have extensive and explicit autonomy in this area, environmental governance has become more of a regional issue. This is reflected in the findings and recommendations.

At the national level, Sudan faces many challenges to meet its international obligations, as set out in the treaties and conventions it has signed over the last thirty years. An additional difficulty in this area is incorporating GOSS-related issues. A range of reforms and significant investment are clearly needed.

The overall technical skill and level of knowledge in the environmental sector are very high and some practical legislation is already in place. However, the regulatory authorities also have critical structural problems, and are under-resourced and ineffective. Further, enforcement is highly variable and there is a fundamental disconnect between the environmental sector, the highest levels of government and the other sectors and ministries responsible for the development of Sudan.

In the conflict- and instability-wracked regions of Darfur and the Three Areas, environmental governance is essentially absent, even though environmental issues are among the causes of the conflict.

In Southern Sudan, finally, environmental governance is in its infancy, but the early signs are positive. High-level political and cross-sector support is visible, and the new structures are considered to be relatively suited to the task. The environment ministry and other authorities presently have negligible capacity and hence require comprehensive capacity-building. Environmental policies, plans and regulations for all sectors need to be developed from first principles. Due to the combination of the lack of environmental governance and the post-conflict development boom, the environment of Southern Sudan is currently extremely vulnerable.

Background to the recommendations

A key theme for the recommendations in this chapter is the need for local ownership and leadership on governance issues. International assistance is needed but must play a supporting role only, particularly with respect to funding. Accordingly, the central recommendation for both GONU and GOSS environmental authorities, and especially for the former, is to work to achieve sustained high-level and mainstreamed political support. This support should then be converted into adequate budgets, appropriate mandates, and assistance in the development, ratification and enforcement of robust legislation.

Recommendations for the Government of National Unity

R13.1 The MEPD should undertake an environmental awareness campaign targeted at GONU senior leadership, ministries and other civil service bodies. This would entail use of materials generated by the NPEM, UNEP and MEPD, and a sustained programme of communication via presentations, bulletins and other tools.

CA: GROL; PB: MEPD; UNP: UNEP and UNDP; CE: 0.2M; DU: 1 year

R13.2 The MEPD Minister should convene the first HCENR meeting with minister-level attendance. This would be an important and symbolic step towards integrating environmental issues into GONU and commencing the reform process.

CA: GROL; PB: MEPD; UNP: UNEP; CE: nil; DU: 3 months

R13.3 Secure funding and mandates, and undertake a comprehensive reform of the GONU core environmental governance structure. This will entail a wide range of activities, as set out in section 13.7, and could take up to two years to complete. The cost estimate covers only the reform process and not the subsequent operational costs of the new structure.

CA: GROL; PB: MEPD; UNP: UNEP and UNDP; CE: 1M; DU: 2 years

R13.4 Undertake a comprehensive and staged legislation development programme. This should start with a revision of the Framework Act, followed by the full suite of supporting statutory guidance, sector and state legislation.

CA: GROL; PB: MEPD; UNP: UNEP; CE:1.5M; DU: 4 years

R13.5 Develop a dedicated environmental data management centre. This centre should focus on the collection, collation and public dissemination of scientifically sound environmental data to support all aspects of environmental governance.

CA: TA; PB: MEPD; UNP: UNEP; CE:1M; DU: 2 years

R13.6 Invest to sustain the operations of the reformed and upgraded environmental governance sector. There is no substitute for sufficient and secured annual funding to allow the MEPD and other related bodies to fulfil their mandates.

CA: GI; PB: MEPD; UNP: UNEP; CE: 5M; DU: per annum minimum

Recommendations for the Government of Southern Sudan

R13.7 Develop interim strategies, plans and directives for environmental governance.

Detailed long-term plans, policies and legislation cannot be rationally developed or implemented due to the current lack of information and governance capacity. Interim measures are clearly needed.

CA: GROL; PB: MEWCT; UNP: UNEP and USAID; CE: 0.3M; DU: 6 months

R13.8 Develop and implement a practical action plan for environmental management in Juba with a range of partners. Practical action programmes are urgently needed in Southern Sudan to demonstrate progress and the benefits of peace. Projects in Juba have added value over other Southern Sudanese cities, in that they are relatively easier to manage, have high visibility and can be used as part of the capacity-building programme.

CA: PA; PB: MEWCT; UNP: UNEP and others; CE: 3M; DU: 3 years

R13.9 Implement a comprehensive capacity-building programme for the MEWCT and other GOSS ministries associated with environment and natural resource management. Development of a skilled and well equipped workforce at the regional and state level is a major multi-year task.

CA: CB; PB: GOSS; UNP: UNEP and USAID; CE: 5M; DU: 3 years

R13.10 Develop the full package of environmental legislation, regulations and implementation plans. Once the basic capacity is in place, longer-term plans and solutions can be developed. This needs to be a multi-sector effort to ensure buy-in and enforceability.

CA: GROL; PB: GOSS; UNP: UNEP and USAID; CE: 1M; DU: 3 years

International Aid
and the Environment

International aid and the environment

14.1 Introduction and assessment activities

Introduction

International aid represents approximately three percent of Sudan's economy, and the humanitarian aid programme in the country is the largest of its kind worldwide. Some 15 percent of the population are completely or largely dependent on international food aid for survival, and the number is rising due to the Darfur crisis.

A core principle for the UN programme in Sudan and elsewhere is to 'do no harm' through the provision of aid. This applies to the environment as well. Indeed, humanitarian, recovery and development aid programmes that inadvertently create or exacerbate local environmental problems may, in the long run, do more harm than good to local communities aspiring to sustainable livelihoods. In this context, a review of the environmental impacts of the international aid programme in Sudan was considered an appropriate component of the UNEP post-conflict assessment.

Furthermore Sudan, like many developing countries, receives international aid from a variety of sources for a number of environmental issues as diverse as biodiversity conservation, climate change adaptation, control of redundant pesticides and transboundary water resources management. In view of UNEP's planned follow-up capacity-building activities in Sudan, an evaluation of the impact of such programmes was also deemed necessary.

Assessment activities

The assessment of the impact of international aid was included in the overall scope of activities carried out by UNEP in Sudan. A significant amount of background information was available on humanitarian, recovery, development, and environmental aid: the UN and Partners Work Plan for 2006 [14.1] provided a detailed basis for a desk-based analysis, and substantial project documentation (including progress and closure reports) was available for virtually all of the environment-specific aid programmes identified, such as those funded by the Global Environment Facility (GEF).

UNEP assessment teams visited dozens of aid projects as they travelled through Sudan, gaining a first-hand impression of impacts in the field. The projects and programmes viewed include:

The influx of large numbers of displaced persons and the associated humanitarian aid has created a 'relief economy' in some Darfurian towns, which in turn drives environmental degradation

- food aid programmes managed by WFP, contractors and partners in several states;

- UN agency and government-managed internally displaced persons camps in Darfur;

- the WFP-managed Southern Sudan roads and Bor dyke projects;

- FAO agricultural projects in Southern Kordofan;

- UN and other agency compound- and facility-building programmes in Southern Sudan;

- return and support programmes managed by WFP, FAO, UNHCR and IOM in Jonglei state;

- EC-sponsored Oxfam agricultural projects in the Tokar delta, in Red Sea state;

- the Dinder National Park GEF project;

- the USAID STEP project training facilities in Southern Sudan;

- the Port Sudan GEF project for the Marine Environmental Protection Authority; and

- the Nile Basin Initiative project offices and sites.

UN Sudan environmental impact grading and integration assessment

The environmental impact of UN aid and peacekeeping programmes is rarely studied, due to the understandable priority of providing urgently needed vital services and commodities such as security, food, drinking water and shelter. In Sudan, however, the humanitarian programme has now been managing a series of crises for over twenty years. The UN and partners spend over USD 2 billion per year in the country (including peacekeeping costs [14.2]) and work in a number of environmentally degraded regions like Northern Darfur, Southern Kordofan and Kassala. UNEP therefore considers that an assessment of the environmental impacts of the UN Sudan programme is warranted.

The international aid community in Sudan operates at least partly outside the national regulatory framework. For environmental issues, such as the potential impact of the programmes it manages, the aid community is effectively fully self-governed. There is no single mandatory or even agreed environmental standard or code of conduct guiding the UN agencies and their partners operating in Sudan and or other post-conflict countries.

To date, the most relevant document is the SPHERE Project Humanitarian Charter and Minimum Standards in Disaster Response [14.3], which includes some guidance notes and limited standards on the environmental impact of specific activities. Several agencies also have internal guidelines, which are generally voluntary and applied at the discretion of the agency country director (or head of mission for peacekeepers).

In the absence of an agreed and appropriate existing standard, UNEP adopted a three-part system for this assessment:

1. Assessing the potential negative environmental impacts of projects using the established UNEP/World Bank 'ABC' project screening system;

2. Searching for evidence of integration of environmental issues into project design and implementation by qualitative review; and

3. Searching for potential positive environmental impacts of projects by qualitative review.

The UNEP/World Bank 'ABC' system for screening the environmental impact of projects is a qualitative process that gives a preliminary rating to projects based on project size, type, and location [14.4]:

- Category A: likely to have significant adverse environmental impacts (on a national scale);

- Category B: likely to have adverse environmental impacts; and

- Category C: likely to have negligible or no environmental impact.

The UN compound in Juba hosts a number of UN and other international agencies providing humanitarian and development assistance in Southern Sudan

14.2 Overview of international aid in Sudan

A major and long-standing aid programme

Foreign aid – which has played a crucial role in the country's development – has had a turbulent history in Sudan, with changes in the political regime and economic crises leading to corresponding modifications in donor country programmes.

Development aid commenced after independence and continues to this day. Sudan first obtained public sector loans for development from a wide variety of international agencies and individual governments. Major lenders included the World Bank (both the International Development Association and the International Finance Corporation), as well as the governments of the United States, China, the United Kingdom and Saudi Arabia. As Sudan defaulted on some of its debts in the late 1970s, however, many of these credit providers have now ceased development loans and provide direct grants or other forms of assistance instead.

Large-scale humanitarian aid, which now constitutes approximately 80 percent of direct international aid to Sudan, started in the 1980s. Operation Lifeline Sudan (OLS) was established in April 1989 as a consortium of two UN agencies, UNICEF and the World Food Programme, as well as more than 35 non-governmental organizations [14.5]. It provided humanitarian assistance to central and south Sudan without a major break for 17 years, and continues today, in modified form. Current large-scale humanitarian assistance operations in Darfur began in 2003 and are ongoing, with over 2,000,000 beneficiaries [14.1].

The aid programme for 2006

Total international aid to Sudan for 2006 was valued at over USD 2 billion, making Sudan the largest recipient of direct aid in Africa. Approximately USD 1.7 billion were received in the form of grants, commodities and services, and other direct assistance monitored by the UN. Other sources of aid, which are less easily quantifiable, included aid managed outside the UN system, aid from Arab states and China, and development loans from a range of international partners.

Given that Sudan's estimated gross domestic product for 2005 was USD 85.5 billion [14.6], international aid in 2006 represented 2 to 4 percent of the economy (depending on the method of measurement and multiplier effect). Table 26 shows the total humanitarian aid requested in the UN Work Plan of January 2006, broken down into twelve themes or sectors. Table 27 shows the same expenditure divided by state and region (with some projects labelled as national in scope).

Table 26. UN and Partners Sudan
Work Plan 2006
Aid projections by sector

Sector	Value (USD)	Number of projects
Basic infrastructure and settlement development	118,138,319	16
Cross-sector support for return	67,287,999	20
Education and vocational training	198,331,275	50
Food aid	603,762,013	44
Food security and livelihood recovery	117,598,136	69
Governance and rule of law	12,706,000	62
Health	142,461,918	140
Mine action	54,819,670	44
NFIs, common services and coordination	157,257,653	28
Nutrition	51,832,047	42
Protection and human rights	72,414,506	80
Water and sanitation	134,954,916	66
Grand total	**1,731,564,452**	**661**

Table 27. UN and Partners Sudan
Work Plan 2006
Aid projections by state and region

Region	Value (USD)
National programmes	144,652,806
Southern Sudan	650,859,700
Darfur	650,422,397
Abyei	23,433,461
Blue Nile	41,122,373
Southern Kordofan	90,017,289
Eastern Sudan	70,042,272
Khartoum and other northern states	61,014,154
Grand total	**1,731,564,452**

In practice, expenditure is further broken down into two major categories: humanitarian (USD 1.519 billion or 88 percent), and recovery and development (USD 211 million or 12 percent).

The strong emphasis on humanitarian projects shows that the majority of international aid to Sudan is currently aimed at saving lives. In line with humanitarian needs, most of the aid goes to Darfur and Southern Sudan. Recovery and development needs are secondary. Projects related to good governance – which is a core issue for environment – received USD 12 million or 0.7 percent of the total amount of aid for 2006.

14.3 Overview of environmental aid programmes in Sudan

Historical programmes related to the environment

Investment in the environment in Sudan began in the form of wildlife-related initiatives in the early 20th century. These were followed in the post-war period by a range of technical studies on soil, flora and fauna, some quite detailed in nature [14.7]. After independence, investment in environmentally beneficial projects continued but on an insignificant scale compared to the environmentally destructive agricultural development projects initiated at the same time. The most significant historical aid projects are probably the forestry and shelter belt projects implemented and managed by FAO from the 1970s to the 1990s, evidence of which UNEP sighted in the course of field reconnaissance in Khartoum state, White Nile state and Northern Kordofan.

Current structure

The current arrangements for the delivery of environmentally oriented aid programmes to Sudan are not structured or formally connected in any way, and are not comprehensively recorded in any management system. Based on the information available, UNEP has categorized environment-related projects and expenditure for 2006 in Table 28 on the following page.

It should be noted that while projects related to water and sanitation do have environmental aspects, they were not categorized as 'environmental projects' in this assessment. The criteria used by UNEP to identify specific 'environmental projects' were those provided by Part 1 of UN Millennium Development Goal no. 7: *integrate the principles of sustainable development into country policies and programmes* and *reverse the loss of environmental resources*. Only projects whose objectives correspond to those criteria were considered as 'targeted environmental projects'. Note that Water and Sanitation is an entire sector of the UN Sudan Work Plan.

Table 28. Summary of environment-related aid activities in Sudan in 2006

Type of programme	Number of projects	2006 Sudan project cost (USD)
Conventional aid programmes		
Total of all UN country programmes – as recorded in the UN 2006 Work Plan (January 2006 version)	661	1,730 million
Targeted environmental projects within conventional humanitarian programmes	3	Approx. 0.30 million
Targeted environmental projects within conventional recovery and development programmes (both inside and outside the Work Plan)	2	Approx. 2.5 million
Conventional humanitarian, recovery and development programmes that have mainstreamed or seriously attempted to mainstream environmental issues into project design and implementation	3	Unknown
Active environmental aid programmes – usually multi-year		
Regional programmes with a major environmental component	7	Unknown – < 10 million
Assistance programmes for implementation of ratified multilateral environmental agreements and conventions (active in 2006)	3	Unknown – < 1 million
Total 2006 active environment-related or integrated projects	18	Unknown

Targeted environmental projects within humanitarian programmes

Using the aforementioned criteria, the assessment identified only three projects in the humanitarian field in 2006 that were specifically targeted at environmental issues; UNEP is involved in two of these:

- the Tearfund Darfur environment study, which began in the third quarter of 2006 [14.8]; this assessment-based project is funded to a total of USD 200,000 by UNICEF, DFID, and UNHCR – UNEP has provided technical assistance;

- the International Red Cross flood preparedness and tree-planting project in IDP settlements in Khartoum; UNEP is funding this project for USD 60,000; and

- the forestation and provision of alternative energy resources (fuel-efficient stoves) project, funded to a total of USD 30,000 by the Fondation Suisse de Déminage (FSD).

Targeted environmental projects within recovery and development programmes

The UNEP assessment found only two projects in the recovery and development field in 2006 that were specifically targeted at environmental issues:

- the UNEP post-conflict environmental assessment for Sudan, funded by Sweden and the United Kingdom; and

- the Sudan Transitional Environment Programme (STEP) funded by USAID for approximately USD 6 million over a period of three years (see Case Study 14.1) [14.9, 14.10].

Mainstreaming environmental issues in conventional country programmes

There are no established criteria within the UN to determine whether an aid project has truly integrated or mainstreamed environmental issues into its design and implementation, or made a serious attempt to do so. Accordingly, the UNEP assessment was based on an ad hoc qualitative analysis using the following checklist of questions:

1. Has any form of environmental impact assessment, even very basic, been carried out?

2. Has the project design been altered significantly on the basis of such an EIA?

3. Have any proactive measures been taken to minimize environmental impacts?

4. Have any opportunities for a positive environmental impact been proactively included in the project?

UNEP screened over 650 country projects for Sudan in 2006 and found that only four could be considered by any reasonable measure to have truly mainstreamed environmental issues or made a serious attempt to do so. None of these were in the 2006 UN Work Plan:

- the USAID-sponsored WFP and GTZ management of the construction-related impacts of the Southern Sudan roads programme [14.11] (see Case Study 14.2);

- the USAID-sponsored construction of the Bor dyke [14.12];

- a camp rehabilitation project managed by UNHCR and IUCN in Kassala state [14.13]; and

- a town planning project sponsored by USAID in Southern Sudan [14.9].

CS 14.1   The USAID Sudan Transitional Environment Programme for Southern Sudan

The USAID Sudan Transitional Environment Programme (STEP), which is focused on stability and the prevention of conflict, was established in August 2005. It aims to address critical environmental issues that constitute potential sources of conflict in Southern Sudan.

The STEP team is currently working with the Directorate of Environmental Affairs in the GOSS Ministry of Environment, Wildlife Conservation and Tourism (MEWCT), to establish an inter-ministerial GOSS Environmental Consultative Group, whose mandate is to bring together representatives of key ministries to discuss and sanction the establishment and implementation of government-wide environmental policies, procedures and guidelines for impact monitoring in selected sectors (transportation and roads, water and sanitation, oil exploration and production, education and health).

To date, the STEP team has trained 120 GOSS officials in environmental impact assessment (EIA) procedures. These trained personnel are expected to conduct EIAs for all projects that are considered to have serious environmental consequences. STEP has also facilitated the establishment of the South Sudan National Environment Association (SSNEA), and contracted a short-term organizational establishment consultant from among the members of the organization to promote early activities within the membership.

In addition, STEP has organized study tours to sub-Saharan African countries for GOSS officials to be exposed to modern environmental and natural resources sustainable management practices.

The Programme's most significant undertaking, in collaboration with the World Food Programme and the GOSS Ministry of Transport and Roads, has been the successful completion of environmental impact assessments for the WFP road project (see Case Study 14.2) and the Bor dyke.

Since late 2003, some 1,400 km of road have been rebuilt under the WFP project

CS 14.2   The Southern Sudan roads project

The Southern Sudan roads project is an example of how the assessment and mitigation of environmental impacts can be built into aid projects, as well as an illustration of how aid-funded development projects can have a significant negative effect on the environment.

Two decades of civil war destroyed the region's road network and most other infrastructure, leaving it isolated and economically crippled. With the signing of the Comprehensive Peace Agreement and the return of peace, the need to connect isolated and remote areas to major towns was deemed a high priority by the Government of Southern Sudan, the United Nations and USAID.

To facilitate the return of internally displaced persons (IDPs) and the delivery of much-needed humanitarian aid to the remote regions of Southern Sudan, USAID contracted the World Food Programme (WFP) to rebuild and maintain the region's dilapidated road network.

The WFP road project aims to rebuild more than 3,000 km of roads in the war-ravaged south, at a cost of USD 183 million. Pending sufficient funding, the entire region will eventually be opened up by improving road links between Kenya, Uganda and Sudan (see Figure 14.1). It will also connect the Nile River to key feeder roads. Once complete, it will be possible, for the first time in a generation, to travel by road from the southern borders of Sudan to Khartoum and onto Egypt. Since late 2003, WFP has rebuilt some 1,400 km of roads, repaired bridges and culverts, and in the process removed and destroyed some 200,000 pieces of unexploded ordnance in Southern Sudan. The project has linked major towns across the south and reopened trade routes with neighbouring countries.

The social and economic benefits of the work completed to date are undeniable: according to a recent WFP survey, the roads built so far have halved travel time to markets, schools and health centres. Bus services now operate on all major routes and the cost of public transport has decreased by 50 to 60 percent. The price of commodities has also fallen. Besides, the roads project employs 1,650 Sudanese nationals, including 250 working in de-mining.

The negative environmental impacts of the project, however, are also clear. According to the USAID-sponsored EIA, these include soil erosion, impacts on local hydrology, negative aspects of abandoned borrow pits, construction camp impacts, road dust, and most importantly, the indirect but real impact of opening up large regions of tropical forest and several protected areas.

UNEP can add one specific issue to this general list: the effect of traffic on wildlife, as seen on the Bor-Padak road in Jonglei state, which cuts directly across the annual migration route of several hundred thousand antelope (tiang and white-eared kob). The road is also likely to attract settlers and make large-scale hunting much easier. Appropriate mitigation measures are needed as a matter of urgency if this road is not to become the root cause of a decline in these wildlife populations.

Figure 14.1   Southern Sudan roads programme

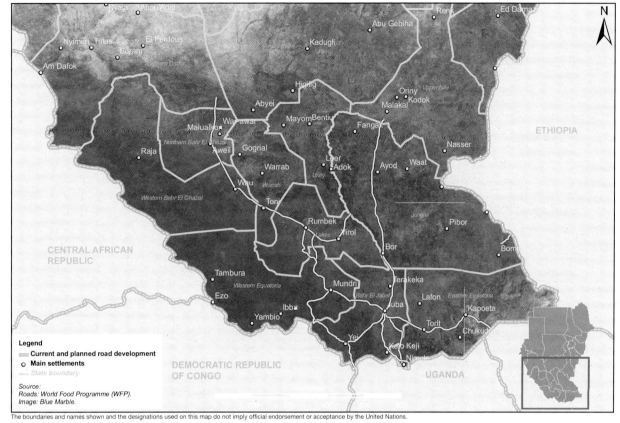

Legend
— Current and planned road development
o Main settlements
--- State boundary

Source:
Roads: World Food Programme (WFP).
Image: Blue Marble.

The boundaries and names shown and the designations used on this map do not imply official endorsement or acceptance by the United Nations.

Proposed and planned environmental programmes for Sudan

A number of projects related to the environment of Sudan have been proposed and are expected to start in 2007, subject to funding and other issues. These include:

- the Africa Parks Foundation-Cousteau Society project (Phase I) for protected area management and integrated coastal zone management (Red Sea state only);

- the Wildlife Conservation Society programme for Southern Sudan wildlife and protected area management;

- the expanded USAID Sudan Transitional Environment Programme (STEP) for Southern Sudan;

- the UNEP-UNICEF Darfur integrated water resource management project;

- the UNEP-UNDP Darfur aid and environment project;

- the UNEP-UNDP Darfur conflict and environment project; and

- the UN Habitat Darfur 'woodless construction' project.

Regional environmental programmes

As set out in Table 29 on the following page, Sudan is a participant in numerous regional programmes that include an element of aid provision on environmental topics, in addition to opportunities for networking and cooperating with surrounding countries. Each programme focuses on the issues related to the management of a major shared natural resource or a shared problem. Note that the total value covers all countries involved in the programme (UNEP efforts to obtain clarity on Sudan's share were unsuccessful due to time constraints).

The majority of the funding for these programmes comes via the Global Environment Facility, and each programme is managed entirely separately. Administration and funds are managed by UNDP Khartoum.

The international aid community in Sudan includes a wide range of actors, as illustrated by this water point established by a partnership of UN and development cooperation agencies and international NGOs

Table 29.   Regional aid-based programmes related to the environment [14.14]

Project title	Total value (million USD)
The Strategic Action Programme for the Red Sea and Gulf of Aden (PERSGA programme)	19,34
The Nile Basin Initiative, the environmental component of which is the Nile Transboundary Environmental Action Project	27,15
Formulation of an action programme for the integrated management of the shared Nubian aquifer	1
Demonstration of sustainable alternatives to DDT and strengthening of national vector control capabilities in the Middle East and North Africa	8,5
Mainstreaming conservation of migratory soaring birds into key productive sectors along the Rift Valley/Red Sea flyway	10,24
Elimination of persistent organic pollutants and adoption of integrated pest management for termites	3,5
Removal of barriers to the introduction of cleaner artisanal gold mining and extraction technologies	7,125
Total	**76,85**

Table 30.   Global Environment Facility projects for Sudan [14.14]

Programme name	Convention	Cost (million USD)
National biodiversity strategies, action plan and the report to the CBD	Biodiversity	0.334
Conservation and management of habitats and species, and sustainable community use of biodiversity in Dinder National Park	Biodiversity	0.75
Clearing-house mechanism enabling activity	Biodiversity	0.014
Assessment of capacity-building needs and country-specific priorities in biodiversity management and conservation in Sudan	Biodiversity	0.102
Community-based rangeland rehabilitation for carbon sequestration	Climate change	1.5
Capacity-building to enable Sudan's response and communication to the UNFCCC	Climate change	0.29
Barrier removal to secure PV market penetration in semi-urban Sudan	Climate change	0.75
Expedited financing of climate change enabling activities (Phase II)	Climate change	0.1
National Adaptation Programme of Action (NAPA)	Climate change	0.2
National Capacity Self-Assessment (NCSA) for Global Environmental Management	Multi-focal areas	0.225
Initial assistance to Sudan to meet its obligations under the Stockholm Convention on Persistent Organic Pollutants (POPs)	Persistent organic pollutants (POPs)	0.5
Total		**4.765**

Global programmes promoting compliance with international conventions

As detailed in Chapter 13, Sudan is a signatory to sixteen multilateral environmental agreements (MEAs). The majority of these MEAs provide aid to developing countries to assist them to work towards compliance with the terms of the agreement. This aid focuses on the years immediately following the signing, to support the signatories in understanding the obligations, collecting data, and planning a country-specific compliance programme. The best funded MEAs are the climate change (UNFCCC) and biodiversity (CBD) conventions, which are funded through the Global Environment Facility.

In the period 2002-2006, Sudan benefited from eleven GEF-funded projects to a total of USD 4.76 million, as detailed in Table 30 above.

Each programme is/was managed entirely separately. Administration and funds are/were managed by

UNDP Khartoum. As of end 2006, Sudan had not yet proposed any projects for GEF funding Tranche 4.

14.4   Overview of impacts and issues for aid and the environment

Unintended impacts and coordination issues

UNEP's assessment revealed a wide range of issues linked to unintended impacts of aid programmes, aid effectiveness and coordination. The key issues were considered to be:

• agricultural substitution by food aid;

• environmental impacts of humanitarian, and recovery and development country programmes;

• lack of issue integration into UN country programmes; and

• environment sector management and effectiveness.

Food distribution in Um Shalaya IDP camp, Western Darfur. Over six million Sudanese depend on food aid provided by the international community

Agricultural substitution by food aid

The dominant but unintended impact of aid on the environment in Sudan is linked to the provision of food aid by the international community to over 6,000,000 destitute people, or approximately 15 percent of the population. Food aid has been supplied to the Sudanese on a large scale since 1989. Its provision has become almost institutionalized and routine, particularly in Southern Sudan and increasingly in Darfur.

Without international or national aid, and in the absence of import purchasing power, this food would have to be produced in Sudan, placing an additional burden on the rural environment, particularly in the northern half of the Sahel. In many of the poorer and arid parts of Sudan such as Northern Darfur, it is clear that this extra load would intensify the observed land degradation to potentially critical levels.

This finding raises the important issue of how the international community proposes to eventually cease large-scale provision of food aid to Sudan. Any exit strategy will need to consider the risk of increased land degradation in the most vulnerable areas, if only to reduce the likelihood of having to remobilize food aid to the same areas as a result of famine arising from desertification.

The option of shifting large return populations to lesser stressed areas in order to reduce food aid is also problematic in the long term, as the assessment has shown that no area in Sudan is immune from the population-linked problems of deforestation and land degradation. Moving people south to higher rainfall areas will not solve the underlying problem.

One potential approach would be to focus on assisting economic development in order to enable more of the population to shift from subsistence agriculture to alternative livelihoods, relying on household purchasing power for food security. Food would be purchased from the domestic market, taking a share of what is currently exported. Such an approach would also have a linked environmental payback.

Environmental impacts of humanitarian, and recovery and development country programmes

Of the 661 projects screened, two projects were classified as Category A (likely to have significant adverse environmental impacts), one project as Category B (likely to have adverse environmental impacts), and 658 projects as Category C (likely to have negligible or no environmental impact).

The two Category A projects are the Southern Sudan roads rehabilitation programme (see Case Study 14.2) and the Bor flood control dyke project in Jonglei state by the Bor-Padak rural trunk road. Both of these major infrastructure initiatives have followed a form of EIA process, and are in this respect considered positive examples for the UN. However – as indicated in the EIA studies themselves [14.11, 14.12] – their negative environmental impacts are likely to be significant on a local scale. The negative environmental impacts of the Bor dyke project, in particular, have a direct link to livelihoods and food security.

While the proactive implementation of an EIA process by USAID is to be commended, the fact that this process was essentially self-managed by USAID and its contractors highlights an evident need for environmental governance at the national level and/or some form of environmental standard for international aid projects of this nature. At present, many bilateral agencies are more advanced than the UN in this respect, as they already have some form of environmental policy, standard and safeguard system in place.

The Category B project-related issue is linked to the operation of health clinics in Southern Sudan. The waste management situation in Southern Sudan is generally problematic, and there are currently no clinical waste management facilities in the region. Disposal options for clinical waste are thus far from optimal, although investments in waste management are underway as of early 2007.

The great majority of projects rated as Category C are considered to have negligible environmental impacts on the national scale, but adverse effects are expected at the local level for all projects, except for purely human resource projects such as training.

However, the cumulative impact of more than 650 projects is expected to be very significant. In this context, environmental best practice or proactive mitigation measures at the local level become more important.

Lack of issue integration into UN country programmes

Not one of the 658 non-environmental projects listed in the 2006 UN Work Plan were judged by UNEP to have fully integrated or 'mainstreamed' environmental issues, though one project had made a serious attempt to do so (the WFP and GTZ management of the construction-related impacts of the Southern Sudan roads programme, see Case Study 14.2).

This finding is surprising in its uniformity and indicates that the UN humanitarian, recovery and development teams in Sudan are clearly not taking environmental issues into account in project planning and implementation in the field, despite some awareness of the importance of environmental issues within the aid community.

UNEP looked for best practice in environmental management in aid projects through a process of project field inspections and desk study reviews, and found that individual examples of good practice stood out against a background of generally indifferent or poor environmental management. Waste management and use of construction materials contributing to deforestation were two key areas of concern.

Environment sector aid management and effectiveness

A range of management issues significantly reduce the environmental aid sector's effectiveness in Sudan. The key problems are fragmentation, lack of coordination, limited prioritization and lack of counterpart funding. These issues are perhaps not unique to Sudan or to the environment sector, but nonetheless need to be addressed if future aid is to be used to the country's best advantage.

The total budget allocated to the environment in Sudan by the international aid community is almost impossible to evaluate accurately, as the sector is extremely fragmented. UNEP identified over twenty ongoing or proposed aid-funded environmental activities for Sudan, through a year-long process of enquiry and discovery; it is likely that a number of additional existing projects were not found.

Coordination is quite limited, and there is no central reporting system. Furthermore, there is no formal or regular forum in which the numerous actors in the environmental field can meet and exchange information – all such events to date have been ad hoc. The MEA and GEF global structure contributes to this confusion, as it results in a number of separate teams and projects running in parallel, with no permanent country presence and multiple reporting lines.

In addition, there is no consistent country-driven prioritization process. Generally speaking, regional programmes appear to be reasonably well aligned with country needs, as they have long consultation and development processes that allow for more meaningful local input. In contrast, global MEA activities in Sudan are presently managed in a formulaic manner, by which a series of standard steps are taken in order to progress eligibility for subsequent funding. This is not conducive to the alignment of future projects with the priorities of the country.

This overall negative review is somewhat offset by the quality of the individual projects. While the UNEP assessment did not extend to a project audit level, the reconnaissance work indicated that individual projects were often very well designed and managed. Many projects had very accurately identified several of the key issues and developed appropriate solutions. Two good examples of this were the programme for Dinder National Park managed by UNDP and HCENR, and the rehabilitation of community rangelands project managed by UNDP. Both have now been completed.

A further defining feature of the environmental aid sector over the last decade has been the very limited extent of government counterpart funding. In many projects, the funding has been 100 percent international, with no financial contribution by the government. This has resulted in aid-generated structural problems and a lack of government ownership and continuity.

The Khartoum-based secretariat of the GONU Higher Council for Environment and Natural Resources was originally conceived as a coordinating body. Now however, most of its funding and activities are focused on the implementation of MEA and GEF-funded projects. As such, it has essentially become an organization sustained by international aid in the form of a series of often unrelated convention projects. Most of the HCENR staff work on a contract basis, and return to academia upon project completion. As a result, there has been negligible capacity-building in the core civil service from these projects.

The lack of government ownership in the environmental sector is also evident in the lack of counterpart funding. In many cases, projects have been shut down when international aid has ceased, and Sudan now has a series of needs assessments, capacity assessments, status reports and management plans that have progressed to final document stage and no further.

This lack of government counterpart funding for environmental issues was relatively understandable in the war economy that prevailed for over two decades. Now however, Sudan should start to contribute significantly to this sector.

Analysis of the findings

In the 2006 Work Plan, environment was designated by the UN as one of four cross-cutting issues for special focus (the other three were HIV/AIDS, gender and capacity-building). UNEP was nominated as the UN focal point for environment, and this assessment is one of its initial activities in attempting to 'mainstream' or integrate environmental issues into the UN aid agenda in Sudan.

The assessment results are overall fairly negative, but not uniformly so, as a number of high quality projects and efforts were noted. Two core problems were identified. First, the impacts of good individual projects and efforts are greatly weakened by a lack of integration into the core government and international aid programmes. Second, the environment and natural resource management sector suffers from a lack of funding and funding continuity. Indeed, the five environment-specific programmes within the UN 2006 Work Plan had a combined budget of approximately USD 2.8 million, representing less than 0.2 percent of the UN country programme expenditure.

In order to direct corrective action, the underlying causes for these problems need to be understood. UNEP has identified the following five factors:

1. Humanitarian focus. Humanitarian responses are typically designed for fast mobilization in emergencies, which allows little time for integration of cross-cutting issues like the environment. Agencies engaged in humanitarian work have mandates and management procedures to focus

The dyke by the Bor-Padak rural trunk road was dug to control flooding in the region, but is now a cause for concern as it is leading the land beyond it to dry out and is thus reducing grazing land for both livestock and wildlife

on supply to beneficiaries without corresponding attention to management of the (natural) resources used for supply. This exacerbates the risk of environmental degradation.

Sudan is unusual in that the emergency has been ongoing for many years, but given that the humanitarian needs are not diminishing on an annual basis, the general approach has not changed. Long-term resident UN programmes are usually development-focused; in Sudan it is the opposite.

2. Lack of a resident agency focal point for the environment. The promotion of environmental issues is a subject at the margins of the mandates of many UN agencies, but only one agency – UNEP – has it as its core mandate. Historically, UNEP has not been present in the field on a residential basis. As a result, the topic of environment is in part orphaned and struggles to compete for attention and funding, given the plethora of other often very urgent issues facing the UN country team.

3. Managerial separation of the global and regional environmental programmes from the UN country programme. At present, the majority of the funding for environment in Sudan comes from the secretariats of the multilateral environmental conventions (MEAs) and the Global Environment Facility (GEF). A very small regional contribution comes directly from UNEP. None of these institutions currently have a residential presence in Sudan and are hence not answerable in any way to the UN country team (bar UNEP starting in 2006). UNDP is often tasked with administering convention and GEF projects, but does not have full discretion on allocation and management issues.

4. Lack of quantification and measurable results. In the general drive for aid effectiveness, it is important that needs and aid programme outputs be measured. This is very well established for the humanitarian sector (food tonnage delivered, number of wells installed etc.). In contrast, work in the environmental sector in Sudan has been largely qualitative. Needs and outputs have not always been clearly defined and stated in the context of the overall goals of the UN response. This tends to work against attracting and retaining aid investment.

5. Lack of high-level government buy-in. The lack of significant and high-level pressure on the UN from GONU regarding environmental issues indicates that the government has not been convinced of the scale and importance of the needs in this sector either.

14.5 Conclusions and recommendations

Conclusion

The assessment of the international aid programme in Sudan has raised a number of issues that need to be resolved to avoid inadvertently doing harm through the provision of aid, and to improve the effectiveness of aid expenditure in the environmental sector.

The dominant impact of aid on the environment in Sudan is the provision of food aid to some 15 percent of the population. Sudan is essentially now caught in a vicious circle of food aid dependence and environmental degradation: if food aid were reduced to encourage a return to agriculture, the result under current circumstances would be an intensification of land degradation, leading to the high likelihood of a return to food insecurity in the long term.

The analysis of the other links between international aid and the environment in Sudan indicates that most aid does not cause significant harm to the environment. However, integration of environmental issues into the current programme is negligible, and the environment-related expenditure that does occur – while it is acknowledged and welcomed – suffers from a range of management problems that reduce its effectiveness.

Background to the recommendations

Given the current environmental situation in Sudan, increased international aid for environmental issues is warranted. All other issues being equal, the level of food security in many parts of Sudan

will gradually drop and rural livelihoods will be increasingly threatened unless problems such as desertification and deforestation are tackled. This in turn will drive conflict, displacement, and further degradation, and as a result increase demands for humanitarian aid and peacekeeping.

At the same time as investment is increased, the effectiveness of all expenditure for environmental issues will need to be significantly improved through better coordination and other structural reforms.

The recommendations below are based on the themes of improved UN coordination and national ownership, which are two of the principles currently driving UN and aid reform in Africa and elsewhere. The majority of the programmes requiring investment are listed in other chapters; the financial investment in this chapter relates solely to coordination and UN agency assistance.

Recommendations for the United Nations in Sudan

R14.1 Implement a focal point and long-term, centralized environmental technical assistance service for aid agencies in Sudan. The long-term goal is the full integration of environmental issues into the UN aid programme in Sudan. This recommendation entails the establishment of UNEP offices in Khartoum and Juba, the provision of a service for environmental advice and rapid assessment for all agencies and NGOs, and a focal point to promote investment and coordination in environmental issues.

CA: TA; PB: UNCT; UNP: UNEP; CE: 3M; DU: 3 years

R14.2 Help mainstream environmental issues into the UN programme through improved structure and monitoring via the UN Work Plan. This would entail measures such as collating and including all ongoing environmental projects from all parties into the annual UN Work Plan process and elevating environment from a 'cross-cutting issue' to an investment sector or sub-sector.

CA: GROL; PB: UN RCHC; UNP: UNEP and UNDP; CE: nil; DU: ongoing

R14.3 Advise future international environmental aid proposals and funding offers to fit within a national management framework presented by the combination of the UN Work Plan, the UNEP assessment and the GONU and GOSS NPEM processes. This would not entail additional fund-raising, but only directing funds towards priority areas and projects as determined by these linked processes, which have already conducted the groundwork to develop a list of priorities and have a high level of ownership at the national level.

CA: GROL; PB: UN RCHC; UNP: UNEP and UNDP; CE: nil; DU: ongoing

R14.4 Set government counterpart funding as a key criterion for funding environmental projects in Sudan. The level of funding provided by the government partner is a litmus test for government commitment and the prospects for sustainable project benefits. The international: national funding ratio should in no case be greater than 4:1, and should ideally be 1:1 or less.

CA: GROL; PB: GONU and GOSS; UNP: UNEP; CE: nil; DU: 3 years then review

Recommendations for the Government of National Unity

R14.5 Officially designate and support the GONU Ministry of Environment and Physical Development as the GONU focal point for liaison for all international aid projects in the environmental sector that require a GONU government partner, including MEAs and GEF projects. This will significantly assist coordination and central planning. Once contact and a framework are established, liaison can be delegated to the appropriate level on a project-specific basis. This initiative needs to include capacity-building (see Chapter 13) to enable the government to participate actively in such projects.

CA: GROL; PB: MEPD; UNP: UNEP; CE: nil; DU: 3 years then review

Recommendations for the Government of Southern Sudan

R14.6 Officially nominate and support the GOSS Ministry of Environment, Wildlife Conservation and Tourism as the GOSS focal point for liaison for all international aid projects in the environmental sector that require a GOSS government partner, including GEF projects. This will significantly assist coordination and central planning. Once contact and a framework are established, liaison can be delegated to the appropriate level on a project-specific basis. This initiative needs to include capacity-building (see Chapter 13) to enable the government to actively participate in such projects.

CA: GROL; PB: MEWCT; UNP: UNEP; CE: nil; DU: 3 years then review

Conclusions

A fish eagle crossing the White Nile flood plain, against a backdrop of seasonal rangeland fires set by pastoralists. Sustainable management and development of natural resources is one of the greatest challenges facing post-conflict Sudan.

Conclusions

15.1 Introduction

The UNEP post-conflict environmental assessment of Sudan has made clear that Sudan is affected by a number of severe environmental issues, which are closely tied to the country's social and political problems with conflict, food insecurity and displacement.

Ignoring these environmental issues will ensure that some political and social problems remain unsolvable and even likely to worsen, as environmental degradation mounts at the same time as population increases. Resolving them will require a cross-cutting effort in the political arena.

Investment in the environmental sector has suffered greatly from the conflicts that have wracked Sudan for most of the last fifty years, and environmental concerns still cannot be adequately addressed in Darfur today. Corrective action, however, can start in much of the rest of the country. Moreover, thanks to the benefits of oil exports, Sudan can for the first time afford to significantly invest its own resources into such action.

Recommendations on each of the various cross-cutting issues and sectors have already been set out in Chapters 3 through 14. These have been viewed and vetted by the Governments of Sudan and other national and international stakeholders. As such, they represent an agreed way forward for each sector.

This chapter summarizes the findings and recommendations of the UNEP post-conflict environmental assessment, and proposes the general way forward for the Governments of Sudan, civil society and the international community, to help ensure that these recommendations are acted upon.

15.2 Key findings

Over 100 environment and governance issues are discussed in Chapters 3 through 14, many of which are closely connected or different aspects of the same problem. These items have been distilled into three positive and seven negative key findings:

Positive findings

1. **The oil-driven economic boom can fund the necessary investment in improved environmental governance.** The total cost of the recommendations listed in this report is approximately USD 120 million over three to five years. With oil exports expected to be in excess of USD 5 billion in 2006, the government clearly has the capacity to pay some if not all of these costs. On this basis, all future international aid projects for environmental governance should have a strong element of matching government funding.

2. **The combination of the natural resources of the south and the resource needs of the north represents a real opportunity for large-scale sustainable trade in raw and added-value natural resources.** Many of the resources of Southern Sudan could be used to drive economic development, but are currently being wasted. For example, Khartoum state imports construction timber even as mahogany trees are burnt to clear land for shifting agriculture in the southern states. While tight controls are obviously needed to avoid over-exploitation, extracting added value from the natural resources of the south is key to both economic development and conservation.

3. **Political support for the environment is strong in the newly formed Government of Southern Sudan, and rising in the Government of National Unity.** Support is both political (in terms of awareness-raising) and practical (in terms of allocating GONU and GOSS core budgets to tackling environmental governance and natural resource management issues).

Negative findings

4. **Environmental degradation in northern, central, eastern and western Sudan is widespread, severe and continuing at a linear rate.** The most common forms of degradation – desertification and deforestation – are long-term problems that may worsen in the future. The northern coastline and marine habitats have been locally damaged near urban areas, but remain in good condition overall.

5. **Environmental degradation in south Sudan is overall moderate but locally severe and generally increasing at a rapid pace.** Ongoing deforestation, which could worsen considerably in the coming years due to the massive refugee and IDP return process underway, represents a significant lost opportunity in sustainable development and economic growth.

6. **Southern Sudan's environment is highly vulnerable to development-induced damage in the post-conflict period.** Given the near complete absence of environmental governance, natural resources such as timber and the remaining wildlife are vulnerable to over-exploitation.

7. **Environmental degradation, as well as regional climate instability and change, are major underlying causes of food insecurity and conflict in Darfur – and potential catalysts for future conflict throughout central and eastern Sudan and other countries in the Sahel belt.** Setting aside all of the social and political aspects of the war in Darfur, the region is beset with a problematic combination of population growth, over-exploitation of resources and an apparent major long-term reduction in rainfall. As a result, much of northern and central Darfur is degraded to the extent that it cannot sustainably support its rural population.

Although not a novel finding to those working in this field in Darfur, it is not commonly understood outside the region. Yet it has major implications for the prospects for peace, recovery and rural development in Darfur and the Sahel. Indeed, the situation in Darfur is uniquely difficult, but many of the same underlying factors exist in other parts of Sudan and in other countries of the Sahel belt. Darfur accordingly holds grim lessons for other countries at risk, and highlights the imperative for change towards a more sustainable approach to rural development.

8. **Long-term peace in Sudan is at risk unless sustainable solutions are found for several environmental issues identified as potential conflict 'flashpoints' in Unity and Upper Nile states, the Three Areas and other north-south border zones.** In general order of priority, these unresolved issues are:

- the environmental impacts of the development of the oil industry;

- the southward migration of northern pastoralists due to land scarcity and degradation;

- tree-felling for the charcoal industry in the north-south boundary zone;

- new and planned dams and major water projects, including any revival of the Jonglei canal project; and

- ivory and bushmeat poaching.

An appreciation and long-term solutions for these environmental issues should be integrated into peacebuilding efforts to reinforce the prospects for sustainable peace.

9. **Environmental governance and policy failures underlie many of the problems observed.** Many of the issues identified cannot be resolved by more aid or investment, but require changes in government policy instead. This is particularly the case for agricultural development. In addition, the basics for good environmental governance are lacking or need substantial strengthening throughout the country. Areas necessitating attention include legislation development, civil service capacity-building and data collection.

10. **United Nations work in the field of environment and aid in Sudan could be much improved by increased efforts in coordination.** At present, environmental issues are not integrated into the larger UN humanitarian programmes, and numerous structural and management problems reduce the effectiveness of environment-specific programmes, such as those funded by the Global Environment Facility. Improved coordination could resolve many of these problems without significantly raising overall aid expenditure.

15.3 Key recommendations and investment requirements

Eighty-five detailed recommendations are provided in Chapters 3 through 14. These have been distilled into four general recommendations:

1. **Invest in environmental management to support lasting peace in Darfur, and to avoid local conflict over natural resources elsewhere in Sudan**. Because environmental degradation and resource scarcity are among the root causes of the current conflict in Darfur, practical measures to alleviate such problems should be considered vital tools for conflict prevention and peacebuilding. Climate change adaptation measures and ecologically sustainable rural development are needed in Darfur and elsewhere to cope with changing environmental conditions and to avoid clashes over declining natural resources.

2. **Build capacity at all levels of government and improve legislation to ensure that reconstruction and economic development do not intensify environmental pressures and threaten the livelihoods of present and future generations.** The new governance context provides a rare opportunity to truly embed the principles of sustainable development and best practices in environmental management into the governance architecture in Sudan.

3. **National and regional governments should assume increasing responsibility for investment in the environment and sustainable development.** The injection of oil revenue has greatly improved the financial resources of both the Government of National Unity and the Government of Southern Sudan, enabling them to translate reform into action.

4. **All UN relief and development projects in Sudan should integrate environmental considerations in order to improve the effectiveness of the UN country programme.** Better coordination and environmental mainstreaming are necessary to ensure that international assistance 'does no harm' to Sudan's environment.

Analysis of chapter recommendations

The recommendations from each chapter have been collated by issue and economic sector in Table 31, and by theme in Table 32.

Table 31. Recommendations by economic sector and geographic region

Issue and economic sector	No.	Cost of recommendation by region/target (USD million)			
		National (including Darfur)	Southern Sudan	International Community	Total
Natural disasters and desertification	3	4.0	–	–	4.0
Conflict	4	–	–	2.9	2.9
Displacement	4	–	–	5.3	5.3
Urban environment and environmental health	6	5.0	2.0	1.0	8.0
Industry	5	2.9	1.0	–	3.9
Agriculture	8	14.6	9.2	–	24.0
Forestry	13	10.6	7.8	0.3	18.7
Water resources	9	11.6	2.0	–	13.6
Wildlife and protected area management	5	3.5	6.0	–	9.5
Marine and coastal resources	8	9.1*	–	–	9.1
Environmental governance and awareness	10	8.7	9.3	–	18.0
International aid and the environment	6	–	–	3.0	3.0
Total	**85**	**70**	**37.3**	**12.5**	**119.8**

*Includes USD 0.7 million by Red Sea state

Table 32. Recommendations by theme and region/target

Recommendation theme	Costs of recommendation by region (USD million)			
	National (including Darfur)	Southern Sudan	International Community	Total
Governance	9.1	6.5	0.3	15.9
Technical assistance	13.0	6.0	6.5	25.5
Capacity-building	7.0	12.0	–	19.0
Government investment	25.1	–	–	25.1
Awareness-raising	0.2	0.1	0.5	0.8
Assessment	9.6	0.7	1.2	11.5
Practical action	6.0	12.0	4.0	22.0
Totals	**70**	**37.3**	**12.5**	**119.8**

Cost of the recommendations

Depending on the approach, the cost of a list of recommendations for the substantial resolution of the major environmental issues in Sudan could run from millions to billions of US dollars. In the context of the competing needs of post-conflict recovery and the ongoing Darfur crisis, it is at present clearly unrealistic to expect such additional expenditure. However, it is critical that expenditure be raised from its current negligible level to one at which a real difference can be made (and measured). Accordingly, the costed recommendations are kept below USD 5 million per government, per sector, and per annum – and address only the most urgent or logical first few items.

The resolution of many of the issues raised will also require considerable time. UNEP estimates that building national capacity and addressing some of the more complex policy, legal and political issues noted in this report will take a minimum of three to five years. Reversing the noted trends of environmental degradation could take much longer.

UNEP does not expect work on all of the listed recommendations to commence in 2007; some indeed may never be taken up. Moreover, the costs listed are only basic estimates that will need to be refined in the project development stage. However, they provide a good indication of the scale of investment required to make a significant difference to the current environmental situation and trends in the country.

It should be noted that in addition to the expenditure discussed above, a major investment in environmental health infrastructure (water supply and treatment, sewage treatment etc.) is unavoidable if GONU and GOSS wish to achieve major improvements in the health sector. In this area, 'soft' approaches like awareness-raising and capacity-building will be of limited benefit in the absence of 'hard' improvements in water supply and sanitation infrastructure.

The total cost of this report's recommendations is estimated at approximately USD 120 million over three to five years: USD 70 million for GONU, USD 37.3 million for GOSS and USD 12.5 million for the international community. These are not large figures compared to the Sudanese GDP in 2005 (USD 85.5 billion), and are hence considered to be relatively affordable for both GONU and GOSS. The recommendations aimed specifically at the international community come to approximately 0.5 percent of annual aid expenditure for Sudan in 2006 – again relatively affordable.

Financing the recommendations

The UNEP proposal is that the Government of National Unity and the Government of Southern Sudan own this list of sector recommendations and contribute the majority of the funds. International aid should make up the difference on a partnership basis, with a view to providing technical assistance and capacity-building rather than just funding. As mentioned in the previous chapter, sole funding by the international aid community is specifically not recommended for three reasons:

1. Prior experience in Sudan and elsewhere has shown that one hundred percent aid-funded recovery and development projects often have

poor sustainability and collapse when donor funds are withdrawn. Part-financing by the government typically results in much better design and national ownership;

2. International aid funding for Sudan has its limits, and urgent humanitarian needs will continue to draw the bulk of the available funds. It will simply not be possible to raise all the required finances from international donors; and

3. Many of the recommendations focus on policy and governance, so the direct costs are limited and internal to government civil services.

Some sectors such as industry, urban development and forestry have a high potential for part-financing by the private sector, but any revenue-generating option, such as license fees and royalty agreements, should be designed and introduced with care to avoid governance problems.

15.4 The way forward

Establishing roles and responsibilities in GONU, GOSS and the UN

UNEP's recommendations envisage a key role for several government ministries within GONU and GOSS, as well as for over ten different UN agencies. Their wholehearted support is required for the implementation of many recommendations.

UNEP and its government counterparts in the GONU and GOSS environment ministries cannot play the roles of the other parties, as they do not have the mandate or the capacity to do so. They can, however, catalyse action from their counterparts to pick up the recommendations and assist them throughout the process. The first stage in the implementation of the recommendations has in fact already occurred, as the respective ministries and UN agencies were asked for their views and support in the report drafting process. The recommendations in this final report reflect that input.

UNEP proposes to maintain a central role through the establishment of a Sudan country programme for the period of at least 2007-2009 (funds permitting). For each recommendation listed, UNEP will have one of three positions:

- a central role as the lead UN agency or one of a small joint agency team;

- a catalysing and supporting role to other UN agencies; or

- a tracking role for recommendations that do not require substantive UN input.

On the government side, the environment and wildlife ministries and authorities will also need to determine their specific role for each recommendation, and engage the appropriate line ministries if required.

UNEP country programme

The UNEP Sudan country programme is still under development as of early 2007, but an outline can be presented.

Funds permitting, UNEP will establish more permanent project offices in Khartoum and Juba, to implement a core programme for the period 2007-2009. In 2009, the possibility of an extension will be reviewed against a set of exit criteria based on the situation in the country and progress on addressing the environmental issues listed in this report. Key themes for the UNEP programme are anticipated to be the same as the recommendation themes:

- governance (with a focus on legislation development);

- technical assistance and capacity-building;

- awareness-raising and advocacy;

- assessment; and

- practical action.

The exception is the recommendation category of government investment, as this is considered to be a role for the GONU and GOSS only.

Advocacy, and awareness- and fund-raising

The funding and political support required to implement the recommendations will need to be found through an organized process of advocacy and awareness-raising. This effort will by default be led in the first instance by UNEP and its government counterparts in GONU and GOSS.

UNEP has developed a range of assessment products to assist this process and will lead fund-raising within the international community. The government counterparts will direct fund-raising within their respective governments, using normal annual budgetary mechanisms and all other avenues for extra-budgetary funding. The existing National Plan for Environmental Management (NPEM) process could be utilized to this end by the GONU Ministry of Environment and Physical Development.

It is anticipated that awareness- and fund-raising will take a minimum of one year to complete substantially. Some projects will start much sooner than this, but major items, such as line ministry policy shifts and infrastructure investments, will probably require one to three years.

Development of national, regional and sectoral plans and action programmes

Once the agreed partners are on board and funds have been allocated, the recommendations list can be converted into a number of national, regional, sectoral and project plans for implementation. Wherever possible, these plans should be integrated into general development and poverty reduction strategies rather than be stand-alone initiatives.

In the water sector, for example, individual states have the responsibility to develop five-year State Water Master Plans; this represents an ideal opportunity to mainstream environment and sustainability issues into concrete policy and investment programmes at the intermediate level. At the international level, UNEP will be working to integrate environmental issues into the UN Development Assistance Framework (UNDAF) process, planned for late 2007, and the joint government-UN Poverty Reduction Strategy Papers (PRSPs).

Annual and three-year progress review

This UNEP assessment project has been a major and relatively costly undertaking. Its first phase has now been successfully completed. The real test, however, will be the rate of implementation of its recommendations, which will only be possible to accurately evaluate some time after the public launch of the report and other assessment products.

It is therefore recommended that UNEP and partners conduct an evaluation of the status of the recommendations at the end of 2009. Interim assessments should be conducted on an annual basis, starting in December 2007.

15.5 Concluding remarks

Sudan is now at a crossroads. While the country clearly faces many severe environmental challenges, the combination of the 2005 Comprehensive Peace Agreement and the oil-driven economic boom represents a major opportunity for positive change.

The sustainable management of the country's natural resources is part of the solution for achieving social stability, sustainable livelihoods and development in the country. For this goal to be reached, however, it will be necessary to deeply embed a comprehensive understanding of environmental issues in the culture, policies, plans and programmes of the Government of Sudan and its international partners, such as the United Nations.

This will require a long-term process and a multi-year commitment from both the Government of Sudan and its international partners. As the environmental expert of the United Nations, UNEP is ready to assist the Government and people of Sudan, as well as their international partners, in taking forward the recommendations developed from this assessment.

Appendices

Appendix I
List of acronyms and abbreviations

AMCEN	African Ministerial Conference on the Environment
AMIS	African Union Mission in Sudan
BOD	Biological Oxygen Demand
°C	Degrees Celsius
CAR	Central African Republic
CBD	Convention on Biological Diversity
CITES	Convention on International Trade in Endangered Species of Wild Fauna and Flora
CPA	Comprehensive Peace Agreement
DEA	Department of Environmental Affairs (GONU MEPD)
DFID	Department for International Development (UK)
DPA	Darfur Peace Agreement
DRC	Democratic Republic of Congo
DSS	Department of Safety and Security (UN)
EC	European Commission
EIA	Environmental Impact Assessment
ERW	Explosive Remnants of War
ESPA	Eastern Sudan Peace Agreement
FAO	Food and Agriculture Organization of the United Nations
FNC	Forests National Corporation
FRA	Forest Resources Assessment
FSD	Fondation Suisse de Déminage
GDP	Gross Domestic Product
GEF	Global Environment Facility
GNP	Gross National Product
GONU	Government of National Unity
GOS	Government of Sudan
GOSS	Government of Southern Sudan
GRASP	Great Apes Survival Project
GRID	Global Resource Information Database (UNEP)
GTZ	Deutsche Gesellschaft für Technische Zusammenarbeit (German Technical Cooperation)
HCE	Higher Council for Environment
HCENR	Higher Council for Environment and Natural Resources
IAEA	International Atomic Energy Agency
ICRAF	International Centre for Research in Agroforestry
ICZM	Integrated Coastal Zone Management
IDP	Internally Displaced Person
IGAD	Inter-government Authority on Drought
INGO	International Non-Governmental Organization
IOM	International Organization for Migration
IUCN	The World Conservation Union
IWRM	Integrated Water Resource Management
JEM	Justice and Equality Movement
km	Kilometre (measurement)
km²	Kilometres squared (area)
km³	Kilometres cubed (volume)
LPG	Liquefied Petroleum Gas
LRA	Lord's Resistance Army
m	Metre (measurement)
m²	Metres squared (area)
m³	Metres cubed (volume)
MAF	Ministry of Agriculture and Forestry (GONU/GOSS)
MAR	Ministry of Animal Resources (GONU)
MARF	Ministry of Animal Resources and Fisheries (GOSS)
MDG	Millennium Development Goal
MEA	Multilateral Environmental Agreement
MEPD	Ministry of Environment and Physical Development (GONU)
MEWCT	Ministry of Environment, Wildlife Conservation and Tourism (GOSS)
MFA	Marine Fisheries Administration (GONU)
MI	Ministry of Interior (GONU)
MIM	Ministry of Industry and Mining (GOSS)

MIWR	Ministry of Irrigation and Water Resources (GONU)
MEM	Ministry of Energy and Mining (GONU)
MEPA	Marine Environment Protection Authority (Red Sea state)
MoF	Ministry of Finance (GONU)
MoI	Ministry of Industry (GONU)
MOSS	Minimum Operating Security Standard
MOU	Memorandum of Understanding
MPA	Marine Protected Area
MTR	Ministry of Transport and Roads (GOSS)
MTW	Ministry of Tourism and Wildlife (GONU)
MWRI	Ministry of Water Resources and Irrigation (GOSS)
NAPA	National Adaptation Programme of Action
NBI	Nile Basin Initiative
NCP	National Congress Party
NDVI	Normalized Difference Vegetative Index
NCSA	National Capacity Self-Assessment
NEPAD	New Partnership for Africa's Development
NFI	Non-Food Item
NGO	Non-Governmental Organization
NPEM	National Plan for Environmental Management
NSAS	Nubian Sandstone Aquifer System
NSWCO	New Sudan Wildlife Conservation Organization
NTEAP	Nile Transboundary Environment Action Project
NWA	Nile Water Agreement
OCHA	United Nations Office for the Coordination of Humanitarian Affairs
OHCHR	Office of the United Nations High Commissioner for Human Rights
OLS	Operation Lifeline Sudan
PCDMB	Post-Conflict and Disaster Management Branch
PCEA	Post-Conflict Environmental Assessment
PERSGA	Regional Organization for the Conservation of the Environment of the Red Sea and the Gulf of Aden
POPs	Persistent Organic Pollutants
PPD	Plant Protection Directorate (GONU MAF)
ppm	Parts per Million
PRSPs	Poverty Reduction Strategy Papers
SCE	State Council for Environment (Red Sea state)
SECS	Sudanese Environment Conservation Society
SPLA	Sudan People's Liberation Army
SPLM	Sudan People's Liberation Movement
SSARP	Southern Sudan Agricultural Revitalization Programme
SSCSE	South Sudan Centre for Statistics and Evaluation
SSNEA	South Sudan National Environment Association
UN	United Nations
UNCCD	United Nations Convention to Combat Desertification
UNCT	United Nations Country Team
UNCTAD	United Nations Conference on Trade and Development
UNDAF	United Nations Development Assistance Framework
UNDG	United Nations Development Group
UNDP	United Nations Development Programme
UNDPKO	United Nations Department of Peacekeeping Operations
UNEP	United Nations Environment Programme
UNESCO	United Nations Educational, Scientific and Cultural Organization
UNFCCC	United Nations Framework Convention on Climate Change
UNFPA	United Nations Population Fund
UNHCR	United Nations High Commissioner for Refugees
UNICEF	United Nations Children's Fund
UNIDO	United Nations Industrial Development Organization
UNMAS	United Nations Emergency Mine Action Programme in Sudan
UNMIS	United Nations Mission in Sudan
UNOPS	United Nations Office for Project Services
UNRCHC	United Nations Resident and Humanitarian Coordinator
USAID	United States Agency for International Development
UXO	Unexploded Ordnance
WFP	World Food Programme
WHC	UNESCO World Heritage Convention
WHO	World Health Organization
WUA	Water Use Associations

Appendix II
List of references

Chapter 2: Country Context

1. *United Nations and Partners Work Plan for Sudan 2007*

2. *United Nations and Partners Work Plan for Sudan 2006*

3. The Economist Intelligence Unit (2006). *Country Report: Sudan (December 2006)*, London: EIU

4. The World Bank 2007 Sudan Country Data Profile (2003 data)
    http://devdata.worldbank.org/external/CPProfile.asp?CCODE=SDN&PTYPE=CP

5. Khartoum Department of Statistics (1993). *1993 Population Census* [in Arabic]

6. CBS/UNFPA 2004 Population Data Sheet

7. South Sudan Centre for Statistics and Evaluation (2004)

8. Sulaiman, S. and A. Ahmed (2006). *Urban Environmental Issues in Khartoum.* Sudanese Environment Conservation Society Report to UNEP

9. Government of Sudan and United Nations Country Team (2004). *Sudan Millennium Development Goals: Interim Unified Report 2004*

10. The World Bank (2003). *Sudan Development Outcomes and Pro-poor Reforms*

11. UNDP (2006). *The Human Development Report 2006*

12. Chapin Metz, H. (1991). *Sudan: A Country Study*. Washington, DC: Library of Congress

13. The International Monetary Fund World Economic Outlook Database 2007
    http://www.imf.org/external/pubs/ft/weo/2007/01/data/index.aspx

14. The World Bank (2001). *Country Economic Memorandum*

15. FAO Aquastat Information System on Water and Agriculture. Sudan Country Profile (2005)
    http://www.fao.org/ag/agl/aglw/aquastat/countries/sudan/index.stm

16. Harrison M.N. and J.K. Jackson (1958). 'Ecological Classification of the Vegetation of the Sudan' in *Forest Bulletin* 2

17. Bashir, M. et al (2001). *Sudan Country Study on Biodiversity*. Khartoum: Ministry of Environment and Tourism

18. FAO Multipurpose Africover Databases on Environmental Resources
    http://www.africover.org/

19. Itto, A. (2001). *Agriculture and Natural Resources of New Sudan. A Report to the SPLM/A*

20. *United Nations Sudan Joint Assessment Mission Report 2005*

Chapter 3: Natural Disasters and Desertification

1. Government of Sudan (2003). Ministry of Environment and Physical Development. *Sudan's First National Communications under the UN Framework Convention on Climate Change*

2. Zeng, N. (2003). 'Drought in the Sahel' in Science, November 2003

3. Sudan Rainfall Station Data Tables, provided by the Sudan Meteorological Department

4. Royal Netherlands Meteorological Institute. Climate Change in West Africa: Eastern Sahel
    http://www.knmi.nl/africa_scenarios/West_Africa/region8/

5. Held I. M et al. (2005). *Simulation of Sahel Drought in the 20th and 21st Century* (Proceedings of the National Academy of Sciences Vol. 102, No. 50)

6. Thornton P.K. et al (2006). *Mapping Climate Vulnerability and Poverty in Africa.* Report to International Livestock Research Institute

7. Government of Sudan (undated). *National Plan for Combating Desertification in the Republic of Sudan*

8. Stebbing E.P. (1953). *The Creeping Desert in the Sudan and Elsewhere in Africa.* Khartoum: McCorquodale & Co

9. Communication with the Northern State Ministry of Water and Irrigation, August 2006

Chapter 4: Conflict and the Environment

1. Johnson, Douglas H. (2003). *The Root Causes of Sudan's Civil Wars.* Bloomington: Indiana University Press

2. Deng, Francis (1995). *War of Visions: Conflict of Identities in Sudan.* Washington, DC: Brookings Institution

3. Wadi, A.I. (1998). *Perspectives on Tribal Conflicts in the Sudan.* University of Khartoum: IAAS

4. *United Nations Sudan Joint Assessment Mission Report 2005*

5. Bashar, Z.M. (2003). *Mechanisms for Peaceful Co-Existence among Tribal Groups in Darfur* [MA Thesis in Arabic]. University of Khartoum

6. *United Nations and Partners Work Plan for Sudan 2006*

7. Christian Aid (2001). *The Scorched Earth: Oil and Water in Sudan*

8. Alier, Abel (1990). *Southern Sudan: Too Many Agreements Dishonoured.* Exeter: Ithaca Press

9. An Naim, Abdullahi and Peter Kok (1991). *Fundamentalism and Militarism: A Report on the Root Causes of Human Rights Violations in the Sudan.* New York: The Fund for Peace

10. Beck, Kent (1996). 'Nomads of Northern Kordofan and the State: From Violence to Pacification' in *Nomadic Peoples* 38

11. Braukamper, Ulrich (2000). 'Management of Conflicts over Pastures and Fields among the Baggara Arabs of the Sudan Belt' in *Nomadic Peoples* Volume 4

12. Lind, J. and K. Sturman (2002). *Scarcity and Surfeit: The Ecology of Africa's Conflicts.* African Centre for Technology Studies and Institute for Security Studies

13. Diehl, P.F. and N.P. Gleditsch (eds.) (2001). *Environmental Conflict.* Boulder: Westview Press

14. Suliman, M. (ed.) (1999). *Ecology, Politics and Violent Conflict.* London: Zed Books

15. Homer-Dixon, T.F. (1999). *Environment, Scarcity and Violence.* Princeton: Princeton University Press

16. Fadul, A.A. (2004). 'Natural Resources Management for Sustainable Peace in Darfur' in *Conference Proceedings: Environmental Degradation as a Cause of Conflict in Darfur* (December 2004)

17. GONU Ministry of Animal Resources Published Statistics, 2005

18. UNEP/FAO/ICRAF (2006). *Post-Conflict Environmental Assessment of the Rural Environment for Sudan* (Draft). Nairobi, Kenya

19. Feinstein International Famine Centre (2005). *Darfur: Livelihoods under Siege.* Cambridge: Tufts University

20. University for Peace (2004). *Conference Proceedings: Environmental Degradation as a Cause of Conflict in Darfur* (December 2004)

21. African Union Mission in Sudan
http://www.amis-sudan.org/

22. Tobiolo, M. L. et al. (2006). *A Report on the Status of the Forest Reserves in Greater Yei County, South Sudan.* Kagelu Forestry Training Centre

23. UN Information Gateway on Sudan
www.unsudanig.org/

24. *USAID Sudan Strategy Paper 2006-2008*
http://www.usaid.gov/locations/sub-saharan_africa/countries/sudan/docs/sudan_strategy.pdf

Chapter 5: Population Displacement and the Environment

1. UNHCR (2006). *2005 Global Refugee Trends*

2. UNHCR (1994). *Populations of Concern to UNHCR: A Statistical Overview*

3. UNHCR (2001). *Statistical Yearbook 2001*

4. UNHCR (2006). *Sudan Country Operations Plan*

5. Norwegian Refugee Council Global IDP Project (2005). *Profile of Internal Displacement: Sudan* (29 October 2005)

6. Internal Displacement Monitoring Centre (2006). *Internal Displacement: Global Overview of Trends and Developments in 2005*

7. *United Nations and Partners Work Plan for Sudan 2005*

8. Kelly, C. (2004). *Summary Report: Darfur Rapid Environmental Assessment.* CARE International and Benfield Hazard Research Centre

9. Women's Commission on Refugee Women and Children (2006). *Beyond Firewood: Fuel Alternatives and Protection Strategies for Displaced Women and Girls*

10. BBC Press Release (21/1/01): 'New Appeal for Drought-hit Sudan'
http://news.bbc.co.uk/2/hi/africa/1129567.stm

11. International Committee for the Red Cross (2004). 'Darfur's Turbulent Times' in *The Red Cross Magazine*, December 2004

12. Malik, S. (2005). 'Sustainable Return Depends on Collaborative Approach' in *Forced Migration Review* 24. Oxford: Refugee Studies Centre

13. McCallum, J. and Willow, G. Y. (2005). 'Challenges Facing Returnees in Sudan' in *Forced Migration Review* 24. Oxford: Refugee Studies Centre

14. OCHA Press Release (23/2/01): 'Under-Secretary-General for Humanitarian Affairs Warns of Sudan Disaster: 600,000 People at Immediate Risk of Starvation'
http://www.reliefweb.int/rw/rwb.nsf/AllDocsByUNID/bf9697ed24f039d2852569fc005efb08

15. OCHA Press Release (17/8/06): 'Sudan: the UN Expresses Concerns at Forced Relocation of 12,000 Displaced People in Greater Khartoum'
http://www.reliefweb.int/rw/RWB.NSF/db900SID/EGUA-6SRMPX?OpenDocument

16. UNHCR (2006). *The State of the World's Refugees: Human Displacement in the New Millennium.* Oxford: Oxford University Press

17. UNHCR (2006). *Global Appeal 2006*

18. Tearfund (2007). *Darfur: Relief in a Vulnerable Environment*

Chapter 6: Urban Environment and Environmental Health

1. Khartoum Department of Statistics (1993). *1993 Population Census* [in Arabic]

2. El Amin Abdel El Rahman, M. and M. Osman El Sammani (2006). *Natural Resources and Socio-Economic Parameters* (Workshop on the Post-Conflict National Plan for Environmental Management in Sudan, Khartoum, July 2006)

3. Creative Associates International, Inc. (2005). *Juba Assessment: Town Planning and Administration* (Draft)

4. Ahmed, A.E.M. (1998). *The Prevailing Situation of Urban Housing in Sudan* [in Arabic]. Conference on Housing in the Arab World (Khartoum, Sudan)

5. Sulaiman, S. and A. Ahmed (2006). *Urban Environmental Issues in Khartoum.* Sudanese Environment Conservation Society Report to UNEP

6. Burhan Eltayeb Bushra Elghazali (2006). *Urban Intensification in Metropolitan Khartoum: Influential Factors, Benefits and Applicability* [Doctoral Thesis]. Stockholm: Royal Institute of Technology

7. Khartoum Department of Statistics (1993). *1993 Population Census* [in Arabic]

8. *United Nations and Partners Work Plan for Sudan 2007*

9. *United Nations and Partners Work Plan for Sudan 2005*

10. Government of Sudan and United Nations Country Team (2004). *Sudan Millennium Development Goals: Interim Unified Report 2004*

11. United Nations Sudan Joint Assessment Mission (2005). *Volume III Cluster Report*

12. Nile Basin Initiative (2005). *National Nile Basin Water Quality Monitoring Baseline Report for Sudan*

13. World Health Organization (2004). *Water, Sanitation and Hygiene Links to Health*

14. World Health Organization Sudan (2006). 'Cholera' in Sudan Update 4 http://www.who.int/csr/don/2006_06_21a/en/index.html

15. World Health Organization verbal report of a Juba UN Country Team Meeting, May 2006

16. Abdelgani M.E. and Z.E. Alabjar (2006). *Environmental Research Capacity in Sudan* (Workshop on the Post-Conflict National Plan for Environmental Management in Sudan, Khartoum, July 2006)

17. Norwegian Refugee Council Global IDP Project (2005). *Profile of Internal Displacement: Sudan* (29 October 2005)

18. Al Adam, E. et al (2001). *Compressed Stabilized Earth Block Manufacture in Sudan.* Paris: UNESCO

19. Al Adam, E. et al (2006). *Urban Environment: Low-Cost Buildings* (Workshop on the Post-Conflict National Plan for Environmental Management in Sudan, Khartoum, July 2006)

Chapter 7: Industry and the Environment

1. Sudan Update (2001). *Raising the Stakes: Oil and Conflict in Sudan* http://www.sudanupdate.org/REPORTS/Oil/Oil.pdf

2. PennWell Petroleum Group (2001 and 2006 editions). *International Petroleum Encyclopedia*

3. Sudan Oil and Gas Conference, London, November 2006 http://www.sudandevelopmentprogram.org/sp/events/oilgas.htm

4. Reuters Press Release (14/8/06): 'Sudan to Ship 400,000 Barrels of Crude'

5. US Energy Information Administration. *Annual Energy Outlook 2006* http://www.eia.doe.gov/oiaf/aeo/index.html

6. The Economist Intelligence Unit (2006). *Country Report: Sudan (December 2006)*, London: EIU

7. United Nations Joint Logistics Centre (2006). *Fuels 2006: A Survey of the Humanitarian Fuels Situation in the Context of Humanitarian and Peacekeeping Operations in the Republic of the Sudan*

8. US Energy Information Administration. *World Proved Reserves of Oil and Natural Gas, Most Recent Estimates* http://www.eia.doe.gov/emeu/international/reserves.html

9. 'Oil Flow Starts at Sudan's Thar Jath Field' in *Oil and Gas Journal*, June 29 2006

10. 'Sudan - Oil and Gas: Crude Petroleum and Natural Gas Extraction'
    http://www.mbendi.co.za/indy/oilg/ogus/af/su/p0005.htm

11. Sudan News Archive. Gulf Oil and Gas 2007
    http://www.gulfoilandgas.com/webpro1/Main/NewsCTRY.asp?nid=SD

12. Switzer, J. (2002). *Oil and Violence in Sudan*. International Institute for Sustainable Development

13. International Crisis Group (2002). *God, War and Oil: Changing the Logic of War in Sudan*

14. Gagnan, G. and J. Ryle (2001). *Report of an Investigation in Oil Development, Conflict and Displacement in Western Upper Nile, Sudan*

15. European Coalition on Oil in Sudan (2006). *Oil Development in Upper Nile*

16. *Sudan National Oil Spill Response Contingency Plan 2000-2004*
    http://www.persga.org/Publications/Technical/pdf/4%20Technical%20Series/TS6%20NOSCP%20Sudan%20Part%20II%20(Eng).pdf

17. Human Rights Watch (2003). *Sudan, Oil, and Human Rights*
    http://www.hrw.org/reports/2003/sudan1103/

Chapter 8: Agriculture and the Environment

1. FAO Aquastat Information System on Water and Agriculture. Sudan Country Profile (2005)
   http://www.fao.org/ag/agl/aglw/aquastat/countries/sudan/index.stm

2. Chapin Metz, H. (1991). *Sudan: A Country Study*. Washington, DC: Library of Congress

3. Government of Sudan (undated). *Sudan National Report to the Conference of Parties on the Implementation of the United Nations Convention to Combat Desertification*

4. Abbadi, K and A. Ahmed (2006). *A Brief Overview of Sudan's Economy and Future Prospects for Agricultural Development*. Khartoum Food Aid Forum (WFP)

5. Mohammed, H. and Hamid (2006). *Range Management and Conservation in Sudan* (Workshop on the Post-Conflict National Plan for Environmental Management in Sudan, Khartoum, July 2006)

6. Agaemi, O. (undated). *Towards a State Environmental Action Plan* (for Gedaref). Gedaref: Gedaref Ministry of Environment and Tourism

7. UNEP (2002.) *Global Environment Outlook 3* (GEO-3)

8. El Faki, A. (undated). *The Problem of Land Use: The Issue and Future Vision* [in Arabic]. Gedaref: Gedaref State Ministry of Agriculture, Animal Resources and Irrigation

9. Abdel Nour, H. 'Gum Arabic in Sudan: Production and Socio-Economic Aspects' in *Medicinal, Culinary and Aromatic Plants in the Near East* (FAO - Proceedings of the International Expert Meeting, Cairo)

10. Harragin, S. (2003). *Nuba Mountains Land and Natural Resources Study* (Part I – Land Study). USAID/UNDP-NMPACT

11. Manger, F. *The Issue of Land in the Nuba Mountains*. UNDP-NMPACT

12. Gezira State Board (undated). *An Introduction to the Gezira Scheme* [Arabic]

13. Hindi, A. et al. (2003). *Management of Public Health Pesticides in Sudan* (Inter-Country Workshop on Public Health Pesticides Management in the Context of the Stockholm Convention of Persistent Organic Pollutants (POPs), Amman, Jordan)

14. Sirag, A. et al. (2004). 'Identification and Concentration of Organochlorine Residues in the Blood of Sudanese Workers at the Gezira Agricultural Scheme' in *International Research on Food Security, Natural Resource Management and Rural Development* (Peter, K. ed.). Berlin: Humboldt-Universität zu Berlin

15. KSC (2006). *Kenana Corporate Environment Strategy*

16. FAO (1995). *Prevention and Disposal of Obsolete and Unwanted Pesticide Stocks in Africa and the Near East*

17. Dukeen, M. et al. (2006). *Vector Control Situation within the Context of Sectoral Coordination in Sudan* (The First Regional Meeting of the Global Environmental Facility-supported countries in the Eastern Mediterranean Region, Muscat, Oman)

18. Gedaref SCENR (undated). *Report of the Environment Secretariat on the Pesticide Warehouse in the City of El Fao* [in Arabic]

19. Northern State Ministry of Agriculture, Animal Resources and Irrigation (undated). *Northern State* [in Arabic]

20. Lahmeyer International (2002). *Environmental Assessment Report for Merowe Dam Project*. Khartoum: Merowe Dam Project Implementation Unit

21. FAO Multipurpose Africover Databases on Environmental Resources http://www.africover.org/

22. UNEP (1998). Mutsambiwa, F. and F. Ali. *Community Forestry Project, Ed Debba, Sudan*

23. UNEP/FAO/ICRAF (2006). *Post-Conflict Environmental Assessment of the Rural Environment for Sudan* (Draft). Nairobi, Kenya

24. UNICEF (2003). *Analysis of Nine Conflicts in Sudan*

25. Zaroug, M. (2000, updated 2002). *FAO Sudan Pasture/Forage Resource Profile*

26. Elsiddig, E. (2002). *Developments in Forestry Education in the Sudan* (VITRI Inauguration Workshop on Tropical Dryland Rehabilitation June 2002, University of Helsinki, Hyytiälä Forestry Field Station). http://www.mm.helsinki.fi/mmeko/VITRI/research/workshops/abdalla.htm

27. Suttie, J. et al (2005). *Grasslands of the World*. Rome: FAO

28. International Fund for Agricultural Development (2004). *Report and Recommendation of the President for the Western Sudan Resources Management Programme*

29. Europa (1964). *The Middle East and North Africa 1964-65*. London: Europa Publications Limited

30. HCENR/WCGA/UNDP (2004). *Management Plan for Dinder National Park*

31. Sudan Ministry of Environment and Tourism/HCENR/SECS/Friedrich Ebert Foundation (1996). *Toward a National Environmental Action Plan for Sudan* [in Arabic]

32. Wallach, B. (2004 revised). 'Improving Traditional Grassland Agriculture in Sudan' in *Geographical Review*

33. Gedaref State Ministry of Agriculture, Animal Resources and Irrigation (undated). *Gedaref State Land Uses* [in Arabic]

34. Northern State Ministry of Agriculture, Animal Resources and Irrigation (undated). *State Preparations for the Winter Season 2006-2007* [in Arabic]

35. Ayoub, A. *Linkages between food security and natural resources conditions*. (Workshop on the Post-Conflict National Plan for Environmental Management in Sudan, Khartoum, July 2006).

36. Government of Sudan (undated). *National Plan for Combating Desertification in the Republic of Sudan* [in Arabic]

37. International Fund for Agricultural Development (2003). *Gash Sustainable Livelihoods Regeneration Project*

38. International Fund for Agricultural Development (2004). *Local Governance to Secure Access to Land and Water in the Lower Gash Watershed*

39. Bashar, K. et al. (2005). *Watershed Erosion and Sediment Transport.* Nile Basin Capacity-Building Network, Khartoum

40. El Rahman, M. and M. El Sammani (2006). *Natural Resources and Socio-Economic Parameters* (Workshop on the Post-Conflict National Plan for Environmental Management in Sudan, Khartoum, July 2006).

41. Musnad, H. and N. Nasr (2004). *Experience-Sharing Tour and Workshop on Shelterbelts and Fuelwood Substitutes in Sudan.* NORAD/NORAGRIC

42. McNeil, M. (1972). 'Lateritic Soils in Distinct Tropical Environments: Southern Sudan and Brazil' in *The Careless Technology: Ecology and International Development* (Farvar, T. and J. Milton, eds.). Garden City: The Natural History Press

Chapter 9: Forest Resources

1. FAO Forestry Department. *Sudan Country Profile*
   http://www.fao.org/forestry/site/countryinfo/en/

2. Gaafar Mohamed, A. (2005). 'Improvement of Traditional Acacia Senegal Agroforestry' in *Tropical Forestry Reports* 26

3. El Taib, A. A. and C. Holding (1988). *Forestry and the Development of a National Forestry Extension Service: A Sudanese Case Study*

4. Sudan Forests National Corporation (1994). *Studies on Consumption of Forest Products*

5. El Warrag, E.A., El Shiekh and A.A. El Feel (2002). *Forest Genetic Resources Conservation in Sudan.* Khartoum: University of Khartoum

6. FAO (2005). *Global Forest Resources Assessment 2005*

7. UNEP/FAO/ICRAF (2006). *Post-Conflict Environmental Assessment of the Rural Environment for Sudan* (Draft). Nairobi, Kenya

8. Tobiolo, M. L. et al. (2006). *A Report on the Status of the Forest Reserves in Greater Yei County, South Sudan.* Kagelu Forestry Training Centre

9. Government of Sudan (2003). Ministry of Environment and Physical Development. *Sudan's First National Communications under the UN Framework Convention on Climate Change*

10. Badi, K. H and A.M. Ibrahim (2006). *Forest Management and Conservation in Sudan* (Workshop on the Post-Conflict National Plan for Environmental Management in Sudan, Khartoum, July 2006)

11. Gorashi, A.R. (2001). *State of Forest Genetics in Sudan.* FAO/IPGRI/ICRAF

12. Ibrahim, A.M. (2004). *National Report to the Fifth Session of the United Nations Forum on Forests.* Khartoum: FNC

13. Van Noordwijk, M. (1984). *The Ecology Textbook for the Sudan.* Khartoum: Khartoum University Press

14. Raddad, E.Y.A. (2006). 'Tropical Dryland Agroforestry on Clay Soils' in *Tropical Forestry Reports* 30

15. FAO Multipurpose Africover Databases on Environmental Resources
    http://www.africover.org/

Chapter 10: Freshwater Resources

1. FAO Aquastat Information System on Water and Agriculture. Sudan Country Profile 2005
   http://www.fao.org/ag/agl/aglw/aquastat/countries/sudan/index.stm

2. GONU Ministry of Irrigation and Water Resources: unpublished data and UNEP interviews during field missions in 2006

3. GONU Ministry of Irrigation and Water Resources (undated). *Irrigation and Water Resources of the Sudan (Past, Present and Future)*

4. Hassan, H. and M. Osman (2006). *Gaps in Natural Resources Management in North Sudan States* (Workshop on the Post-Conflict National Plan for Environmental Management in Sudan, Khartoum, July 2006)

5. Mahboub, E. and K. Riak (2006). *Wetland Management* (Workshop on the Post-Conflict National Plan for Environmental Management in Sudan, Khartoum, July 2006)

6. Lahmeyer International (2002). *Environmental Assessment Report for Merowe Dam Project.* Khartoum: Merowe Dam Project Implementation Unit

7. Teodoru, C., Wüest, A., and B. Wehrli (2006). *Independent Review of the Environmental Impact Assessment for the Merowe Dam Project.* Kastanienbaum: EAWAG Aquatic Research Group

8. Giles, J. (23/3/06). 'Tide of censure for African dams'. Nature Publishing Group News Release

9. International Rivers Network Merowe Campaign http://www.irn.org/programs/merowe/

10. Mohammed, Y (2005). *The Nile Hydroclimatology: Impact of the Sudd Wetland* [PhD Thesis]. Delft: Delft University of Technology/UNESCO-IHE, Balkelma Publishers

11. Liabwel, I. (2006). *Water Management in Southern Sudan* (Workshop on the Post-Conflict National Plan for Environmental Management in Sudan, Khartoum, July 2006)

12. Mefit, Babtie (1983). 'Development Studies in the Jonglei Canal Area: Final Report, Vol. 5' in *Wildlife Studies*, April 1983

13. Winter, P. (1997). 'Southern Sudan' in *Antelope Survey Update* 5

14. Moghray, A. et al (1982). 'The Jonglei Canal: A needed Development or Potential Ecodisaster' in *Environmental Conservation* 9

15. World Commission on Dams (2000). *Dams and Development: A New Framework for Decision-Making.* London: Earthscan Publications

16. UN Sudan Transition and Recovery Database (July 2004 data). Southern Kordofan State

17. Ramsar Convention (2006). *Ramsar Information Sheet for the Sudd Wetlands*

18. Thieme, M. et al. (2005). *Freshwater Ecoregions of Africa and Madagascar: A Conservation Assessment.* Washington, DC: Island Press

19. PERSGA (2003). *Status of Mangroves in the Red Sea and Gulf of Aden*, Technical Series No.11

20. Bashir, M. et al (2001). *Sudan Country Study on Biodiversity.* Khartoum: Ministry of Environment and Tourism

21. Blower, J.R. (1977). *Wildlife Conservation and management in the Southern Sudan* (UNDP/ FAO Sudan Project Findings and Recommendations)

22. Navarro, Luis A. and George Phiri (eds.) (2000). *Water Hyacinth in Africa and the Middle East. A Survey of Problems and Solutions.* International Development Research Centre

23. Agaemi, O. (undated). *Towards a State Environmental Action Plan (for Gedaref).* Gedaref: Gedaref Ministry of Environment and Tourism

24. Al Sunut Development Company http://www.alsunut.com/index.php

25. International Fund for Agricultural Development Exploitation of Groundwater Resources Website http://www.International Fund for Agricultural Development.org/evaluation/public_html/eksyst/doc/lle/pn/l103nrme.htm

26. Tearfund (2007). *Darfur: Relief in a Vulnerable Environment*

27. Abu Zeid, K. (2002). *The Transboundary Nubian Groundwater Basin*. CEDARE

28. South Valley Development Project (Toshka and East Oweinat)
http://www.amcham.org.eg/BSAC/StudiesSeries/report20.asp

29. Action Programme for the Integrated Management of the Shared Nubian Aquifer
http://www.gefonline.org/projectDetails.cfm?projID=2020

30. *Report of the Symposium on Aquaculture in Africa in Accra, Ghana, 30 September – 2 October 1975*. Reviews and Experience Papers: CIFA Technical Paper No. 4 (Supplement 1)
http://www.fao.org/docrep/005/AC672B/AC672B31.htm

31. Murakami, Masahiro (1995). *Managing Water for Peace in the Middle East: Alternative Strategies*. Tokyo: United Nations University Press

32. UNICEF Sudan Country Statistics (2002)
http://www.unicef.org/infobycountry/sudan_statistics.html

33. HCENR/WCGA/UNDP (2004). *Management Plan for Dinder National Park*

34. Sudan Ministry of Environment and Tourism/HCENR/SECS/Friedrich Ebert Foundation (1996). *Toward a National Environmental Action Plan for Sudan* [in Arabic]

35. The Juba Post Press Release (10/8/06): 'Digging Jonglei Canal Resumes with Egypt'

Chapter 11: Wildlife and Protected Area Management

1. Sudan Ministry of Environment and Physical Development (2003). *Assessment of Capacity-Building Needs and Country-Specific Priorities in Biodiversity Management and Conservation in Sudan*

2. Boma Wildlife Training Centre (2006). *UNEP PCEA Wildlife Conservation and Protected Areas* (Draft)

3. Sudanese Environmental Conservation Society (2006). *Sudan Wildlife Status Report* (Draft)

4. Hillman, J.C. (1985). *Wildlife Research in the Sudan in Relation to Conservation and Management* (Proceedings of the Seminar on Wildlife Conservation in the Sudan, Khartoum, March 1985)

5. Kemp, R. and J. (1984). *Survival Anglia. The Mysterious Journey* [Film Documentary]

6. Hillman, J.C. (1982). *Wildlife Information Booklet*. Sudan Ministry of Wildlife Conservation and Tourism, Department of Wildlife Management

7. Bashir, M. et al (2001). *Sudan Country Study on Biodiversity*. Khartoum: Ministry of Environment and Tourism

8. Blower, J.R. (1977). *Wildlife Conservation and management in the Southern Sudan* (UNDP/ FAO Sudan Project Findings and Recommendations)

9. Cave et al (1958). *Birds of Sudan, their Identification and Distribution*. UK: Oliver and Boy

10. Robertson, P. (2001). 'Sudan' in *Important Bird Areas in Africa and its Associated Islands: Priority Sites for Conservation*. Fishpool, L.D.C. and M.I. Evans (eds). Birdlife International Series 11

11. Setzer, H.W. (1956). *Mammals of the Anglo-Egyptian Sudan* (Proceedings US Natural History Museum 106)

12. IUCN Red List of Threatened Species 2006
http://www.iucnredlist.org/

13. Abdel Salam, Mohammed Younis (2006). *Marine and Costal Environmental Conservation in Sudan: The Role of Marine Protected Areas* (Workshop on the Post-Conflict National Plan for Environmental Management in Sudan, Khartoum, July 2006)

14. Mackinnon J. and K. (1986). *Reviews of the Protected Areas System in the Afrotropical Realm*. IUCN/UNEP

15. World Resources Institute (1995). *Twelfth Annual Report*

16. Birdlife International
www.birdlife.org/datazone/sites/

17. Seymour, C. (2001). *Saharan Flooded Grassland*. WWF
www.worldwildlife.org/wildworld/profiles/terrestrial/at/at0905_full.html

18. Stuart, S.N. and R.J. Adams (1990). *Biodiversity in Sub-Saharan Africa and its Islands: Conservation, Management and Sustainable Use*

19. Winter, P. (1997). 'Southern Sudan' in *Antelope Survey Update* 5

20. East, R. (1998). *African Antelope Database 1998*. IUCN/SSC Antelope Specialist Group

21. Food and Agriculture Organization of the United Nations (1981). *Forest Resources of Tropical Africa* (Part II: Country briefs – Sudan). FAO Tropical Forest resources Assessment project

22. Hammerton, D. (1964). *Hydrological Research in Sudan* (Sudan Phl. Soc. 12th Annual Symposium, Khartoum)

23. Hughes, R.H. and J.S. (1991). *Directory of African Wetlands*. IUCN/UNEP

24. Moghray, A. et al (1982). 'The Jonglei Canal: A needed Development or Potential Ecodisaster' in *Environmental Conservation* 9

25. Nikolaus, G. (1985). *Necessary Conservation and Education Programmes, Protection of Wetlands of International Importance and Migratory Birds in Sudan* (Proceedings of the Seminar on Wildlife Conservation and Management in the Sudan, March 1985, Khartoum)

26. Ojok, L.I., Morjan, M.D., and B.B. Nicholas (2001). *The Impact of Conflict on Wildlife and Food Security in South Sudan: the Survey of Boma National Park, South Sudan* (Draft)

27. Ramsar Convention. *Sudd Nomination Document*
www.ramsar.org

Chapter 12: Marine Environments and Resources

1. Hassan, M. (2006). *Sudan Marine and Coastal Environment* (Draft)

2. Abdel Salam, Mohammed Younis (2006). *Marine and Costal Environmental Conservation in Sudan: The Role of Marine Protected Areas* (Workshop on the Post-Conflict National Plan for Environmental Management in Sudan, Khartoum, July 2006)

3. UNEP/IUCN (1988). *Coral Reefs of the World Vol. 2: Indian Ocean, Red Sea and the Gulf*. UNEP Regional Seas Directories and Bibliographies

4. PERSGA/GEF (1998). *Strategic Action Programme for the Red Sea and Gulf of Aden. Country Report: Republic of the Sudan*

5. UNEP-WCMC (2006). World Database on Protected Areas

6. African Parks Foundation (2006). *Expedition to Sanganeb and Dongonab National Parks*
http://www.africanparks-conservation.com/sudan-expedition.php

7. PERSGA (2003). *Status of Mangroves in the Red Sea and Gulf of Aden*, Technical Series No.11

8. Government of Sudan. *2004 Marine Fisheries Statistics*

Chapter 13: Environmental Governance and Awareness

1. Government of Sudan (2005). *The Comprehensive Peace Agreement between the Government of the Republic of Sudan and the Sudan People's Liberation Movement/Army*, Nairobi, 9 January 2005

2. Government of Sudan (2005). *The Interim National Constitution of Sudan*

3. Government of Southern Sudan (2005). *The Interim Constitution of Southern Sudan*

4. UNEP (2006). *Assessment of Environmental Policy, Institutions and Legal Framework in North and South Sudan* (Draft)

5. Sudan National Environmental Action Plan (2007) (Draft)

6. Scholte, P. and M. Babiker (2005). *Terminal Evaluation for the Conservation, Management of Habitat, Species and Sustainable Community Use of Biodiversity in Dinder National Park* (Report to UNDP-GEF)

7. UNEP (2002). *Capacity-Building for Sustainable Development. An Overview of UNEP Environmental Capacity Development Activities*

8. Abdel Ati, Hassan A. (ed.) (2002). *Sustainable Development in Sudan, Ten Years after the Rio Summit. A Civil Society Perspective.* Environmentalists Society, EDGE for Consultancy and Research and Heinrich Boll Foundation Regional Office Horn of Africa

9. Bashir, M. et al (2001). *Sudan Country Study on Biodiversity.* Khartoum: Ministry of Environment and Tourism

10. Government of Sudan, Secretariat for Wildlife Conservation and Tourism (2004). Draft Frame Document, The New Site – South Sudan, January 2004

11. UNEP (2005). *Ridding the World of POPs: A Guide to the Stockholm Convention on Persistent Organic Pollutants*

12. IUCN (2004). *An Introduction to the African Convention on the Conservation of Nature and Natural Resources.* Environmental Policy and Law Paper No. 56

13. Government of Sudan (2003). Ministry of Environment and Physical Development. *Sudan's First National Communications under the UN Framework Convention on Climate Change*

14. Convention on Biological Diversity. Three Sudan National Reports to the Conference of the Parties http://www.cbd.int/reports/list.aspx?type=all

15. Government of Sudan (undated). *Sudan National Report to the Conference of Parties on the Implementation of the United Nations Convention to Combat Desertification*

16. Government of Sudan. *Sudan's Ninth Report to the Conference of the Parties of the Ramsar Convention*

17. The Cartagena Protocol on Biosafety http://www.unep.ch/biosafety/development/countryreports/SDNBFrep.pdf

18. Ramsar Convention www.ramsar.org

19. The Bamako Convention on the Ban of the Import into Africa and the Control of Transboundary Movement and Management of Hazardous Wastes within Africa http://www.ban.org/Library/bamako_treaty.html

20. The Basel Convention on the Control of Transboundary Movements of Hazardous Wastes and their Disposal http://www.basel.int/

Chapter 14: International Aid and the Environment

1. *United Nations and Partners Work Plan for Sudan 2006*

2. UN Mission in Sudan http://www.unmis.org/english/en-main.htm

3. Sphere Standards, Sphere Project Website http://www.sphereproject.org/

4. UNEP-ETB (2002). *Environmental Impact Assessment Resource Manual*

5. Chapin Metz, H. (1991). *Sudan: A Country Study.* Washington, DC: Library of Congress

6. The Economist Intelligence Unit (2006). *Country Report: Sudan (December 2006),* London: EIU

7. Andrews F.W. (1950). *The Flowering Plants of the Anglo-Egyptian Sudan.* Arbroath, T. Buncle and Co. Ltd for the Sudan Government

8. Tearfund (2007). *Darfur: Relief in a Vulnerable Environment*

9. *USAID Interim Strategic Plan for Sudan 2004-2006*

10. *USAID Sudan Strategy Paper 2006-2008*
    http://www.usaid.gov/locations/sub-saharan_africa/countries/sudan/docs/sudan_strategy.pdf

11. USAID Sudan Transitional Environment Programme (2006). *Programmatic Environmental Assessment of Road Rehabilitation Activities in Southern Sudan*

12. The Centre for Environmental Economics and Policy in Africa, Faculty of Natural and Agricultural Sciences, University of Pretoria (2006). *Environmental Impact Assessment of the Bor Counties' Dyke Rehabilitation Project, South Sudan: Integrated Assessment Report*

13. Sustainable Options for Livelihood Security in Eastern Sudan (SOLSES) Project Website
    http://www.unhabitat.org/content.asp?cid=4544&catid=271&typeid=13&subMenuId=0

14. Global Environmental Facility Projects Website
    http://www.gefweb.org/Projects/Focal_Areas/focal_areas.html

Special note: In the course of the assessment in 2006, UNEP sponsored two major environmental workshops, one held in Khartoum in July and the other in Juba in November. At these events, technical papers on the environmental issues of Sudan were presented as input to both the UNEP report and the forthcoming government-owned National Plan for Environmental Management. The full list of these papers is offered below. Individual papers have been included in the chapter references where appropriate.

Proceedings of the Khartoum Workshop on the Post-Conflict National Plan for Environmental Management in Sudan, July 2006

- Abdelgani, M.E. and Z.E. Alabjar. *Environmental Research Capacity in Sudan.*

- Kitundo, M. *Environmental Education and Public Participation in Sudan.*

- Mohamed, Y.A. *Public Participation in Natural Resources Management in Sudan.*

- El Tayeb, G. and N. Kuku. *The Role of Environmental Societies in Post-Conflict Sudan.*

- Bashir, M. and F. Tong. *Sudan Protected Areas*

- Mahgoub, E.F.E.T and K. M. Riak. *Wetland Management in Sudan*

- Ibrahim, A.M and K. H. Badi. *Forest Management and Conservation*

- Mohamed, H.M. and A. R.M. Hamid. *Range Management and Conservation in Sudan*

- Abdel Rahman, M.E.A. and M. O. El Sammani. *Natural Resources and Socio-economic Parameters*

- Abdel Salam, M.Y. *Marine and Coastal Environment Conservation in Sudan.*

- Omwenga, J.M. *Global Management of Freshwater Resources – The Nile Basin – A Perspective*

- Liabwel, I. *Water Management in Southern Sudan*

- Ayoub, A.T. *Linkages between Food Security and Natural Resource Conditions*
- Awad, N.M. *International and Regional Agreements*
- Satti, M. *Partnership for Sustainable Development on the Red Sea Coast*
- El Hassan, H.M. and M. Osman. *Gaps in Natural Resources Management in North Sudan States*
- El Hassan, B.A. Resource-Based Conflicts and Land Use Systems
- Abdelbagi, A.O., Mohamed, A.A., El Hindi, A.M. and A.M. Ali. *Impact of Pesticides and Other Chemicals on the Environment*
- Murkaz Ali, E.T. *Overview of Relevant Policies, Strategies and Legislation Related to Environment and their Relevancy under the CPA*
- Ibnoaf, M. *A Pro-Poor Post-Conflict Participatory Approach*
- El Moghraby, A.I. *Management of Natural Resources in the Sudan*
- Desertification Control and Mitigation of Drought Effects in Sudan

Proceedings of the Juba Workshop on the Post-Conflict National Plan for Environmental Management in Sudan, November 2006

- Hassan, K.I. *The Impact of Climate Change on Food Security*
- Bojoi, M. *Wildlife Tourism and Poverty. Present State and Strategy for Development in South Sudan*
- Dima, S.J. *Land Use Systems in South Sudan and their Impact on Land Degradation*
- Wurda, V. *The Current Development of Instructional and Regulatory Framework for Environmental Management in South Sudan*
- Badawi Bashir M. K. *Management of the Environment in the Sudan's Oil Industry*
- Dhol, J.C. *Sustainable Agricultural Development in Sudan*
- Abate, A.L. *Livestock Production Challenges in the Rangelands Ecosystem of South Sudan*
- Udo, M.G. *Sustainable Livestock/Range Management Systems – A Way Forward to Progressive Development of South Sudan*
- Riak, K.K. *Sudd Area as a Ramsar Site: Biophysical Features*
- Liabwel, I. *Water Resources in Southern Sudan*
- Tier, A.M. *The State and Capacity of Environment Institutions: Legal and Structural*
- Gore, P. *A Demographic Profile of Southern Sudan*

Appendix III
List of contributors

Members of the UNEP Assessment Team

UNEP Post-Conflict and Disaster Management Branch – Senior Management
Mr.   Henrik Slotte, Chief
Mr.   Muralee Thummarukudy, Operations Manager
Mr.   David Jensen, Policy and Planning Coordinator

UNEP Assessment Team
Mr.   Andrew Morton, Sudan Project Coordinator
Mr.   Hassan Partow, Senior Environment Expert
Mr.   Grant Wroe-Street, Project Coordinator
Mr.   Joseph Bartel, Natural Resources Expert
Mr.   David Meadows, Programme Officer
Mr.   Edward Wilson, Wildlife Consultant
Mr.   David Stone, Consultant
Mr.   John Carstensen, Environmental Law Expert
Mr.   Mahgoub Hassan, Marine Expert
Ms.   Silja Halle, Report Editor

UNEP Regional Office for Africa and Headquarters
Mr.   Sekou Toure, Director, Regional Office for Africa
Mr.   Nehemiah Rotich, Programme Officer (Biodiversity), Regional Office for Africa
Mr.   Mohammed Abdel Monem, Programme Officer (Natural Resources), Regional Office for Africa
Mr.   Serge Bounda, Chief Librarian
Mr.   Steve Jackson, Head of Audiovisual

Special Thanks

Ministry of Environment and Physical Development (GONU)
H.E.   Ahmed Babikir Nahar, Minister
H.E.   Ms. Teresa Siricio Iro, State Minister
Mr.   El Fadil Ali Adam, Undersecretary
Mr.   Saadeldin Izzeldin, Secretary General, Higher Council for Environment and Natural Resources
Mr.   Mamoun Abdel Kader, Director, Directorate of Environment
Mr.   Mahgoub Hassan, Deputy Secretary General, Higher Council for Environment and Natural Resources
Ms.   Mona Abdel Hafeez, Directorate of Environment
Mr.   Bashir Omar, Directorate of Environment
Ms.   Samyah Ibrahim, Secretary, Environment Council Secretariat, Gedaref State
Mr.   Ahmed El Rashid Said, Secretary General, State Council for Environment and Natural Resources, Nile State
Mr.   Yacoub Salih, Secretary General, State Council for Environment and Natural Resources, Northern State
Mr.   Ghassan Ahmed, Marine Environment Protection Authority

Ministry of Environment, Wildlife Conservation and Tourism (GOSS)
H.E.   James Loro Siricio, Minister
Major General Alfred Akwoch Omoli, Permanent Undersecretary
Mr.   Victor Wurda LoTombe, Director General for Environment
Mr.   George Modi, Environment Information Centre
Ms.   Kapuki Tognun, Librarian, Environment Information Centre
Mr.   Moses Gogonya, Environmental Inspector
Mr.   Alex Gubek, Environmental Inspector

Government of Southern Sudan Cabinet and Line Ministries

H.E.   Luka Biong Deng, Minister of the Office of the President
Hon.   Gabriel Matur Malek, Chairman of the Committee for Land, Natural Resources and Environment
Mr.   Waragak Gatluak Fequir, Undersecretary, Agriculture
Mr.   Jaden Tongun Emilio, Undersecretary, Forestry
Mr.   Raymond Pitia, Undersecretary, Housing, Land and Public Utilities
Mr.   Francis Latio, Undersecretary, Economic Planning
Dr.   Cirino Hiteng Ofuho, Undersecretary, Regional Cooperation
Dr.   Majok Yak, Undersecretary, Health
Mr.   Chour Deng Mareng, Undersecretary, Industry and Mining
Dr.   Daniel Wani, Undersecretary, Transport and Roads
Dr.   Makuei Malual Kaang, Undersecretary, Animal Resources and Fisheries
Mr.   Bortel Mori Nyombe, Undersecretary, Cooperatives and Rural Development
Mr.   Isaac Liabwel, Undersecretary, Water Resources and Irrigation

Other Sudan Government Agencies

Mr.   Ismail Jelab, Governor, Southern Kordofan State
Mr.   Ahmed Saad, Governor, Sennar State
Mr.   Azhari Abdel Rahman, Minister of Agriculture and Natural Resources, Gezira State
Mr.   Mustapha El Khalil, Minister of Health, Gedaref State
Mr.   Jaafer Salih, Minister of Planning and Public Works, White Nile State
Mr.   Ahmed Gamal Dawood, Minister of Agriculture, Animal Resources and Irrigation, Northern State
Mr.   Abdallah Mohammed Edam, General Director, Ministry of Health, Northern Kordofan State
Mr.   Saoud Mohammed, Director, Office of the Minister of Agriculture and Natural Resources, Gezira State
Mr.   Abdel Adhim Tayfoor, Deputy Director, Ministry of Agriculture, Nile State
Ms.   Amna Hamid, Director, Remote Sensing Authority
Dr.   Salwa Abdel Hameed, Director, Wildlife Research Centre and Ramsar Focal Point, Ministry of Science
            and Technology
Mr.   Mohammed Ballal, Director, Gum Arabic Research Station Office, Agricultural Research Corporation
Mr.   Salah El Din, General Director, Ministry of Energy and Mining
Mr.   Mukhtar Ali Mutkhtar, Environment and Sustainable Development Advisor, Dams Implementation Unit
Mr.   Muawia Salih Elbager, Environmental Affairs Director, Dams Implementation Unit
Mr.   Haidar Bekhit, Director, Nile Water Directorate, Ministry of Irrigation and Water Resources
Mr.   Muatism Al Awadh, Director, Sennar Dam
Mr.   Ahmed Abbas, Assistant Director, Roseires Dam
Mr.   Sameer Ahmed, Director, Khashm El Girba Dam
Mr.   El Hadi Adam, Research Engineer, Ministry of Irrigation and Water Resources, Sennar
Mr.   El Tayib El Alam, Director, Agricultural Directorate, Gezira Scheme
Mr.   Abel Adhim Banaga, Director, Occupational Health and Saftey, Gezira Scheme
Mr.   Hassan Kambal, Director, Directorate of Planning and Social and Economic Research, Gezira Scheme
Mr.   El Tayib El Feel, Director, Irrigation Unit, Gezira Scheme
Mr.   Amr Hassan, Deputy Director, New Halfa Agricultural Scheme
Mr.   Tabayq Tabayq, Acting Director, Ministry of Agriculture, Sennar State
Mr.   El Nour El Nour, Director, Forest National Corporation, Southern Kordofan State
Mr.   Mohammed El Jaak, Director, Forestry Directorate, Gezira State
Mr.   Ibrahim Daoka, Director, Forestry Directorate, Gezira State
Mr.   Youssif Obeid, Director, Forestry Directorate, Nile State
Mr.   Awadh Adam, Forestry Director, El Shuwak
Mr.   El Tijani Hussein Abdallah, Forestry Inspector, Talodi
Mr.   Jamal El Deen Mohammed, Forestry Inspector, Kadugli
Mr.   Muhayi Adam Othman, Assistant Forestry Director, Abu Jubayhah
Mr.   Adam Jadallah Ardeeb, Forestry Inspector, Dilling
Mr.   El Sheikh Dein Hussein, Forestry Inspector, Umm Rawaba
Mr.   Abdullah Hamid, Director, Marine Fisheries Association, Ministry of Animal and Fish Resources
Mr.   Maknoon Othman, Ministry of Agriculture, El Hasahesa

Mr.   Merghani El Sayid, Director, Plant Protection Directorate, Gedaref State
Ms.   Samiha Ishaq, Director, Rangelands and Fodder Directorate, Gedaref State
Mr.   Ousama Ibrahim, Deputy Director, Forestry Department, Khartoum
Mr.   Abdellah Harun, Wildlife Conservation General Administration, Ministry of Interior, Red Sea State
Mr.   Asam Qassem, Manager, Suba Wastewater Treatment Station, Khartoum
Colonel Sanad Bin Suleiman, Dinder National Park, Wildlife Conservation General Administration, Ministry of Interior
Mr.   Mubarak Ibrahim, Wildlife Research Centre

Sudan Civil Society and Private Sector

Mr.   Muawia Shadad, Chairman, Sudan Environment Conservation Society
Ms.   Suad M. Sulaiman, Director, Sudan Environment Conservation Society
Ms.   Huda Khogali, Environment Expert, Sudan Environment Conservation Society
Mr.   Taalat Abd El Majed, Environment Expert, Sudan Environment Conservation Society
Mr.   Sumaia M. Elsayed, Sudan Environment Conservation Society
Ms.   Salma El Tayb, Deputy Director, Sudan Environment Conservation Society, Kosti
Mr.   Hussein Musa, Director, Sudan Environment Conservation Society, Wad Medani
Mr.   Abel Latif, Jawdan, Sudan Environment Conservation Society, Wad Medani
Mr.   El Nayir Suleiman, Sudan Environment Conservation Society, Gedaref
Mr.   Izat Taha, Consultant
Ms.   Susan Ayot, Consultant
Mr.   Malik Marjan, Principal, Boma Wildlife Training Centre
Captain Abdel Helim bin Abdel Helim, Red Seas Enterprise
Ms.   Somaya Mohammed, Head, Department of Biological Oceanography, Red Sea University
Ms.   Nahid Osman, Faculty of Marine Sciences and Fisheries, Red Sea University
Mr.   Suliman Suliman, HSE Consultant, Shell Sudan
Mr.   Hamza Ibrahim, Deputy General Manager, Nile Cement Co.
Mr.   Samuel Mule, Concern, Southern Kordofan State
Ms.   Nidal Ibrahim, Jamiyat Ro'ait El Kheir
Mr.   Ahmed El Bashir, Deputy Dean, Wadi El Neel University
Mr.   Alex Murray, Field Coordinator, ADRA, Um Jawasir
Mr.   Nasser Bur, Section Head Environment, HSE Department, GNPOC
Mr.   Fatih Youssif, HSE Supervisor, Heglig, GNPOC
Mr.   Mohammed Abdullah, Director, Administration Department, Gezira Tanneries
Mr.   Hamza Fath El Rahman, Deputy Director General, El Rabak Cement Factory
Mr.   Siddiq Abdul Rahman, Chemical Engineer, El Rabak Cement Factory
Mr.   Muawia Ali, General Manager, Kenana Sugar Company
Mr.   Mohammed El Sheikh, General Director, Assalaya Sugar Factory
Mr.   Mohammed Abou Raouf, President of the Pastoralist Union and Leader of the Al Rifaa Tribe, Sennar State
Mr.   Mahmood Khalid, President, Farmers Union, Nile State

United Nations in Sudan

Mr.   Omer Egemi, Head of Environmental Section, UNDP
Ms.   Hanan Mutwakil, Senior Programme Associate, UNDP
Mr.   Thomas Carter, UNDP Juba
Mr.   John Fox, UNDP Juba
Mr.   Sadig Ibrahim Elamin, Sudan Interagency Mapping, OCHA
Mr.   George Okech, Head of Office, FAO Juba
Mr.   Mohammed Hussein, FAO Khartoum
Mr.   John Smith, Livestock Officer, FAO
Mr.   Greg Wilson, Country Director, UNOPS
Mr.   Akuila Buadromo, Project Manager, UNOPS
Mr.   Steve Crosskey, Roads Programme Manager, WFP
Ms.   Malar Smith, Head of Office, UNHCR Bor
Mr.   Tom Hockley, Deputy Head, RCO Khartoum
Mr.   Marcus Culley, UNDSS Juba
Mr.   Kakuca Mladen, OIM Juba

International Organizations and Individuals

Mr.   Yves Barthélemy, Remote Sensing Expert
Mr.   Dominique Del Pietro, UNEP DEWA GRID-Europe
Mr.   Brendan Bromwich, Tearfund
Mr.   Azene Tesemma Bekele, Project Manager, ICRAF
Mr.   Sean White, Winrock International
Mr.   Douglas Varchol, DZAP Productions
Mr.   Philip Winter, Rift Valley Institute
Mr.   Thomas Catterson, USAID
Mr.   Paul Symonds, European Commission
Mr.   Gedion Asfaw, NTEAP Manager, Nile Basin Initiative
Ms.   Astrid Hillers, World Bank
Mr.   Jörn Laxén, University of Helsinki
Mr.   Steve McCann, MMackintosh
Mr.   Jon Bennett, Oxford Development Consultants
Mr.   Evert Van Walsum, Consultant
Ms.   Jane Upperton, Consultant
Ms.   Mette Møglestue, Consultant
Mr.   Laks Akella, Consultant

UNEP Post-Conflict and Disaster Management Branch

Mr. Henrik Slotte, Chief
Mr. Muralee Thummarukudy, Operations Manager
Mr. Andrew Morton, Country Operations Coordinator
Mr. David Jensen, Policy and Planning Coordinator
Mr. Joseph Bartel, Natural Resources Expert
Mr. Mario Burger, Senior Scientific Advisor
Ms. Rachel Dolores, Project Assistant
Ms. Silja Halle, Communications Advisor
Mr. David Meadows, Programme Officer
Ms. Cecilia Morales, Advisor
Ms. Mani Nair, Project Assistant
Ms. Satu Ojaluoma, Administrative Officer
Ms. Elena Orlyk, Project Assistant
Mr. Hassan Partow, Senior Environment Expert
Mr. Matija Potočnik, Media Assistant
Mr. Gabriel Rocha, Systems Administrator
Ms. Joanne Stutz, Programme Assistant
Mr. Koen Toonen, Project Coordinator
Ms. Maliza van Eeden, Associate Programme Officer
Ms. Anne-Cécile Vialle, Operations and Research Assistant
Mr. Richard Wood, Technical Coordinator
Mr. Grant Wroe-Street, Project Coordinator
Mr. Dawit Yared, Project Assistant